LONDON MATHEMATICAL SOCIETY LECTURE NOTE SERIES

Managing Editor: Professor M. Reid, Mathematics Institute,
University of Warwick, Coventry CV4 7AL, United Kingdom

The titles below are available from booksellers, or from Cambridge University Press at
http://www.cambridge.org/mathematics

London Mathematical Society Lecture Note Series: 413

Optimal Transportation
Theory and Applications

Edited by

YANN OLLIVIER
Université Paris-Sud, France

HERVÉ PAJOT
Université de Grenoble, France

CÉDRIC VILLANI
Université de Lyon, France

CAMBRIDGE
UNIVERSITY PRESS

CAMBRIDGE
UNIVERSITY PRESS

University Printing House, Cambridge CB2 8BS, United Kingdom

One Liberty Plaza, 20th Floor, New York, NY 10006, USA

477 Williamstown Road, Port Melbourne, VIC 3207, Australia

314-321, 3rd Floor, Plot 3, Splendor Forum, Jasola District Centre, New Delhi - 110025, India

79 Anson Road, #06-04/06, Singapore 079906

Cambridge University Press is part of the University of Cambridge.

It furthers the University's mission by disseminating knowledge in the pursuit of
education, learning and research at the highest international levels of excellence.

www.cambridge.org
Information on this title: www.cambridge.org/9781107689497

First published 2014

A catalogue record for this publication is available from the British Library

ISBN 978-1-107-68949-7 Paperback

Contents

Contributors

Filippo Santambrogio, Laboratoire de Mathématiques, Université Paris-Sud, Orsay Cedex 91405, France.

Ivan Gentil, Institut Camille Jordan, Université Lyon I, Villeurbanne Cedex 69622, France.

Luigi Ambrosio, Faculty of Sciences, Scuola Normale Superiore di Pisa, Pisa 56126, Italy.

Alessio Figalli, Department of Mathematics, University of Texas, Austin, TX 78712, USA.

Peter Topping, Mathematics Institute, University of Warwick, Coventry CV4 7AL, England.

Sara Danieri, S.I.S.S.A., Trieste 34151, Italy.

Guiseppe Savare, Departimento di Matematica, Università di Pavia, Pavia 27100, Italy.

Shin-ichi Ohta, Department of Mathematics, Kyoto University, Kyoto 606-8502, Japan.

Olivier Besson, Institut de Mathématiques, Université de Neuchatel, Neuchatel 2009, Switzerland.

Martine Picq, Université de Lyon CNRS, INSA-Lyon ICJ UMR, Villeurbanne Cedex F-69100, France.

Jérome Pousin, Université de Lyon CNRS, INSA-Lyon ICJ UMR, Villeurbanne Cedex F-69100, France.

Mathias Beiglbock, Faculty of Mathematics, University of Vienna, Vienna 1090, Austria.

Christian Léonard, Laboratoire de Mathématiques, Université Paris Ouest, Nanterre Cedex 92001, France.

Walter Schachermayer, Faculty of Mathematics, University of Vienna, Vienna 1090, Austria.

François Bolley, Université de Paris IX (Paris-Dauphine), Paris Cedex 16 F-75775, France.

Patrick Cattiaux, Institut de Mathématiques de Toulouse, Université Paul Sabatier, Toulouse Cedex 09, France.

Arnaud Guillin, Laboratoire de Mathématiques, Université Blaise Pascal, Aubière 63170, France.

Quentin Mérigot, Laboratoire Jean Kuntzmann, Université Grenoble Alpes and CNRS, Grenoble, France.

Preface

This book contains the proceedings of the summer school "Optimal transportation: Theory and Applications" held at the Fourier Institute (University of Grenoble I, France). The first 2 weeks were devoted to courses that described the main properties of optimal transportation and discussed its applications to analysis, differential geometry, dynamical systems, partial differential equations and probability theory. Courses were addressed both to students and researchers. A workshop took place during the last week. The aim of this conference was to present very recent developments of optimal transportation and also its applications in biology, mathematical physics, game theory and financial mathematics.

The first part of the book contains (expanded) versions of the courses. There are two sets of notes by F. Santambrogio. The first one gives a short introduction to optimal transport theory. In particular, the Kantorovich duality, the structure of Wasserstein spaces and the Monge–Ampère equations related to optimal transport are presented to the readers. These notes could be seen as an introduction for the other papers of the book. The second one describes applications to economics, game theory and urban planning.

The notes of I. Gentil, P. Topping and S.-I. Ohta describe (with different flavours) the connections between optimal transport and the notion of Ricci curvature, which is a very important tool in classical Riemannian geometry. A notion of curvature-dimension condition was defined by D. Bakry and M. Émery to study geometric properties of diffusions and to get functional inequalities. I. Gentil's notes study the Bakry–Émery condition in the case of the Ornstein–Uhlenbeck semigroup. A quite different approach using optimal transport theory to obtain logarithmic Sobolev-type inequalities is also discussed. A definition of metric measure spaces with lower Ricci curvature bound (which coincides with the classical definition in the case of Riemannian manifolds) was proposed very recently by K.-T. Sturm and J. Lott–C.

Villani independently. S.-I. Ohta's long paper discusses in detail the geometry of such spaces. For instance, versions of the Brunn–Minkowski inequality, of the Lichnerowicz inequality, of Bishop–Gromov volume comparison, of the Bonnet–Myers diameter bound and also the stability under Hausdorff–Gromov convergence are proved in this general setting. The theory of Ricci flow as developed by R. Hamilton and others since 1982 is an essential element in the proofs by G. Perelman of the Poincaré conjecture and Thurston's geometrisation conjecture. The objective of P. Topping's lectures is to explain this theory from the point of view of optimal transport.

The fundamental work of Y. Brenier related to the Euler equation played an important role in the renewal of optimal transport in the 1980s. The notes of L. Ambrosio and A. Figalli describe some recent results on Brenier's variational models for the incompressible Euler equations.

The paper by S. Daneri and G. Savaré gives an overview of the theory of gradient flows in Euclidean spaces and then in metric spaces. Applications to evolution equations in the Wasserstein spaces of probability measures are also discussed.

Apart from these mini-courses, this book also contains five research/survey papers. O. Besson, M. Picq and J. Pousin present an algorithm for a computing mass transport problem inspired from optimal transport and whose origin lies in hearts' images tracking. M. Beigblöck, C. Léonard and W. Schachermayer discuss the duality theory for the Monge–Kantorovich transport problem. In particular, they give a version of Fenchel's perturbation method. The paper of F. Bolley reviews recent quantitative results on the approximation of mean field diffusion equations by large systems of interacting particles, obtained by optimal coupling methods. P. Cattiaux and A. Guillin describe some recent results on Poincaré-type inequalities, transportation-information inequalities or logarithmic Sobolev inequality obtained via Lyapounov conditions. Q. Mérigot proves the stability of the Federer curvature measures with respect to the Wasserstein distance. This was motivated by problems of reconstruction of curves and surfaces from point cloud approximation that come from image analysis for instance. These five contributions illustrate the variety of possible applications of optimal transport to pure and applied mathematics, and also to computer science.

PART ONE

Short Courses

1

Introduction to optimal transport theory

FILIPPO SANTAMBROGIO

Abstract

These notes constitute a sort of crash course in optimal transport theory. The different features of the problem of Monge–Kantorovitch are treated, starting from convex duality issues. The main properties of space of probability measures endowed with the distances W_p induced by optimal transport are detailed. The key tools connecting optimal transport and partial differential equations are provided.

Contents

AMS Subject Classification (2010): 00-02, 49J45, 49Q20, 35J60, 49M29, 90C46, 54E35
Keywords: Monge problem, linear programming, Kantorovich potential, existence, Wasserstein distances, transport equation, Monge–Ampère, regularity

3

1.1 Introduction

These very short lecture notes are not intended to be an exhaustive presentation of the topic, but only a short list of results, concepts and ideas which are useful when dealing for the first time with the theory of optimal transport. Several of these ideas have been used, and explained in greater detail, during the other classes of the Summer School "Optimal Transportation: Theory and Applications" which were the occasion for the redaction of these notes. The style that was chosen when preparing them, in view of their use during the Summer School, was highly informal, and this revised version will respect the same style.

The main references for the whole topic are the two books on the subject by C. Villani [15, 16]. For what concerns curves in the space of probability measures, the best specifically focused reference is [2]. Moreover, I am also very indebted to the approach that L. Ambrosio used in a course at SNS Pisa in 2001–02 and I want to cite this as another possible reference [1].

The motivation for the whole subject is the following problem proposed by Monge in 1781 [14]: given two densities of mass f, $g \geq 0$ on $\mathbb{R}^d$, with $\int f = \int g = 1$, find a map $T : \mathbb{R}^d \to \mathbb{R}^d$ pushing the first one onto the other, i.e. such that

$$\int_A g(x)dx = \int_{T^{-1}(A)} f(y)dy \quad \text{for any Borel subset } A \subset \mathbb{R}^d \quad (1.1)$$

and minimizing the quantity

$$\int_{\mathbb{R}^d} |T(x) - x| f(x)dx$$

among all the maps satisfying this condition. This means that we have a collection of particles, distributed with density f on $\mathbb{R}^d$, that have to be moved, so that they arrange according to a new distribution, whose density is prescribed and is g. The movement has to be chosen so as to minimize the average displacement. The map T describes the movement (that we must choose in an optimal way), and $T(x)$ represents the destination of the particle originally located at x. The constraint on T precisely accounts for the fact that we need to reconstruct the density g. In the following, we will always define, similarly to (1.1), the image measure of a measure μ on X (measures will indeed replace the densities f and g in the most general formulation of the problem) through a measurable map $T : X \to Y$: it is the measure denoted by $T_{\#}\mu$ on Y and characterized by

$$T_\# \mu(A) = \mu(T^{-1}(A)) \quad \text{for every measurable set } A,$$

$$\text{or } \int_Y \phi \, d \, (T_\# \mu) = \int_X \phi \circ T \, d\mu \quad \text{for every measurable function } \phi.$$

The problem of Monge has stayed with no solution (Does a minimizer exist? How to characterize it? . . .) until the progress made in the 1940s. Indeed, only with the work by Kantorovich in 1942 has it been inserted into a suitable framework which gave the possibility to approach it and, later, to find that solutions actually exist and to study them. The problem has been widely generalized, with very general cost functions $c(x, y)$ instead of the Euclidean distance $|x - y|$ and more general measures and spaces. For simplicity, here we will not try to present a very wide theory on generic metric spaces, manifolds and so on, but we will deal only with the Euclidean case.

1.2 Primal and dual problems

In what follows we will suppose Ω to be a (very often compact) domain of $\mathbb{R}^d$ and the cost function $c : \Omega \times \Omega \to [0, +\infty[$ will be supposed continuous and symmetric (i.e. $c(x, y) = c(y, x)$).

1.2.1 Kantorovich and Monge problems

The generalization that appears as natural from the work of Kantorovich [12] of the problem raised by Monge is the following:

Problem 1. Given two probability measures μ and ν on Ω and a cost function $c : \Omega \times \Omega \to [0, +\infty]$ we consider the problem

$$(K) \quad \min \left\{ \int_{\Omega \times \Omega} c \, d\gamma \, | \, \gamma \in \Pi(\mu, \nu) \right\}, \tag{1.2}$$

where $\Pi(\mu, \nu)$ is the set of the so-called *transport plans*, i.e. $\Pi(\mu, \nu) = \{\gamma \in \mathcal{P}(\Omega \times \Omega) : (p^+)_\# \gamma = \mu, (p^-)_\# \gamma = \nu\}$, where p^+ and p^- are the two projections of $\Omega \times \Omega$ onto Ω. These probability measures over $\Omega \times \Omega$ are an alternative way to describe the displacement of the particles of μ: instead of saying, for each x, which is the destination $T(x)$ of the particle originally located at x, we say for each pair (x, y) how many particles go from x to y. It is clear that this description allows for more general movements, since from a single point x particles can a priori move to different destinations y. If multiple destinations really occur, then this movement cannot be described through a map T. Notice that the constraints on $(p^{\pm})_{\#}\gamma$ exactly mean that we restrict our attention to the movements that really take particles distributed according to the distribution μ and move them onto the distribution ν.

The minimizers for this problem are called *optimal transport plans* between μ and ν. Should γ be of the form $(id \times T)_{\#}\mu$ for a measurable map $T : \Omega \to \Omega$ (i.e. when no splitting of the mass occurs), the map T would be called an *optimal transport map* from μ to ν.

Remark 1. It can be easily checked that if $(id \times T)_{\#}\mu$ belongs to $\Pi(\mu, \nu)$ then T pushes μ onto ν (i.e. $\nu(A) = \mu(T^{-1}(A))$ for any Borel set A) and the functional takes the form $\int c(x, T(x))\mu(dx)$, thus generalizing Monge's problem.

This generalized problem by Kantorovich is much easier to handle than the original one proposed by Monge, for instance, in the Monge case we would need existence of at least a map T satisfying the constraints. This is not verified when $\mu = \delta_0$, if ν is not a single Dirac mass. On the contrary, there always exists a transport plan in $\Pi(\mu, \nu)$ (for instance, $\mu \otimes \nu \in \Pi(\mu, \nu)$). Moreover, one can state that (K) is the relaxation of the original problem by Monge: if one considers the problem in the same setting, where the competitors are transport plans, but sets the functional at $+\infty$ on all the plans that are not of the form $(id \times T)_{\#}\mu$, then one has a functional on $\Pi(\mu, \nu)$ whose relaxation is the functional in (K) (see [3]).

Anyway, it is important to notice that an easy use of the direct method of calculus of variations (i.e. taking a minimizing sequence, saying that it is compact in some topology – here it is the weak convergence of probability measures – finding a limit, and proving semicontinuity (or continuity) of the functional we minimize, so that the limit is a minimizer) proves that a minimum does exist.

As a consequence, if one is interested in the problem of Monge, the question may become "Does this minimum come from a transport map T?" Actually, if the answer to this question is yes, then it is evident that the problem of Monge has a solution, which also solves a wider problem, that of minimizing among transport plans. In some cases, proving that the optimal transport plan comes from a transport map (or proving that there exists at least one optimal plan coming from a map) is equivalent to proving that the problem of Monge has a solution, since very often the infimum among transport plans and among transport maps is the same. Yet, in the presence of atoms, this is not always the case, but we will not insist any more on this degenerate case.

1.2.2 Duality

Since the problem (K) is a linear optimization under linear constraints, an important tool will be duality theory, which is typically used for convex problems. We will find a dual problem (D) for (K) and exploit the relations between dual and primal.

The first thing we will do is find a formal dual problem, by means of an inf–sup exchange.

First, express the constraint $\gamma \in \Pi(\mu, \nu)$ in the following way: notice that if γ is a non-negative measure on $\Omega \times \Omega$, then we have

$$\sup_{\phi, \psi} \int \phi \, d\mu + \int \psi \, d\nu - \int (\phi(x) + \psi(y)) \, d\gamma = \begin{cases} 0 & \text{if } \gamma \in \Pi(\mu, \nu) \\ +\infty & \text{otherwise} \end{cases}.$$

Hence, one can remove the constraints on γ if one adds the previous sup, since if they are satisfied nothing has been added and if they are not one gets $+\infty$, and this will be avoided by the minimization. Hence, we may look at the problem we get and interchange the inf in γ and the sup in ϕ, ψ:

$$\min_{\gamma} \int c \, d\gamma + \sup_{\phi, \psi} \left(\int \phi \, d\mu + \int \psi \, d\nu - \int (\phi(x) + \psi(y)) \, d\gamma \right)$$

$$= \sup_{\phi, \psi} \int \phi \, d\mu + \int \psi \, d\nu + \inf_{\gamma} \int (c(x, y) - (\phi(x) + \psi(y))) \, d\gamma.$$

Obviously it is not always possible to exchange inf and sup, and the main tool to do it is a theorem by Rockafellar requiring concavity in one variable, convexity in the other one, and some compactness assumption. We will not investigate anymore whether in this case these assumptions are satisfied or not. But the result is true.

Afterwards, one can rewrite the inf in γ as a constraint on ϕ and ψ, since one has

$$\inf_{\gamma \geq 0} \int (c(x, y) - (\phi(x) + \psi(y))) \, d\gamma$$

$$= \begin{cases} 0 & \text{if } \phi(x) + \psi(y) \leq c(x, y) \text{ for all } (x, y) \in \Omega \times \Omega \\ -\infty & \text{otherwise} \end{cases}.$$

This leads to the following dual optimization problem:

Problem 2. Given the two probabilities μ and ν on Ω and the cost function $c : \Omega \times \Omega \to [0, +\infty]$, we consider the problem

$$(D) \quad \max \left\{ \int_\Omega \phi \, d\mu + \int_\Omega \psi \, d\nu \;\middle|\; \phi \in L^1(\mu), \, \psi \in L^1(\nu) \, : \, \phi(x) + \psi(y) \right.$$
$$\left. \leq c(x, y) \text{ for all } (x, y) \in \Omega \times \Omega \right\}. \quad (1.3)$$

This problem does not admit a straightforward existence result, since the class of admissible functions lacks compactness. Yet, we can better understand this problem and find existence once we have introduced the notion of c-transform (a kind of generalization of the well-known Legendre transform).

Definition 1. Given a function $\chi : \Omega \to \overline{\mathbb{R}}$ we define its *c-transform* (or *c*-conjugate function) by

$$\chi^c(y) = \inf_{x \in \Omega} c(x, y) - \chi(x).$$

Moreover, we say that a function ψ is *c-concave* if there exists χ such that $\psi = \chi^c$ and we denote by $\Psi_c(\Omega)$ the set of *c*-concave functions.

It is quite easy to realize that, given a pair (ϕ, ψ) in the maximization problem (D), one can always replace it with (ϕ, ϕ^c), and then with (ϕ^{cc}, ϕ^c), and the constraints are preserved and the integrals increased. Actually, one could go on, but it is possible to prove that $\phi^{ccc} = \phi^c$ for any function ϕ. This is the same as saying that $\psi^{cc} = \psi$ for any *c*-concave function ψ, and this perfectly recalls what happens for the Legendre transform of convex funtions (which corresponds to the particular case $c(x, y) = x \cdot y$).

A consequence of these considerations is the following well-known result:

Proposition 1.1. *We have*

$$\min(K) = \max_{\psi \in \Psi_c(\Omega)} \int_\Omega \psi \, d\mu + \int_\Omega \psi^c \, d\nu, \quad (1.4)$$

where the max on the right-hand side is realized. In particular, the minimum value of (K) is a convex function of (μ, ν), as it is a supremum of linear functionals.

Definition 2. The functions ψ realizing the maximum in (1.4) are called *Kantorovich potentials* for the transport from μ to ν. This is in fact a small abuse, because usually this term is used only in the case $c(x, y) = |x - y|$, but it is usually understood in the general case as well.

Notice that any *c*-concave function shares the same modulus of continuity of the cost c. This is the reason why one can prove existence for (D) (which is the same of the right-hand side problem in Proposition 1.1), by applying Ascoli–Arzelà's theorem.

In, particular, in the case $c(x, y) = |x - y|^p$, if Ω is bounded with diameter D, any $\psi \in \Psi_c(\Omega)$ is pD^{p-1}-Lipschitz continuous. Notice that the case where c is a power of the distance is actually of particular interest, and two values of the exponent p are remarkable: the cases $p = 1$ and $p = 2$. In these two cases we provide characterizations for the set of c-concave functions. Let us denote by $\Psi_{(p)}(\Omega)$ the set of c-concave functions with respect to the cost $c(x, y) = |x - y|^p/p$. It is not difficult to check that

$$\psi \in \Psi_{(1)}(\Omega) \Longleftrightarrow \psi \text{ is a 1-Lipschitz function;}$$

$$\psi \in \Psi_{(2)}(\Omega) \Longrightarrow x \mapsto \frac{x^2}{2} - \psi(x) \text{ is a convex function;}$$

$$\text{if } \Omega = \mathbb{R}^d \text{ this is an equivalence.}$$

1.2.3 The case $c(x, y) = |x - y|$

The case $c(x, y) = |x - y|$ shows a lot of interesting features, even if from the point of the existence of an optimal map T it is one of the most difficult. A first interesting property is the following:

Proposition 1.2. *For any 1-Lipschitz function ψ we have $\psi^c = -\psi$. In particular, (1.4) may be rewritten as*

$$\min(K) = \max(D) = \max_{\psi \in \text{Lip}_1} \int_\Omega \psi \, d(\mu - \nu).$$

The key point of Proposition 1.2 is proving $\psi^c = -\psi$. This is easy if one considers that $\psi^c(y) = \inf_x |x - y| - \psi(x) \le -\psi(x)$ (taking $x = y$), but also $\psi^c(y) = \inf_x |x - y| - \psi(x) \ge \inf_x |x - y| - |x - y| + \psi(y) = \psi(y)$ (making use of the Lipschitz behavior of ψ).

Another peculiar feature of this case is the following:

Proposition 1.3. *Consider the problem*

$$(B) \quad \min\{M(\lambda) \mid \lambda \in \mathcal{M}^d(\Omega); \ \nabla \cdot \lambda = \mu - \nu\}, \tag{1.5}$$

where $M(\lambda)$ denotes the mass of the vector measure λ and the divergence condition is to be read in the weak sense, with Neumann boundary conditions, i.e. $-\int \nabla \phi \cdot d\lambda = \int \phi \, d(\mu - \nu)$ for any $\phi \in C^1(\overline{\Omega})$. If Ω is convex then it holds

$$\min(K) = \min(B).$$

This proposition links the Monge–Kantorovich problem to a minimal flow problem which was first proposed by Beckmann [5], under the name of *continuous transportation model*. He did not know this link, as Kantorovich's theory was being developed independently almost in the same years. In Section 2.1 we will see some more details on this model and on the possibility of generalizing it to the case of distances $c(x, y)$ coming from Riemannian metrics. In particular, in the case of a nonconvex Ω, (B) would be equivalent to a Monge–Kantorovich problem where c is the geodesic distance on Ω.

To have an idea of why these equivalences between (B) and (K) hold true, one can look at the following considerations.

First, a formal computation. We take the problem (B) and rewrite the constraint on λ by means of the quantity

$$\sup_{\phi} \int -\nabla\phi \cdot d\lambda + \int \phi \, d(\mu - \nu) = \begin{cases} 0 & \text{if } \nabla \cdot \lambda = \mu - \nu \\ +\infty & \text{otherwise} \end{cases}.$$

Hence, one can write (B) as

$$\min_{\lambda} M(\lambda) + \sup_{\phi} \int -\nabla\phi \cdot d\lambda + \int \phi \, d(\mu - \nu)$$

$$= \sup_{\phi} \int \phi \, d(\mu - \nu) + \inf_{\lambda} M(\lambda) - \int \nabla\phi \cdot d\lambda,$$

where inf and sup have been exchanged formally as in the previous computations. After that, one notices that

$$\inf_{\lambda} M(\lambda) - \int \nabla\phi \cdot d\lambda = \inf_{\lambda} \int d|\lambda| \left(1 - \nabla\phi \cdot \frac{d\lambda}{d|\lambda|}\right) = \begin{cases} 0 & \text{if } |\nabla\phi| \le 1 \\ -\infty & \text{otherwise,} \end{cases}$$

and this leads to the dual formulation for (B), which gives

$$\sup_{\phi : |\nabla\phi| \le 1} \int_{\Omega} \phi \, d(\mu - \nu).$$

Since this problem is exactly the same as (D) (a consequence of the fact that Lip_1 functions are exactly those functions whose gradient is smaller than 1), this gives the equivalence between (B) and (K).

Most of the considerations above, especially those on the problem (B), do not hold for costs other than the distance $|x - y|$. The only possible generalizations I know concern either a cost c which comes from a Riemannian distance $k(x)$ (i.e. $c(x, y) = \inf\{\int_0^1 k(\sigma(t))|\sigma'(t)|dt : \sigma(0) = x, \sigma(1) = y\}$, which gives a problem (B) with $\int k(x)d|\lambda|$ instead of $M(\lambda)$) or the fact that p-homogeneous costs may become 1-homogeneous through the introduction of time as an extra variable (see [11]). Some more details on the problem (B) can be found in the lectures notes in Chapter 2.

1.2.4 $c(x, y) = h(x - y)$ with h strictly convex and the existence of an optimal T

We summarize here some useful results for the case where the cost c is of the form $c(x, y) = h(x - y)$, for a strictly convex function h.

The main tool is the duality result. If we have equality between the minimum of (K) and the maximum of (D) and both extremal values are realized, one can consider an optimal transport plan γ and a Kantorovich potential ψ and write

$$\psi(x) + \psi^c(y) \leq c(x, y) \text{ on } \Omega \times \Omega \text{ and } \psi(x) + \psi^c(y) = c(x, y) \text{ on } \operatorname{spt} \gamma.$$

The equality on $\operatorname{spt} \gamma$ is a consequence of the inequality which is valid everywhere and of

$$\int c \, d\gamma = \int \psi \, d\mu + \int \psi^c \, dv = \int (\psi(x) + \psi^c(y)) \, d\gamma,$$

which implies equality γ-a.e. These functions being continuous, the equality passes to the support of the measure.

Once we have that, let us fix a point $(x_0, y_0) \in \operatorname{spt} \gamma$. One may deduce from the previous computations that

$$x \mapsto \psi(x) - h(x - y_0) \quad \text{is minimal at } x = x_0$$

and, if ψ is differentiable at x_0, one gets $\nabla \psi(x_0) \in \partial h(x_0 - y_0)$. For a strictly convex function h one may inverse the relation passing to ∇h^*, thus getting

$$x_0 - y_0 = \nabla h^*(x_0) = (\partial h)^{-1}(x_0).$$

This solves several questions concerning the transport problem with this cost, provided ψ is differentiable a.e. with respect to μ. This is usually guaranteed by requiring μ to be absolutely continuous with respect to the Lebesgue measure, and using the fact that ψ may be proven to be Lipschitz. Then, one may use the previous computation to deduce that, for every x_0, the point y_0 such that $(x_0, y_0) \in \operatorname{spt} \gamma$ is unique (i.e. γ is of the form $(id \times T)_{\#}\mu$, where $T(x_0) = y_0$). Moreover, this also gives uniqueness of the optimal transport plan and of the gradient of the Kantorovich potential.

We may summarize everything in the following theorem:

Theorem 1.4. *Given μ and v probability measures on a domain $\Omega \subset \mathbb{R}^d$ there exists an optimal transport plan π. It is unique and of the form $(id \times T)_{\#}\mu$, provided μ is absolutely continuous. Moreover, there exists also at least a*

Kantorovich potential ψ, and the gradient $\nabla\psi$ is uniquely determined μ-a.e. (in particular, ψ is unique up to additive constants, provided the density of μ is positive a.e. on Ω). The optimal transport map T and the potential ψ are linked by $T(x) = x - (\nabla h^*)(\nabla\psi(x))$. Moreover, we have $\psi(x) + \psi^c(T(x)) = c(x, T(x))$ for μ-a.e. x. Conversely, every map T which is of the form $T(x) = x - (\nabla h^*)(\nabla\psi(x))$ for a function $\psi \in \Psi_c(\Omega)$ is an optimal transport plan from μ to $T_{\#}\mu$.

Remark 2. Actually, the existence of an optimal transport map is true under weaker assumptions: we can replace the condition of being absolutely continuous with the condition "$\mu(A) = 0$ for any $A \subset \mathbb{R}^d$ such that $\mathcal{H}^{d-1}(A) < +\infty$" or with any condition which ensures that the non-differentiability set of ψ is negligible. In the theorem we used the Lipschitz behavior of $\psi \in \Psi_c$ and applied the Rademacher theorem, but c-concave functions are often more regular than only Lipschitz.

Remark 3. In Theorem 1.4 only the part concerning the optimal map T is not symmetric in μ and ν; hence, the uniqueness of the Kantorovich potential is true even if ν (and not μ) has positive density a.e. (since one can retrieve ψ from ψ^c and vice versa).

Remark 4. Theorem 1.4 may be particularized to the quadratic case $c(x, y) = |x - y|^2/2$, thus getting the existence of an optimal transport map

$$T(x) = x - \nabla\psi(x) = \nabla\left(\frac{x^2}{2} - \psi(x)\right) = \nabla\phi(x)$$

for a convex function ϕ. By using the converse implication (sufficient optimality conditions), this also proves the existence and the uniqueness of a gradient of a convex function transporting μ onto ν. This well-known fact was investigated first by Brenier [6] and is often known as Brenier's theorem.

Let us, moreover, notice that a specific approach for the case $|x - y|^2$, based on the fact that we can withdraw the parts of the cost depending on x or y only and maximize $\int x \cdot y \, d\gamma$, gives the same result in an easier way: we actually get $\phi(x_0) + \phi^*(y_0) = x_0 \cdot y_0$ for a convex function ϕ and its Legendre transform ϕ^* and we deduce $y_0 \in \partial\phi(x_0)$.

All the costs of the form $c(x, y) = |x - y|^p$ with $p > 1$ fall under Theorem 1.4.

We finish the part dedicated to positive results by noticing that the same method may not be used if h is only convex, or at least does not give results as strong as what it does if h is strictly convex. Yet, there is anyway something which is known for the case $c(x, y) = |x - y|$. The results are a bit weaker (and

much harder) and are summarized below (this is the classical Monge case and
we refer to [3], even if several different proofs have been provided by different
methods). Notice that a lot of literature is currently being dedicated to the case
of other norms than the Euclidean one and other distance functions.

Theorem 1.5. *Given μ and v probability measures on a domain $\Omega \subset \mathbb{R}^d$
there exists at least an optimal transport plan π for the cost $c(x, y) = |x - y|$.
Moreover, one such plan is of the form $(id \times T)_{\#}\mu$ provided μ is absolutely
continuous. There exists a Kantorovich potential ψ, and its gradient is unique
μ-a.e. and we have $\psi(x) - \psi(T(x)) = |x - T(x)|$ for μ-a.e. x, for any choice
of optimal T and ψ.*

Here, the absolute continuity assumption is essential to have existence of
an optimal transport map, in the sense that in general it cannot be replaced by
weaker assumptions as in the strictly convex case.

Moreover, we can provide a counter-example showing that, in general, it is
necessary that μ does not give mass to "small" sets.

Example 1. Set

$$\mu = \mathcal{H}^1 \llcorner A \text{ and } v = \frac{\mathcal{H}^1 \llcorner B + \mathcal{H}^1 \llcorner C}{2}$$

where A, B and C are three vertical parallel segments in $\mathbb{R}^2$ whose vertexes
lie on the two lines $y = 0$ and $y = 1$ and the abscissas are 0, 1 and -1,
respectively, and $\mathcal{H}^1$ is the 1-dimensional Haudorff measure. It is clear that no
transport plan may realize a cost better than 1 since, horizontally, every point
needs to be displaced by a distance 1. Moreover, one can get a sequence of maps
$T_n : A \to B \cup C$ by dividing A into $2n$ equal segments $(A_i)_{i=1,...,2n}$ and B and
C into n segments each, $(B_i)_{i=1,...,n}$ and $(C_i)_{i=1,...,n}$ (all ordered downwards).
Then define T_n as a piecewise affine map which sends A_{2i-1} onto B_i and A_{2i}
onto C_i. In this way the cost of the map T_n is less than $1 + 1/n$, which implies
that the infimum of the Kantorovich problem is 1, as well as the infimum on
transport maps only. Yet, no map T may obtain a cost 1, as this would imply
that all points are sent horizontally, but this cannot respect the push-forward
constraint. On the other hand, the transport plan associated with T_n weakly
converges to the transport plan $\frac{1}{2}T_{\#}^+\mu + \frac{1}{2}T_{\#}^-\mu$, where $T^{\pm}(x) = x \pm e$ and
$e = (1, 0)$. This transport plan turns out to be the only optimal transport plan,
and its cost is 1.

Notice that the same construction also provides an example of the relaxation
procedure leading from Monge to Kantorovich.

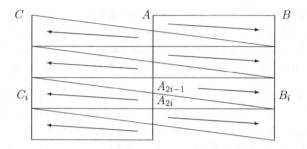

1.3 Wasserstein distances and spaces

Starting from the values of the problem (K) in (1.2) we can define a set of
distances over $\mathcal{P}(\Omega)$. For any $p \geq 1$ we can define

$$W_p(\mu, \nu) = \left(\min(K) \text{ with } c(x, y) = |x - y|^p\right)^{1/p}.$$

We recall that, by the duality formula, we have

$$\frac{1}{p} W_p^p(\mu, \nu) = \sup_{\psi \in \Psi_{(p)}(\Omega)} \int_\Omega \psi \, d\nu + \int_\Omega \psi^c \, d\mu. \tag{1.6}$$

Theorem 1.6. *If Ω is compact, for any $p \geq 1$ the function W_p is in fact
a distance over $\mathcal{P}(\Omega)$ and the convergence with respect to this distance is
equivalent to the weak convergence of probability measures. In particular, any
functional $\mu \mapsto W_p(\mu, \nu)$ is continuous with respect to weak topology.*

To prove that the convergence according to W_p is equivalent to weak convergence one first establishes this result for $p = 1$, through the use of the duality
with the functions in Lip_1. Then it is possible to use the inequalities between
the distances W_p (see below) to extend the result to a general p.

The case of a noncompact Ω is a little more difficult. First, the distance
must be defined only on a subset of the whole space of probability measures,
to avoid infinite values. We will use the space of probabilities with finite p-th
momentum:

$$\mathcal{W}_p(\Omega) = \{\mu \in \mathcal{P}(\Omega) : m_p(\mu) := \int_\Omega |x|^p \mu(dx) < +\infty\}.$$

Theorem 1.7. *For any $p \geq 1$ the function W_p is a distance over $\mathcal{W}_p(\Omega)$
and, given a measure μ and a sequence $(\mu_n)_n$ in $\mathcal{W}_p(\Omega)$, the following are
equivalent:*

- *$\mu_n \to \mu$ according to W_p;*
- *$\mu_n \rightharpoonup \mu$ and $m_p(\mu_n) \to m_p(\mu)$;*

- $\int_\Omega \phi \, d\mu_n \to \int_\Omega \phi \, d\mu$ *for any* $\phi \in C^0(\Omega)$ *whose growth is at most of order* p *(i.e. there exist constants A and B depending on ϕ such that $|\phi(x)| \leq A + B|x|^p$ for any x).*

Notice that, as a consequence of Hölder (or Jensen) inequalities, the Wasserstein distances are always ordered, i.e. $W_{p_1} \leq W_{p_2}$ if $p_1 \leq p_2$. Reversed inequalities are possible only if Ω is bounded, and in this case we have, if set $D = \mathrm{diam}(\Omega)$, for $p_1 \leq p_2$,

$$W_{p_1} \leq W_{p_2} \leq D^{1-p_1/p_2} W_{p_1}^{p_1/p_2}.$$

From the monotone behavior of Wasserstein distances with respect to p it is natural to introduce the distance W_∞: set $\mathcal{W}_\infty(\Omega) = \{\mu \in \mathcal{P}(\Omega) : \mathrm{spt}(\mu)$ is bounded$\}$ (obviously if Ω itself is bounded one has $\mathcal{W}_\infty(\Omega) = \mathcal{P}(\Omega)$) and then

$$W_\infty(\mu, \nu) = \inf \left\{ \gamma\text{-esssup}_{x,y \in \Omega \times \Omega} |x - y| \ : \ \gamma \in \Pi(\mu, \nu) \right\}.$$

Here, γ-esssup denotes the essential sup with respect to γ, i.e. the norm in the space $L^\infty(\Omega \times \Omega; \gamma)$, which is the same, for continuous functions such as $|x - y|$, as the maximal value on the support of γ. It is easy to check that $W_p \nearrow W_\infty$, and it is interesting to study the metric space $\mathcal{W}_\infty(\Omega)$. Curiously enough, this supremal problem in optimal transport theory, even if quite natural, has not deserved much attention, up to the very recent paper [10].

The W_∞ convergence is stronger than any W_p convergence, and hence also than the weak convergence of probability measures. The converse is not true, and the convergence in W_∞ turns out to be actually rare; consequently, there is a great lack of compactness in $\mathcal{W}_\infty$. For instance, it is not difficult to check that, if we set $\mu_t = t\delta_{x_0} + (1 - t)\delta_{x_1}$, where $x_0 \neq x_1 \in \Omega$, we have $W_\infty(\mu_t, \mu_s) = |x_0 - x_1|$ if $t \neq s$. This implies that the balls $B(\mu_t, |x_0 - x_1|/2)$ are infinitely many disjoint balls in $\mathcal{W}_\infty$ and prevent compactness.

The following statement summarizes the compactness properties of the spaces W_p for $1 \leq p \leq \infty$ and its proof is a direct application of the considerations above and of Theorem 1.7.

Proposition 1.8. *For $1 \leq p < \infty$ the space $W_p(\Omega)$ is compact if and only if Ω itself is compact. Moreover, for an unbounded Ω the space $\mathcal{W}_p(\Omega)$ is not even locally compact. The space $W_\infty(\Omega)$ is neither compact nor locally compact for any choice of Ω with $\#\Omega > 1$.*

1.4 Geodesics, continuity equation, and displacement convexity

1.4.1 Metric derivatives in Wasserstein spaces

We are concerned in this section with several properties linked to the curves in the Wasserstein space W_p. For this subject the main reference is [2]. Before giving the main result we are interested in, we recall the definition of metric derivative, which is a concept that may be useful when studying curves which are valued in generic metric spaces.

Definition 3. Given a metric space (X, d) and a curve $\gamma : [0, 1] \to X$ we define the *metric derivative* of the curve γ at time t as the quantity

$$|\gamma'|(t) = \lim_{s \to t} \frac{d(\gamma(s), \gamma(t))}{|s - t|}, \tag{1.7}$$

provided the limit exists.

As a consequence of Rademacher theorem, it can be seen (see [4]) that for any Lipschitz curve the metric derivative exists at almost every point $t \in [0, 1]$. We will be concerned quite often with metric derivatives of curves which are valued in the space $\mathcal{W}_p(\Omega)$.

Definition 4. If we are given a Lipschitz curve $\mu : [0, 1] \to \mathcal{W}_p(\Omega)$, we define the velocity field of the curve of any vector field $v : [0, 1] \times \Omega \to \mathbb{R}^d$ such that for a.e. $t \in [0, 1]$ the vector field $v_t = v(t, \cdot)$ belongs to $[L^p(\mu_t)]^d$ and the *continuity equation*

$$\frac{d}{dt}\mu_t + \nabla \cdot (v \cdot \mu_t) = 0$$

is satisfied in the sense of distributions; this means that for all $\phi \in C_c^1(\Omega)$ and any $t_1 < t_2 \in [0, 1]$ it holds that

$$\int_\Omega \phi \, d\mu_{t_2} - \int_\Omega \phi \, d\mu_{t_1} = \int_{t_1}^{t_2} ds \int_\Omega \nabla \phi \cdot v_s \, d\mu_s,$$

or, equivalently, in differential form:

$$\frac{d}{dt} \int_\Omega \phi \, d\mu_t = \int_\Omega \nabla \phi \cdot v_t \, d\mu_t \qquad \text{for a.e. } t \in [0, 1].$$

We say that v is the *tangent* field to the curve μ_t if, for a.e. t, v_t has minimal $[L^p(\mu_t)]^d$ norm for any t among all the velocity fields (actually, this is not the true definition of a tangent vector field, since this would involve the definition of a tangent space for the "manifold" $\mathcal{W}_p$, but it is in this case the same).

The following theorem is concerned with the existence of tangent fields and comes from Theorem 8.3.1 and Proposition 8.4.5 in [2]:

Theorem 1.9. *If $p > 1$ and $\mu = (\mu_t)_t$ is a curve in* $\text{Lip}([0, 1]; W_p(\Omega))$ *then there exists a unique vector field v characterized by*

$$\frac{\partial}{\partial t}\mu + \nabla \cdot (v \cdot \mu) = 0, \tag{1.8}$$

$$||v_t||_{L^p(\mu_t)} \leq |\mu'|(t) \text{ for a.e. } t, \tag{1.9}$$

where the continuity equation is satisfied in the sense of distributions as previously explained. Moreover, if (1.8) holds for a family of vector fields $(v_t)_t$ with $||v_t||_{L^p(\mu_t)} \leq C$, then $\mu \in \text{Lip}([0, 1]; W_p(\Omega))$ and $|\mu'|(t) \leq ||v_t||_{L^p(\mu_t)}$ for a.e. t.

To have an idea of the meaning of this theorem and of the relationship between curves of measures and the continuity equation, some considerations could be useful.

Actually, at least when the vector fields v_t are regular enough, the solution of the continuity equation $\partial \mu / \partial t + \nabla \cdot (v \cdot \mu) = 0$ is obtained by taking the images of the initial measure μ_0 through the maps $\sigma(t, \cdot)$ obtained by taking the solution of

$$\begin{cases} \sigma'(t, x) = v_t(\sigma(t, x)), \\ \sigma(0, x) = x. \end{cases}$$

This explains why the vector field v_t is called the "velocity field" of the curve μ_t: if every particle follows at each time t the velocity field v_t, then the position of all the particles at time t reconstructs exactly the measure μ_t that appears in the continuity equation together with v_t!

Think for a while to the case of two time steps only: there are two measures μ_t and μ_{t+h} and there are several ways for moving the particles so as to reconstruct the latter from the former. It is exactly as when we look for a transport. One of these transports is optimal in the sense that it minimizes $\int |T(x) - x|^p \mu_t(dx)$ and the value of this integral equals $W_p^p(\mu_t, \mu_{t+h})$. If we call $v_t(x)$ the "discrete" velocity of the particle located at x at time t, i.e. $v_t(x) = (T(x) - x)/h$, one has $||v_t||_{L^p(\mu_t)} = \frac{1}{h}W_p(\mu_t, \mu_{t+h})$. The result of Theorem 1.9 may be easily guessed as obtainable as a limit as $h \to 0$.

1.4.2 Geodesics and geodesic convexity

Once we know about curves in their generality, it is interesting to think about geodesics. The following result is a characterization of geodesics in $W_p(\Omega)$

when Ω is a convex domain in $\mathbb{R}^d$. This procedure is also known as *McCann's linear interpolation*.

Theorem 1.10. *All the spaces $\mathcal{W}_p(\Omega)$ are length spaces, and if μ and v belong to $\mathcal{W}_p(\Omega)$, and γ is an optimal transport plan from μ to v for the cost $c_p(x, y) = |x - y|^p$, then the curve*

$$\mu^\gamma(s) = (p_s)_\# \gamma,$$

where $p_s : \Omega \times \Omega \to \Omega$ is given by $p_s(x, y) = x + s(y - x)$, is a constant-speed geodesic from μ to v. In the case $p > 1$, all the constant-speed geodesics are of this form, and if μ is absolutely continuous, then there is only one geodesic and it has the form

$$\mu(s) = [(1 - s)id + sT]_\# \mu,$$

where T is the optimal transport map from μ to v.

By means of this characterization of geodesics we can also define the useful concept of displacement convexity introduced by McCann [13].

Definition 5. Given a functional $F : \mathcal{W}_p(\Omega) \cap L^1 \to [0, +\infty]$, we say that it is *displacement convex* if all the maps $t \mapsto F(\mu^\gamma(t))$ are convex on $[0, 1]$ for every choice of μ and v in $\mathcal{W}_p(\Omega)$ and γ optimal transport plan from μ to v with respect to $c(x, y) = |x - y|^p$.

The following well-known result provides a wide set of displacement convex functionals. In the case $p = 2$ this result is due to McCann [13], while the generalization to any p can be found in [2].

Theorem 1.11. *Consider the following functionals on the space $\mathcal{W}_p(\Omega)$, where Ω is any convex subset of $\mathbb{R}^N$:*

$$J^1(\mu) = \begin{cases} \int_\Omega f(u(x))\,dx & \text{if } \mu = u \cdot \mathcal{L}^d \\ +\infty & \text{if } \mu \text{ is not absolutely continuous;} \end{cases}$$

$$J^2(\mu) = \int_\Omega V(x)\,\mu(dx);$$

$$J^3(\mu) = \int_\Omega \int_\Omega w(x - y)\mu(dx)\mu(dy).$$

Suppose that $f : [0, +\infty] \to [0, +\infty]$ is a convex and superlinear lower semicontinuous function with $f(0) = 0$, and that $V : \Omega \to [0, +\infty]$ and $w : \mathbb{R}^d \to [0, +\infty]$ are convex functions. Then the functionals J^2 and J^3 are displacement convex in $\mathcal{W}_p(\Omega)$ and the functional J^1 is displacement convex *provided the map*

$$r \mapsto r^d f(r^{-d})$$

is convex and nonincreasing on $]0, +\infty[$.

1.5 Monge–Ampère equation and regularity

The final issue that we will approach in these lecture notes will be concerned with some regularity properties of T and ψ (the optimal transport map and the Kantorovich potential, respectively) and their relations with the densities of μ and ν. We will consider only the quadratic case $c(x, y) = |x - y|^2/2$, because it is the one where more results have been proven. Very recent results for generic costs have been developed by Ma, Trudinger, Wang, Loeper, Figalli, etc. They require some very rigid assumptions on the costs, so that, surprinsingly enough, the quadratic cost is one of the few powers that satisfies the suitable hypotheses.

It is easy – just by a change-of-variables formula – to transform the equality $\nu = T_{\#}\mu$ into the partial differential equation $v(T(x)) = u(x)/|JT|(x)$, where u and v are the densities of μ and ν (which have to be supposed regular enough) and J denotes the determinant of the Jacobian matrix. Recalling that we may write $T = \nabla\phi$ with ϕ convex (Remark 4), we get the Monge–Ampère equation

$$M\phi = \frac{u}{v(\nabla\phi)}, \tag{1.10}$$

where M denotes the determinant of the Hessian

$$M\phi = \det H\phi = \det \left[\frac{\partial^2 \phi}{\partial x_i \, \partial x_j} \right]_{i,j}.$$

This equation up to now is satisfied by $\phi = \frac{x^2}{2} - \psi$ in a formal way only. We define various notions of solutions for (1.10):

- we say that ϕ satisfies (1.10) in the Brenier sense if $(\nabla\phi)_{\#}u \cdot \mathcal{L}^d = v \cdot \mathcal{L}^d$ (and this is actually the sense to be given to this equation);
- we say that ϕ satisfies (1.10) in the Alexandroff sense if $H\phi$, which is always a positive measure for ϕ convex, is absolutely continuous and its density satisfies (1.10) a.e.;
- we say that ϕ satisfies (1.10) in the viscosity sense if it satisfies the usual comparison properties required by viscosity theory but restricting the comparisons to regular convex test functions (since M is in fact monotone just when restricted to positively definite matrices);
- we say that ϕ satisfies (1.10) in the classical sense if it is of class C^2 and the equation holds pointwise.

Notice that any notion except the first may also be applied to the more general equation $M\phi = f$, while the first one just applies to this specific transportation case. The results we want to use are well summarized in Theorem 50 of [15]:

Theorem 1.12. *If u and v are $C^{0,\alpha}(\Omega)$ and are both bounded from above and from below on the whole Ω by positive constants and Ω is a convex open set, then the unique Brenier solution ϕ of (1.10) belongs to $C^{2,\alpha}(\Omega) \cap C^{1,\alpha}(\overline{\Omega})$ and ϕ satisfies the equation in the classical sense (hence also in the Alexandroff and viscosity senses).*

Even if this precise statement is taken from [15], we just detail a possible bibliographical path to arrive at this result. It is not easy to deal with Brenier solutions, so the idea is to consider viscosity solutions, for which it is in general easy to prove existence by Perron's method. Then prove some regularity result on viscosity solutions, up to getting a classical solution. After that, once we have a classical convex solution to the Monge–Ampère equation, this will be a Brenier solution too. Since this is unique (up to additive constants), we have got a regularity statement for Brenier solutions. We can find results on viscosity solutions in [7], [9], and [8]. In [7], some conditions to ensure strict convexity of the solution of $M\phi = f$ when f is bounded from above and below are given. In [9], for the same equation, $C^{1,\alpha}$ regularity is proved provided we have strict convexity. In this way the term $u/v(\nabla\phi)$ becomes a $C^{0,\alpha}$ function, and in [8] $C^{2,\alpha}$ regularity is proved for solutions of $M\phi = f$ with $f \in C^{0,\alpha}$.

References

[1] L. Ambrosio, Lecture notes on optimal transport problems, *Mathematical Aspects of Evolving Interfaces,* Springer Verlag, Berlin, Lecture Notes in Mathematics (1812), 1–52, 2003.

[2] L. Ambrosio, N. Gigli and G. Savaré, *Gradient Flows in Metric Spaces and in the Spaces of Probability Measures.* Lectures in Mathematics, ETH Zurich, Birkhäuser, 2005.

[3] L. Ambrosio and A. Pratelli. Existence and stability results in the L^1 theory of optimal transportation, in *Optimal Transportation and Applications,* Lecture Notes in Mathematics (CIME Series, Martina Franca, 2001) 1813, L.A. Caffarelli and S. Salsa (eds), 123–160, 2003.

[4] L. Ambrosio and P. Tilli, *Topics on Analysis in Metric Spaces.* Oxford Lecture Series in Mathematics and its Applications (25). Oxford University Press, Oxford, 2004.

[5] M. Beckmann, A continuous model of transportation, *Econometrica* (20), 643–660, 1952.

[6] Y. Brenier, Dücomposition polaire et rüarrangement monotone des champs de vecteurs. *C. R. Acad. Sci. Paris Sér. I Math.* (305), no. 19, 805–808, 1987.

[7] L. Caffarelli, A localization property of viscosity solutions to the Monge–Ampère equation and their strict convexity. *Ann. Math.* (131), no. 1, 129–134, 1990.

[8] L. Caffarelli, Interior $W^{2,p}$ estimates for solutions of the Monge–Ampère equation. *Ann. Math.* (131), no. 1, 135–150, 1990.

[9] L. Caffarelli, Some regularity properties of solutions of Monge Ampère equation. *Comm. Pure Appl. Math.* (44), no. 8–9, 965–969, 1991.

[10] T. Champion, L. De Pascale, P. Juutinen, The ∞-Wasserstein distance: local solutions and existence of optimal transport maps, *SIAM J. Math. An.* (40), no. 1, 1–20, 2008.

[11] C. Jimenez, Optimisation de problèmes de transport, PhD thesis, Université du Sud-Toulon-Var, 2005.

[12] L. Kantorovich, On the transfer of masses. *Dokl. Acad. Nauk. USSR*, (37), 7–8, 1942.

[13] R. J. McCann, A convexity principle for interacting gases. *Adv. Math.* (128), no. 1, 153–159, 1997.

[14] G. Monge, Mümoire sur la thüorie des düblais et des remblais, *Histoire de l'Académie Royale des Sciences de Paris*, 666–704, 1781.

[15] C. Villani. *Topics in Optimal Transportation*. Graduate Studies in Mathematics, AMS, 2003.

[16] C. Villani, *Optimal Transport: Old and New*, Springer Verlag (Grundlehren der mathematischen Wissenschaften), 2008.

2

Models and applications of optimal transport in economics, traffic, and urban planning

FILIPPO SANTAMBROGIO

Abstract

Some optimization or equilibrium problems involving the concept of optimal transport are presented in these notes, mainly devoted to applications to economics and game theory settings. A variant model of transport, taking into account traffic congestion effects, is the first topic, and it shows various links with Monge–Kantorovich theory and partial differential equations. Then, two models for urban planning are introduced. The last section is devoted to two problems from economics and their translation into the language of optimal transport.

Contents

AMS Subject Classification (2010): 00-02, 90B06, 49J45, 90C25, 91B40, 91B50, 91D10, 35Q91
Keywords: traffic congestion, Wardrop equilibrium, optimal flows, equilibrium problems, Kantorovich potential, contract theory, Nash equilibrium

2.1 Introduction

These lecture notes will present the main issues and ideas of some variational problems that use or touch the theory of optimal transportation. They all come from economic-oriented applications. Problems will be presented through the main ideas, with almost no proofs. As I did during the class in Grenoble and for the other lecture notes (Chapter 1), I will try to keep a very informal level of presentation.

The first topic I will present is a variant of the usual Monge problem, which takes into account congestion effects in the transportation. It has a more dynamical taste since it asks for looking at the trajectories followed by the particles instead of simple pairs $(x, T(x))$. The starting point will be the equivalent formulation by Beckmann, modified in order to take into account nonuniform metrics and then traffic intensity. In this way one obtains a model which is linked to game theory and Nash equilibria and has a well-known discrete counterpart on networks. Moreover, the optimality conditions for the minimization let a classical transport problem appear (for a metric which is not known a priori).

After this second section, two models for the distribution of some fundamental elements of the structure of urban regions (residents, jobs, industrial areas, services, etc.) are discussed. The first is purely variational: we suppose that these distribution optimize a total welfare functional, as if a benevolent planner could control the city; the functional lets a transport cost (say, a Wasserstein distance) appear explicitly. The second deals with much more delicate equilibrium issues on the agents' choices and it does not address explicitly any transport problem in its formulation; yet, the theory of Monge–Kantorovitch is very useful in its resolution. In both cases the Kantorovitch potential play an essential role.

By the end of this fourth section, the reader who is not familiar with economic theories should have got more accustomed to the some typical concepts of the rational behavior of the consumers, and will be ready for Section 2.5. This section presents two classical problems in economics (in situations of competition and of monopoly) and it shows how to translate them into problems involving transport costs and Kantorovitch potentials.

2.2 Traffic congestion

2.2.1 Generalizations of Beckmann's problem

We saw in the introductory lecture notes (Chapter 1) the problem (B):

$$\min\{M(\lambda) : \lambda \in \mathcal{M}^d(\Omega); \nabla \cdot \lambda = \mu - \nu\},$$

where $M(\lambda)$ denotes the mass of the vector measure λ and we said that it is equivalent to the original problem of Monge. Actually, one way to produce a solution to this divergence-constrained problem is the following: take an optimal transport plan γ and build a vector measure v_γ defined through

$$\langle v_\gamma, \phi \rangle := \int_{\Omega \times \Omega} \int_0^1 \omega'_{x,y}(t) \cdot \phi(\omega_{x,y}(t)) dt \, d\gamma,$$

for every $\phi \in C^0(\Omega; \mathbb{R}^d)$, $\omega_{x,y}$ being a parameterization of the segment $[x, y]$.

It is not difficult to check that this measure satisfies the divergence constraint, since if one takes $\phi = \nabla \psi$ then

$$\int_0^1 \omega'_{x,y}(t) \cdot \phi(\omega_{x,y}(t)) = \int_0^1 \frac{d}{dt} \left(\psi(\omega_{x,y}(t)) \right) dt = \psi(y) - \psi(x),$$

and hence $\langle v_\gamma, \nabla \psi \rangle = \int \psi \, d(v - \mu)$.

To estimate its mass we can see that $|v_\gamma| \leq \sigma_\gamma$, where the scalar measure σ_γ is defined through

$$\langle \sigma_\gamma, \phi \rangle := \int_{\Omega \times \Omega} \int_0^1 |\omega'_{x,y}(t)| \phi(\omega_{x,y}(t)) dt \, d\gamma, \quad \forall \phi \in C^0(\Omega; \mathbb{R})$$

and it is called *transport density*. The mass of σ_γ is obviously

$$\int d\sigma_\gamma = \int \int_0^1 |\omega'_{x,y}(t)| dt \, d\gamma = \int |x - y| d\gamma = W_1(\mu, \nu),$$

which proves the optimality of v_γ.

It is interesting to investigate whether $\sigma_\gamma \ll \mathcal{L}^d$, since this would imply that problem (B) is well posed in L^1 instead of the space of vector measure. For the sake of the variants that we will see later on, it would be interesting to give conditions so that $\sigma_\gamma \in L^p$ as well. All these subjects have been widely studied by De Pascale and Pratelli (see [16–18]), but there is a more recent (and shorter) proof of the same estimates in [24]. It is in particular true that $\mu, \nu \in L^p$ implies that $\sigma_\gamma \in L^p$ and that it is sufficient that one of the two measures is absolutely continuous in order to get the same on σ_γ.

The simplest possible generalization of problem (B) is the following:

$$\min \int k(x)|v(x)|dx \ : \ \nabla \cdot v = \mu - v,$$

which corresponds, by duality with the functions u such that $|\nabla u| \leq k$, to

$$\min \int d_k(x, y)d\gamma \ : \ \gamma \in \Pi(\mu, v),$$

where $d_k(x, y) = \inf_{\omega(0)=x, \omega(1)=y} L_k(\omega) := \int_0^1 k(\omega(t))|\omega'(t)|dt$ is the distance associated with the Riemannian metric k. It would be possible to build in this case an optimal v_γ by replacing the curves $\omega_{x,y}$ with the k-geodesics (instead of the segments).

This generalization above comes from the modelization of a nonuniform cost for the movement (due to geographical obstacles or configurations). It can be applied to several situations, but it is anyway evident that one should look for more realistic models, at least in the case of urban transport. In this case the metric k is usually not known a priori, but it depends on the traffic distribution itself.

The simplest model could be considering a metric $k(x) = g(|v(x)|)$ depending through an increasing function g on the traffic itself (represented by the intensity of v). In this case a very naive model would be obtained by setting $H(t) = tg(t)$ and then solving

$$\min \int H(|v(x)|)dx \ : \ \nabla \cdot v = \mu - v.$$

In most cases H is strictly convex, and this is a strictly convex counterpart to the problem by Beckmann (which was suggested by Beckmann himself in [1] and [3]). Notice that this model is not completely realistic either, since it allows for "cancellation" effects: several flows in opposite directions at the same point x may give a total vector $v(x) = 0$, even if the number of travellers at x is high. Yet, in Section 2.2.3 we will see that this simplifed model will turn out to be equivalent to a more precise one.

We just mention that there exist concave variants too, which are known under the name of *branched transport*. This name is used for addressing all the transport problems where the cost for a mass m moving on a distance l is proportional to l but subadditive w.r.t. m. Typically, it is proportional to a power m^α $(0 < \alpha < 1)$. The adjective "branched" in the name stands for one of the main features of the optimal solutions: they gather mass together, masses tend to move jointly as long as possible, and then they branch towards different destinations, thus giving rise to a tree-shaped structure.

This problem comes from a discrete problem on graphs, where the cost of a graph G whose edges e_h are weighted with coefficients w_h is of the form $\sum_h p_h^\alpha \mathcal{H}^1(e_h)$. It has a continuous generalization where the energy to be minimized is

$$M^\alpha(v) = \begin{cases} \int_M \theta^\alpha d\mathcal{H}^1 & \text{if } v = U(M, \theta, \xi) \\ +\infty & \text{otherwise} \end{cases},$$

where $v = U(M, \theta, \xi)$ means that v is a rectifiable measure supported on the set M, with orientation ξ and density (multiplicity) θ. The energy M^α is then minimized under the constraint $\nabla \cdot v = \mu - \nu$.

These notes will not develop this alternative problem any further and the reader may find the whole theory of branched transport in the recent book by Bernot et al. [4].

2.2.2 Wardrop equilibria, the discrete case

We will describe in this section a traffic problem which has some interesting issues on equilibria and some interesting relations with optimal transport theory. We will start from the discrete case on networks and then generalize to the continuous case. The network case was introduced in [25] and then studied in [2].

In the discrete framework, one considers

- a finite graph with edges $e \in E$ and a set of sources S and destinations D;
- the set $C(s, d) = \{\omega \text{ from } s \text{ to } d\}$ of possible paths from s to d;
- a demand input $\gamma = (\gamma(s, d))_{s,d}$ denoting the quantity of commuters from each $s \in S$ to each $d \in D$, or a set Γ of possible γ (for instance, this could be the set of all demands where the total number of commuters leaving each point s and the total number arriving to each point d are prescribed, but the coupling, i.e. how many commuters for each pair (s, d), is not);
- an unknown repartition strategy (to be looked for) $q = (q_\omega)_\omega$ such that $\sum_{\omega \in C(s,d)} q_\omega = \gamma(s, d)$;
- a consequent traffic intensity on each edge e (depending on q) $i_q = (i_q(e))_e$ given by $i_q(e) = \sum_{e \in \omega} q_\omega$;
- an increasing function $g : \mathbb{R}^+ \to \mathbb{R}^+$ such that $g(i_q(e))$ represents the congestioned cost of e;
- the cost for each path ω, given by $c(\omega) = \sum_{e \in \omega} g(i_q(e))$.

The global strategy q represents the overall distribution on choices of commuters' paths. Imposing a *Nash equilibrium* condition (no single commuter wants to change their choice, provided all the others keep the same strategy) gives the following condition:

$$\omega \in C(s, d), \, q_\omega > 0 \implies c(\omega) = \min\{c(\tilde\omega) \, : \, \tilde\omega \in C(s, d)\}.$$

This condition is well known among geographical economists as the *Wardrop equilibrium*.

The existence of at least an equilibrium comes from the following variational principle.

Optimizing an overall congestion cost means minimizing a quantity $J(q) := \sum_e H(i_q(e))$ (where $H : \mathbb{R}^+ \to \mathbb{R}^+$ is an increasing function; for instance, if one takes $H(t) = tg(t)$, the value of $J(q)$ gives the total cost for all commuters) among all possible strategies q.

The minimization of J obviously has a solution and one can look for optimality conditions. Suppose that H and Γ are convex, so that the necessary conditions will also be sufficient: it is easy to see that q minimizes if and only if, for every other admissible $\tilde q$, one has

$$\sum_e H'(i_q(e))(i_{\tilde q}(e) - i_q(e)) \geq 0.$$

Set $\xi(e) := H'(i_q(e))$ and rewrite the right-hand side as

$$\sum_e \xi(e)(i_{\tilde q}(e) - i_q(e)) = \sum_e \sum_{\omega \ni e} \xi(e)(\tilde q(\omega) - q(\omega))$$

$$= \sum_\omega \left(\sum_{e \in \omega} \xi(e)\right)(\tilde q(\omega) - q(\omega)).$$

This says that, if one sets $L_\xi(\omega) := \sum_{e \in \omega} \xi(e)$, the optimal q must minimize $\sum_\omega L_\xi(\omega)q(\omega)$, since we got $\sum_\omega L_\xi(\omega)\tilde q(\omega) \geq \sum_\omega L_\xi(\omega)q(\omega)$.

This means two facts. First, since the conditions of admissibility on q only look at starting and arrival points, it is pointless to put some mass on those curves ω from a source s to a destination d such that the value $L_\xi(\omega)$ is strictly larger than $d_\xi(s, d) := \min_{\omega \in C(s,d)} L_\xi(\omega)$. This means that $q(\omega) > 0$ and $\omega \in C(s, d)$ imply $L_\xi(\omega) = d_\xi(s, d)$.

Second, another condition occurs when the demand γ is not fixed. We said that to optimize $\sum_\omega L_\xi(\omega)q(\omega)$ we only use curves where $L_\xi = d_\xi$, and this gives

$$\sum_\omega L_\xi(\omega)q(\omega) = \sum_{s,d} d_\xi(s, d)\left(\sum_{\omega \in C(s,d)} q_\omega\right) = \sum_{s,d} d_\xi(s, d)\gamma(s, d).$$

In particular, one also needs to choose $\gamma \in \Gamma$ so as to minimize

$$\sum_{s,d} d_\xi(s,d)\gamma(s,d), \quad \gamma \in \Gamma.$$

This second condition is empty if Γ only contains one γ, but it is of particular interest when $\Gamma = \Pi(\mu, \nu)$, since it says that γ must solve a Kantorovitch problem for the cost d_ξ.

The first condition, on the other hand, always gives some information on q and exactly says: if q is optimal, then it is a Wardrop equilibrium for $g = H'$.

2.2.3 Wardrop equilibria, the continuous case and equivalences

It is possible to give a continuous formulation and prove analogous results (see [13]). In a domain $\Omega \subset \mathbb{R}^n$ the demands are represented by probabilities $\gamma \in \mathcal{P}(\Omega \times \Omega)$. We are given a set $\Gamma \subset \mathcal{P}(\Omega \times \Omega)$ as the set of admissible demand couplings: usually $\Gamma = \{\bar\gamma\}$ or

$$\Gamma = \Pi(\mu, \nu) = \{\gamma \in \mathcal{P}(\Omega \times \Omega) : (\pi_X)_\sharp \gamma = \mu, \ (\pi_Y)_\sharp \gamma = \nu\}.$$

Let us also set

$$C = \{\text{Lipschitz paths } \omega : [0,1] \to \Omega\}$$
$$C(s,d) = \{\omega \in C : \omega(0) = s, \, \omega(1) = d\}.$$

We look for a probability $Q \in \mathcal{P}(C)$ such that $(\pi_{0,1})_\sharp Q \in \Gamma$.

We want to define a traffic intensity $i_Q \in \mathcal{M}^+(\Omega)$ such that the quantity $i_Q(A)$ stands for "how much" the movement takes place in A ... For $\phi \in C^0(\Omega)$ and $\omega \in C$, set $L_\phi(\omega) = \int_0^1 \phi(\omega(t))|\omega'(t)|dt$.

Then we define i_Q by

$$\langle i_Q, \phi \rangle = \int_C L_\phi(\omega)Q(d\omega) = \int_C \left(\int_0^1 \varphi(\omega(t))|\omega'(t)|dt \right) Q(d\omega).$$

Notice that this is exactly what happens for the transport density! The traffic intensity i_Q is a generalization of the transport density, since it deals with the case where Q is any measure on C, while the transport density only looks at the measure concentrated on the segments $[x, y]$ for (x, y) in the support of an optimal γ.

In this continuous framework, it is more convenient to start from the optimization point of view (instead of looking at the equilibrium as a starting point):

we minimize the convex functional

$$J(Q) = \begin{cases} \int H(i_Q(x))dx & \text{if } i_Q \ll \mathcal{L}^n \\ +\infty & \text{otherwise} \end{cases}$$

among all admissible strategies Q, H being a convex, increasing and superlinear function. Typically, $H(t) = t^p$, or $H(t) = t + t^p$ (which is more reasonable, since in general we have $g(0) = H'(0)$ and we do not want $g(0) = 0$; this would mean that moving on an empty road costs nothing, which is usually not the case).

First, one should prove finiteness of the minimum, which is not evident since in the continuous case one needs to prove the existence of a Q such that $i_Q \in L^p$. This is, in the case of $\mu, \nu \in L^p$, a consequence of the summability results on the transport density, since the transport density is, as we said, a particular choice for i_Q. This is why we explicitly cited the L^p result fo De Pascale and Pratelli (besides its interest in itself).

It is possible to look for optimality conditions and to reobtain the same Wardrop equilibrium + optimization of d_ξ. Here, ξ will be the metric $\xi(x) = H'(i_Q(x))$. Yet, this function is not continuous nor l.s.c., and some effort should be spent to give a meaning to the concept of geodesic distance in the case $\xi \in L^q$.

It is also interesting to notice that this problem looks at the movement of some players whose individual goal is fixed but whose utility also looks at the density of all the other players (i.e. their movement is more expensive if they pass where the density is higher); this seems to be a particular case of the so-called *mean field games* introduced by Lasry and Lions [19].

All the results we cited are valid for the case $\Gamma = \{\bar{\gamma}\}$ as well as for $\Gamma = \Pi(\mu, \nu)$ (all the transport plans).

Yet, in this second case, something more may be said. Instead of defining a scalar traffic intensity, one can define a vector measure v_Q by

$$\int_{\overline{\Omega}} \varphi(x) \, dv_Q(x) := \int_{C([0,1];\overline{\Omega})} \left(\int_0^1 \varphi(\omega(t)) \cdot \omega'(t)dt \right) dQ(\omega), \ \forall \varphi \in C(\overline{\Omega}, \mathbb{R}^N),$$

i.e. a sort of vector version of i_Q. It is immediate to check that $|v_Q| \le i_Q$, and that

$$\nabla \cdot v_Q = \mu - \nu, \quad v_Q \cdot \nu = 0 \text{ on } \partial\Omega.$$

Since H is increasing, this implies that the infimum of the previous problem with i_Q is larger than that of the minimal flow problem:

$$\inf\left\{\int_\Omega \mathcal{H}(v)\,dx \;:\; \nabla \cdot v = \mu - \nu, \; v \cdot v = 0 \text{ on } \partial\Omega\right\}, \qquad (2.1)$$

where $\mathcal{H}(v) := H(|v|)$.

A natural question, arising for instance from a comparison with the Monge case, where looking for the vector or the scalar transport density was the same, is the possible equivalence of the two problems.

One can see that a minimizer of the scalar problem can be built formally from a minimizer of the vector one in the following way: if v is the unique solution of the vector problem (2.1) and μ and ν are absolutely continuous (so that we will write μ and ν for their densities as well), we consider the non-autonomous Cauchy problem

$$\begin{cases} \omega'(s) = w(s, \omega(s)) \\ \omega(0) = \quad x \end{cases} \qquad (2.2)$$

for the non-autonomous vector field

$$w(t, x) = \frac{v(x)}{(1-t)f_0(x) + t f_1(x)}, \qquad (t, x) \in [0, 1] \times \Omega. \qquad (2.3)$$

The latter will not have any Lipschitz continuity property in general, unless the optimizer v of (2.1) is regular; anyway, if we assume that one can prove $v \in \mathrm{Lip}(\Omega)$, then the flow $X : [0, 1] \times \Omega \to \Omega$ of v is well defined as the solution of (2.2) and we can take μ_t as the image of μ through the map $X(t, \cdot)$. One can see that μ_t must coincide with the linear interpolating curve $(1-t)\mu + t\nu$ (because this curve solves the continuity equation thanks to the divergence condition). This yields that $(X(1, \cdot))_\sharp f_0 = f_1$, which ensures that $X(1, \cdot)$ transports μ on ν. If we now consider the probability measure concentrated on the flow, i.e.

$$Q = \delta_{X(\cdot, x)} \otimes \mu,$$

then Q is admissible and it is not difficult to see that $i_Q = |\sigma|$, since

$$\int \phi \, di_Q = \int \int_0^1 \phi(\omega_x(t))|\omega_x'(t)|\,dt\,d\mu = \int_0^1 dt \int \phi(\omega_x(t)) \frac{|v(\omega_x(t))|}{\mu_t(\omega_x(t))}\,d\mu$$

$$= \int_0^1 dt \int \phi \frac{|v|}{\mu_t}\,d\mu_t = \int \phi|v|.$$

This finally implies that the minima of the two problems coincide. Moreover, this construction provides a transport map (that is $X(1, \cdot)$) from μ to ν, whose transport "rays" evidently do not cross and which is monotone on transport "rays" (as a consequence of the Cauchy–Lipschitz theorem).

Notice that if one wanted to prove rigorously what we stated, one should investigate a little bit the regularity of the optimal v. This may be done if one writes optimality conditions for v and sees that one has $v = \nabla \mathcal{H}^*(\nabla u)$ where u solves

$$\begin{cases} \nabla \cdot \nabla \mathcal{H}^*(\nabla u) = \mu - v, & \text{in } \Omega \\ \nabla \mathcal{H}^*(\nabla u) \cdot \hat{n} = 0, & \text{on } \partial\Omega \end{cases} \tag{2.4}$$

For $H(t) = t^2$ this is a simple Laplace equation, and regularity theory is well known. For $H(t) = t^p$ this gives a p'-Laplace equation, and here lots of studies have been done as well. Yet, for modeling reasons, we said that it is important to look at the case $H'(0) > 0$, and we suggested as a typical case

$$\mathcal{H}(\sigma) = \frac{1}{p}|v|^p + a|v|, \ v \in \mathbb{R}^N, \tag{2.5}$$

which leads to a function $\mathcal{H}^*$ which vanishes on $\overline{B_1}$. In particular, the corresponding equation for u is very very degenerate and regularity results are less studied (see [7], both for the equivalence with the Wardrop problem and for some regularity proofs).

2.3 The urban planning of residents and services

A very simplified model that has been proposed for studying the distribution of residents and services in a given urban region Ω passes through the minimization of a total quantity $\mathcal{F}(\mu, v)$ concerning two unknown densities μ and v.

The two measures μ and v will be searched among probabilities on Ω. This means that the total amounts of population and production are fixed as problem data. The definition of the total cost functional to optimize takes into account some criteria we want the two densities μ and v to satisfy:

(i) there is a transportation cost T for moving from the residential areas to the services areas;
(ii) people do not want to live in areas where the density of population is too high;
(iii) services need to be concentrated as much as possible in order to increase efficiency and decrease management costs.

Fact (i) is described, in its easiest version, through a p-Wasserstein distance ($p \geq 1$). We will look at $T(\mu, v) = W_p^p(\mu, v)$.

Fact (ii) will be described by a penalization functional, a kind of total unhappiness of citizens due to a high density of population, obtained by integrating with respect to the citizens' density their personal unhappiness.

Fact (iii) is modeled by a third term representing costs for managing services once they are located according to the distribution v, taking into account that efficiency depends strongly on how much v is concentrated.

The cost functional to be considered is then

$$\mathcal{F}(\mu, v) = T(\mu, v) + F(\mu) + G(v), \tag{2.6}$$

where $F, G : \mathcal{P}(\Omega) \to [0, +\infty]$ are functionals chosen so that the first one favors spread measures and the second one concentrated measures, in suitable senses.

We stress that this model is a very naive one, since it disregards equilibrium issues and several other parameters, and that it could be applied only in those cases where a planner could control the whole behavior of the region. We refer to [6, 9, 10, 22, 23] for the study of this model and of similar ones.

As far as particular choices for the functionals F and G are concerned, we may consider

$$F(\mu) = \begin{cases} \int_\Omega f(u) \, d\mathcal{L}^d & \text{if } \mu = u \cdot \mathcal{L}^d \\ +\infty & \text{otherwise} \end{cases},$$

$$G(v) = \begin{cases} \sum_{k \in \mathcal{N}} g(a_k) & \text{if } v = \sum_{k \in \mathcal{N}} a_k \delta_{x_k} \\ +\infty & \text{otherwise} \end{cases},$$

where the integrand $f : [0, +\infty] \to [0, +\infty]$ is assumed to be lower semicontinuous and convex, with $f(0) = 0$ and superlinear at infinity; that is,

$$\lim_{t \to +\infty} \frac{f(t)}{t} = +\infty,$$

and the function g is required to be subadditive, lower semicontinuous, and such that

$$g(0) = 0 \quad \text{and} \quad \lim_{t \to 0} \frac{g(t)}{t} = +\infty.$$

In this form we have two *local* lower semicontinuous functional on measures (see [5]: a functional on measures is said to be local if it is additive on mutually singular measures). This is a useful class of functionals over measures including both concentration-preferring functionals and functionals favoring spread measures.

Without loss of generality, by subtracting constants to the functional F, we can suppose $f'(0) = 0$. Owing to the assumption $f(0) = 0$, the ratio $f(t)/t$

is an incremental ratio of the convex function f, and thus it is increasing in t. Then, if we write the functional F as

$$\int_\Omega \frac{f(u(x))}{u(x)} u(x)\,dx,$$

we can see the quantity $f(u)/u$, which is increasing in u, as the unhappiness of a single citizen when they live in a place where the population density is u. Integrating it with respect to $\mu = u \cdot \mathcal{L}^n$ gives a quantity to be seen as the total unhappiness of the population.

Concerning G, we can think that we are requiring v to be concentrated on a limited number of service poles and that the effects of the managing costs and of the production of a pole whose size is a are summarized in a cost function $g(a)$.

For G, there are other interesting choices among functionals which favor concentration. One of them could be

$$G(v) = \int_\Omega \int_\Omega h(|x - y|)\, v(dx) v(dy),$$

where h is an increasing function and $h(|x - y|)$ stands for the cost of managing the interactions between services located at x and at y. This new choice for G is more concerned with the positions of the services, and not only with the size of each pole.

These two choices and other possible models give different interesting results when one looks at the minimizers. In the first case several atoms occur in v, and μ is concentrated on balls around these centers, which corresponds to sub-cities; in the other, a single-center city is obtained. The mathematical properties which are obtainable thanks to what we know from the theory of optimal transport are remarkable.

As a simple example, we will mention that the solutions of

$$(P_v) \quad \min_\mu W_p^p(\mu, v) + F(\mu); \qquad \text{for fixed } v \in \mathcal{P}(\Omega)$$

are characterized by

$$\mu = u \cdot \mathcal{L}^n; \quad u = (f')^{-1}\left(const - \psi_{\mu,v}\right)_+,$$

where $\psi_{\mu,v}$ is a Kantorovitch potential for the transport from μ to v and the cost $c(x, y) = |x - y|^p$.

Moreover, in the whole minimization with respect to μ and v, in the particular case $T(\mu, v) = W_2^2(\mu, v)$ and $G(v) = \lambda \int_{\Omega \times \Omega} |x - y|^2 v(dx) v(dy)$, $F(\mu) = ||\mu||^2_{L^2(\Omega)}$, any pair of minimizers (μ, v) is shaped as follows:

- μ is concentrated on a ball $B(x_0, r_\lambda)$ (intersected with Ω) and has a density u given by

$$u(x) = \frac{\lambda}{2\lambda + 1}(r_\lambda^2 - |x - x_0|^2);$$

- v is concentrated on the ball $B(x_0, r_\lambda/(2\lambda + 1))$ and it is the image of μ under the homothety of ratio $(2\lambda + 1)^{-1}$ and center x_0;
- x_0 is the barycenter of both μ and v.

The main tool for all these results is the following computation: if $\mu_\varepsilon = (1 - \varepsilon)\mu + \varepsilon\mu_1$, then

$$\lim_{\varepsilon \to 0} \frac{W_p^p(\mu_\varepsilon, v) - W_p^p(\mu, v)}{\varepsilon} = \int \psi_{\mu,v} \, d(\mu_1 - \mu),$$

where $\psi_{\mu,v}$ is, again, a Kantorovitch potential for the transport from μ to v and the cost $|x - y|^p$. This formula says that the Kantorovitch potentials stand for Gateaux derivatives of the functional $W_p^p(\cdot, v)$. Thanks to standard convex analysis, it is not difficult to guess it, and to apply it to variational problems, if one thinks that the duality formula $W_p^p(\mu, v) = \sup \int \psi \, d\mu + \psi^c \, dv$ exactly says that this functional is convex and the optimal ψ are the elements of its subdifferential.

Notice that this kind of technique for finding optimality conditions of problems such as (P_v) is useful in other contexts as well, and in particular, for $p = 2$, when *gradient flows* for functional on the Wasserstein space $\mathcal{W}_2$ are concerned. Actually, the standard *minimizing movement* procedure for the gradient flow of a functional F passes through discrete minimization steps for a quantity like

$$\mu \mapsto \frac{W_2^2(\mu, v)}{2\tau} + F(\mu),$$

where τ is a time step and a discrete sequence $(\mu_k)_k$ is built taking for μ_{k+1} the solution of (P_{μ_k}).

We finish the section by saying that other models with different costs, for instance when T is no more a Wasserstein distance but comes from a congested or branched transport problem (see Section 2.4), have been investigated as well (see [15]).

2.4 Equilibrium structure of a city

This second part is devoted to a much more detailed model on the structure of a city which looks at an equilibrium configuration for the behavior of residents, firms, and landowners. This has a much more economical taste and it has been studied by G. Carlier and I. Ekeland [11, 12].

The elements in this description of the city are the following:

- A domain $\Omega \subset \mathbb{R}^d$, which stands for the urban region we consider.
- A measure $\mu = N(x)dx$ on Ω standing for the residents; this is unknown as well as its mass.
- A measure $\nu = n(x)dx$ standing for jobs, which is unknown too.
- A transportation cost $c(x, y)$ for commuting inside Ω; this is given.
- A wage function $\psi : \Omega \to \mathbb{R}$, where $\psi(x)$ stands for the salary that workers employed by the firm located at x receive from the firm (this is an unknown of the problem).
- A revenue function $\phi : \Omega \to \mathbb{R}$ (unknown) standing for the revenues net of commuting cost that residents earn; people living at x will solve $\max_y \psi(y) - c(x, y) := \phi(x)$ so as to choose where to work according to this optimization problem and, conversely, firms located at y will decide whom to hire solving $\min_x \phi(x) + c(x, y)$ and getting again $\psi(y)$ as a (minimal) wage to be assured at y so that there are workers who do accept to work at y.
- A same utility function for all the residents $U(C, S)$ depending on their consumption level C and on the quantity S of land they use, as well as a fixed utility level $\bar{u}$ that every agent wants to realize. These are given as exogenous (fixed), and $\bar{u}$ may be thought of as the utility level outside Ω; i.e., the utility realized if one decides to move out of the city.
- A price for residential rent $Q(x)$; at every point x the residents want to choose a consumption C and a land surface S so that they obtain at least the utility $\bar{u}$; i.e., if $Q(x)$ is known, they solve $\min\{C + Q(x)S : U(C, S) \geq \bar{u}\}$ and they get the minimal amount of money they need. At the equilibrium this amount will necessarily be $\phi(x)$ (i.e., the money they actually have). This gives a relation between Q and ϕ and finds the optimal value $S(x)$ as well. One obviously has $N(x) = 1/S(x)$.
- A productivity function $z : \Omega \to \mathbb{R}$ which is supposed to depend increasingly on ν (say, $z(x) = \nu(B(x, r))$ or z is obtained through a more general convolution of ν; the idea is that the productivity is higher where there is a higher concentration of workers).
- A production $f(z, n)$ which gives the output of a firm employing n workers in a zone where the productivity is z.
- A price for industrial rent $q(x)$ which is obtained, at the equilibrium, by imposing that all the surplus of the firm may be absorbed by the landlord, so that $q(x) = \max_n f(z(x), n) - \psi(x)n$. This also gives the optimal value $n(x)$.

An equilibrium is given by a pair of measures (μ, ν), some continuous functions z, ϕ, ψ, and a transport plan $\gamma \in \Pi(\mu, \nu)$ such that

- μ and ν have the same mass;
- γ and the pair (ϕ, ψ) are compatible in the sense that γ is concentrated on the set $\{(x, y) : \phi(x) = \psi(y) - c(x, y)\}$ and the inequality $\phi(x) \geq \psi(y) - c(x, y)$ holds for every (x, y);
- z is obtained from ν through the productivity relation that we mentioned above;
- once Q and q are computed (depending on ϕ, ψ, and z), one finds the optimal N and n to be equal to the densities of μ and ν, respectively;
- μ is concentrated on $Q \geq q$ and ν on $q \geq Q$ (this depends on the landlords' behavior: they would not rent to residents if renting to firms is more profitable, nor vice versa).

The application of the optimal transport theory is straightforward and it allows one to pose the problem as a fixed-point issue on μ and ν: once μ and ν are given, one only needs to take for γ the optimal transport plan for the cost c and for ϕ and ψ the Kantorovitch potentials. This is what Carlier and Ekeland did, proving well-posedness results in a framework which was much more general than what was studied before in the literature (mainly one-dimensional or radially symmetric cases).

2.5 Application to economics: Kantorovitch potential as prices or utilities

2.5.1 Hotelling

The Hotelling problem is a double-step equilibrium problem for the strategic location of N firms trying to maximize their incomes from a given distribution μ of consumers in a domain Ω, according to the following criterion. Notice that the domain may be interpreted in a geographical way, or represent the different features of the goods the firms sell.

If we know the positions x_i of the firms and the prices p_i that they choose, the consumer located at x will choose where to buy their good by minimizing the sum $c(x, x_i) + p_i$ over $i = 1, \ldots, N$ (the cost $c(x, y)$ representing for instance the distance from x to y or taking into account the utility that x has when they buy a product of type y). In this way some influence regions

$$A_i = \{x : x_i \text{ minimizes } c(x, x_i) + p_i\}$$

and some demands $d_i = \mu(A_i)$ are obtained. Every firm wants to maximize the profit $p_i d_i$ and a *Nash equilibrium* configuration for prices is a choice of the N prices so that no firm wants to change its mind (i.e., changing its price p_i, supposing that all the others do not change their own prices). Supposing that, for every configuration of the positions of the firms, there is a unique equilibrium, every firm knows the function associating the profits to positions. An equilibrium configuration is hence a configuration where no firm wants to move in order to enhance its profit, provided the others do not move (once again, a Nash equilibrium). The Hotelling problem looks exactly at finding such an equilibrium (see [20]).

An easy but interesting link with optimal transport is the following and concerns the first step (i.e., price equilibria). The idea is that instead of taking the prices p_i, look at the demands d_i. It will be possible to reconstruct the p_i from the d_i; in order to do that, just consider the measure $\nu = \sum_{i=1}^{N} d_i \delta_{x_i}$ and prove that the function $p : \{x_1, \ldots, x_N\} \to \mathbb{R}$ is a Kantorovitch potential for the transport from ν to μ for the cost c. Once this is done, the problem may be translated into a condition on ν which involves its Kantorovitch potential. Notice that writing down the precise conditions on ν involves understanding how the Kantorovitch potential depends on ν, which is a very delicate issue that we will meet again.

2.5.2 Rochet–Choné

There are different models on the prices that a monopolist firm may impose for the goods it produces. One of the mathematically most interesting is the Rochet–Choné model (see [21]), which is an optimization problem under convexity constraint. The convex structure comes from the simplifying assumption that the space of goods y and the space of consumers x are subsets of $\mathbb{R}^N$ and they are coupled through the function $(x, y) \mapsto x \cdot y$ representing the utility that a consumer of type x has in buying y. Once the distribution μ of consumers is known, the firm may choose the price for its good; i.e., a function $p : Y \to [0, \infty[$, defined on the goods space Y. Then every consumer x choses what to buy by solving

$$\max_{y} \; x \cdot y - p(y)$$

and getting a utility $u(x) := \max_y x \cdot y - p(y)$, realized by a good y_x. The firm may reconstruct its total gain by integrating $p(y(x)) - C(y(x))$ (if $C(y)$ is the cost for producing y). The total profit is hence given by

$$\int_X (p(y(x)) - C(y(x))) \, \mu(dx) = \int_X (x \cdot y(x) - u(x) - C(y(x))) \, \mu(dx).$$

One can also notice that $y_x = \nabla u(x)$ (differentiating the expression of u), and hence the maximization of the profit is a problem that may be stated in terms of u:

$$\max F(u) = \int_X (x \cdot \nabla u(x) - u(x) - C(\nabla u(x)))\,\mu(dx),$$

where the constraints on u are convexity (from its defintion) and positivity ($u \geq 0$ is a consequence of the fact that consumers do not buy if they get a negative utility; it may be stated saying that a certain "empty" good called 0 belongs to Y and that we impose $p(0) = 0$; the firm is not allowed to charge for buying this empty good, but this good interests nobody) and a constraint on the gradient: $\nabla u \in Y$. This is the minimization problem under convexity constraint we referred to. It falls into the framework of the convexity-constrained problems studied for instance by Carlier and Lachand-Robert (see [14]), where some C^1 regularity results are also proven. The same class of problems also includes the well-known Newton problem of minimal resistance. For both the problems, some numerical insights in particular cases exist, but lots of information is lacking.

An interesting change of variable, using the image measure $\nu = (\nabla u)_{\sharp}\rho$, is possible, since every measure is the image of ρ through the gradient of a convex function (which is exactly the well-known result by Brenier in transport theory; see [8]). It is interesting to link this reformulation to optimal transport. The most natural cost to be considered would be the scalar product, but we know that considering $-x \cdot y$ or $\frac{1}{2}|x - y|^2$ is the same. Hence, we may rewrite the previous problem as

$$\min_{u \text{ convex}} \tilde{F}(u) = \int_X \left(\frac{|x|^2}{2} - x \cdot \nabla u(x) + \frac{|\nabla u|^2}{2} + u(x) + \tilde{C}(\nabla u(x)) \right) \mu(dx),$$

where $\tilde{C}(z) = C(z) - |z|^2/2$ and we are allowed to add the term in $|x|^2/2$ since it does not depend on u.

We can in the end rewrite the problem in terms of ν as

$$\min_{\nu \in \mathcal{P}(Y)} G(\nu) = \frac{1}{2} W_2^2(\mu, \nu) + \int_Y \tilde{C}\,d\nu + \int_X u_\nu\,d\mu,$$

u_ν being for a measure ν the unique convex function satisfying $\nabla u_{\sharp}\mu = \nu$ and $\min u = 0$ (which is obtained as a Kantorovitch potential for the cost $-x \cdot y$ or $\frac{1}{2}|x - y|^2$).

This kind of functional may be considered via the transport theory. The existence of a minimizer is easy, and the interesting point is finding optimality conditions. The difficult part is handling the term

$$v \mapsto \int_X u_v \, d\mu.$$

For getting optimality conditions, it would be useful to differentiate this term with respect to variations of v. Yet, computing

$$\lim_{\varepsilon \to 0} \frac{u_{v_\varepsilon} - u_v}{\varepsilon}, \qquad v_\varepsilon = v + \varepsilon(\tilde{v} - v),$$

is a challenging issue; possible strategies include the linearization of the Monge–Ampère equation, but lots of questions are open.

References

[1] M. Beckmann, A continuous model of transportation, *Econometrica* (20), 643–660, 1952.

[2] M. Beckmann, C. McGuire, and C. Winsten, C., *Studies in Economics of Transportation*, Yale University Press, New Haven, 1956.

[3] M. Beckmann and T. Puu, *Spatial Economics: Density, Potential and Flow*, North-Holland, Amsterdam, 1985.

[4] M. Bernot, V. Caselles, and J.-M. Morel, *Optimal Transportation Networks, Models and Theory*, Lecture Notes in Mathematics, Springer, Vol. 1955, 2008.

[5] G. Bouchitté and G. Buttazzo, New lower semicontinuity results for nonconvex functionals defined on measures, *Nonlinear Anal.* (15), 679–692, 1990.

[6] G. Buttazzo, Three optimization problems in mass transportation theory. *Nonsmooth Mechanics and Analysis*, pp. 13–23, Adv. Mech. Math., 12, Springer, New York, 2006.

[7] L. Brasco, G. Carlier, and F. Santambrogio, Congested traffic dynamics, weak flows and very degenerate elliptic equations, *J. Math. Pures Appl.* (93), no. 6, 652–671, 2010.

[8] Y. Brenier, Polar factorization and monotone rearrangement of vector-valued functions. *Comm. Pure Appl. Math.* (44), no. 4, 375–417, 1991.

[9] G. Buttazzo and F. Santambrogio, A model for the optimal planning of an urban area. *SIAM J. Math. Anal.* (37), no. 2, 514–530, 2005.

[10] G. Buttazzo and F. Santambrogio, A mass transportation model for the optimal planning of an urban region. *SIAM Rev.* (51), no. 3, 593–610, 2009.

[11] G. Carlier and I. Ekeland, The structure of cities, *J. Global Optim.* (29), 371–376, 2004.

[12] G. Carlier and I. Ekeland, Equilibrium structure of a bidimensional asymmetric city. *Nonlinear Anal. Real World Appl.* (8), no. 3, 725–748, 2007.

[13] G. Carlier, C. Jimenez, and F. Santambrogio, Optimal transportation with traffic congestion and Wardrop equilibria, *SIAM J. Control Optim.* (47), 1330–1350, 2008.

[14] G. Carlier and T. Lachand-Robert, Régularité des solutions d'un problème variationnel sous contrainte de convexité. *C. R. Acad. Sci. Paris Sér. I Math.* (332), no. 1, 79–83, 2001.

[15] G. Carlier and F. Santambrogio, A variational model for urban planning with traffic congestion, *ESAIM Control Optim. Calc. Var.* (11), 595–613, 2007.

[16] L. De Pascale and A. Pratelli, Regularity properties for Monge transport Density and for Solutions of some Shape Optimization Problem, *Calc. Var. Par. Diff. Eq.* (14), no. 3, 249–274, 2002.

[17] L. De Pascale, L. C. Evans and A. Pratelli, Integral estimates for transport densities, *Bull. London Math. Soc.* (36), no. 3, 383–385, 2004.

[18] L. De Pascale and A. Pratelli, Sharp summability for Monge transport density via interpolation, *ESAIM Control Optim. Calc. Var.* (10), no. 4, 549–552, 2004.

[19] J.-M. Lasry and P.-L. Lions, Mean-field games, *Japan. J. Math.* (2), 229–260, 2007.

[20] H. Hotelling, Stability in competition, *Econ. J.* (39), 41–57, 1929.

[21] J.-C. Rochet and P. Choné, Ironing, sweeping and multidimensional screening, *Econometrica* (66), no. 4, 783–826, 1998.

[22] F. Santambrogio, Transport and concentration problems with interaction effects, *J. Global Optim.* (38), no. 1, 129–141, 2007.

[23] F. Santambrogio, *Variational problems in transport theory with mass concentration*, PhD thesis, Edizioni della Normale, Birkhäuser, 2007.

[24] F. Santambrogio, Absolute continuity and summability of transport densities: simpler proofs and new estimates, *Calc. Var. Par. Diff. Eq.* (36), no. 3, 343–354, 2009.

[25] J.G. Wardrop, Some theoretical aspects of road traffic research, *Proc. Inst. Civ. Eng.*, (2), no. 2, 325–378, 1952.

3

Logarithmic Sobolev inequality for diffusion semigroups

IVAN GENTIL

Abstract

Through the main example of the Ornstein–Uhlenbeck semigroup, the Bakry–Emery criterion is presented as a main tool to get functional inequalities as Poincaré or logarithmic Sobolev inequalities. Moreover, an alternative method using the optimal mass transportation is also given to obtain the logarithmic Sobolev inequality.

3.1 Introduction

The goal of this course (given in 2009 in Grenoble) is to introduce inequalities as Poincaré or logarithmic Sobolev for diffusion semigroups. We will focus more on examples than on the general theory. A main tool to obtain those inequalities is the so-called Bakry–Emery Γ_2-criterion. This criterion is well known to prove such inequalities and has also been used many times for other problems; see, for instance, [BÉ85, Bak06]. We will focus on the example of the Ornstein–Uhlenbeck semigroup and on the Γ_2-criterion.

In Section 3.2 we investigate the main example of the Ornstein–Uhlenbeck semigroup, whereas in Section 3.3 we show how the Γ_2-criterion implies such inequalities. In Section 3.4 we will explain an alternative method to get a logarithmic Sobolev inequality under curvature assumption. It is called the *mass transportation method* and has been introduced recently; see [CE02, OV00, CENV04, Vil09]. In this way we will also obtain another inequality called the *Talagrand inequality* or $\mathcal{T}_2$ *inequality*.

Mathematics Subject Classification (2000): Primary 35B40, 35K10, 60J60
Keywords: logarithmic Sobolev inequality, Poincaré inequality, Ornstein–Uhlenbeck semigroup, Bakry–Emery criterion

3.2 The Ornstein–Uhlenbeck semigroup and the Gaussian measure

In the general setting, if $(X_t)_{t \geq 0}$ is a Markov process on $\mathbb{R}^n$, then the family of operators

$$\mathbf{P_t}(f)(x) = E(f(X_t)),$$

where $X_0 = x$ and a smooth function f, is defined as a Markov semigroup on $\mathbb{R}^n$. There are two main examples. The first one is the heat semigroup, which is associated with the Brownian motion on $\mathbb{R}^n$. In this course we will study the second one which is the Ornstein–Uhlenbeck semigroup. We will see that the Ornstein–Uhlenbeck semigroup is associated with a linear stochastic differential equation driven by a Brownian motion.

In this note, a smooth function f in $\mathbb{R}^n$ is a function such that all computations done as integration by parts are justified; for example, $\mathcal{C}_c^\infty(\mathbb{R}^n)$.

3.2.1 Definition and general properties

Definition 1. Let us define the family of operators $(\mathbf{P_t})_{t \geq 0}$, for any $f \in \mathcal{C}_b(\mathbb{R}^n)$, then

$$\mathbf{P_t} f(x) = \int f(e^{-t}x + \sqrt{1 - e^{-2t}}\, y) d\gamma(y), \tag{3.1}$$

where

$$d\gamma(y) = \frac{e^{-|y|^2/2}}{(2\pi n)^{n/2}} dy$$

is the standard Gaussian distribution in $\mathbb{R}^n$ and $|\cdot|$ is the Euclidean norm on $\mathbb{R}^n$. The family of operator $(\mathbf{P_t})_{t \geq 0}$ is called the Ornstein–Uhlenbeck semigroup.

Remark 1. Let $(X_t)_{t \geq 0}$ be a Markov process solution of the stochastic differential equation

$$\begin{cases} dX_t = \sqrt{2}\,dB_t - X_t dt \\ X_0 = 0. \end{cases} \tag{3.2}$$

Since the stochastic differential equation is linear, there is an explicit solution

$$X_t = e^{-t}X_0 + \int_0^t \sqrt{2}e^{s-t}dB_s,$$

and Equation (3.1) is known as the Mehler formula. Moreover, Itô's formula gives that for all continuous and bounded functions f on $\mathbb{R}^n$

$$\mathbf{P_t} f(x) = E_x(f(X_t)).$$

Proposition 3.1. *The Ornstein–Uhlenbeck semigroup is a linear operator satisfying the following properties:*

(i) $P_0 = Id$

(ii) *For all functions* $f \in C_b(\mathbb{R}^n)$, *the map* $t \mapsto P_t f$ *is continuous from* $\mathbb{R}^+$ *to* $L^2(d\gamma)$.

(iii) *For all* $s, t \geqslant 0$ *one has* $P_t \circ P_s = P_{s+t}$.

(iv) $P_t 1 = 1$ *and* $P_t f \geqslant 0$ *if* $f \geqslant 0$.

(v) $\|P_t f\|_\infty \leqslant \|f\|_\infty$.

We say that the Ornstein–Uhlenbeck semigroup is a Markov semigroup on $(C_b(\mathbb{R}^n), \|\cdot\|_\infty)$.

Proof. We will give only some indications of the proof. First, it is easy to prove items (i), (ii), (iv) and (v).

For item (iii), you just have to compute the Ornstein–Uhlenbeck as follows: $P_t f(x) = E(f(e^{-t}x + \sqrt{1 - e^{-2t}}Y))$ where Y is a random variable with a Gaussian distribution. Then compute $P_t(P_s f)$ to obtain $P_{t+s} f$. In fact, since the solution of stochastic differential equation (3.2) is a Markov process, then (iii) is a natural property of the Ornstein–Uhlenbeck semigroup. $\square$

Proposition 3.2. *For all smooth functions* f *one has*

$$\forall x \in \mathbb{R}^n, \ \forall t \geqslant 0, \ \frac{\partial}{\partial t}P_t f(x) = L(P_t f)(x) = P_t(Lf)(x),$$

where for all smooth functions f, $Lf = \Delta f - x \cdot \nabla f$.

The linear operator L *is known as the infinitesimal generator of the Ornstein–Uhlenbeck semigroup.*

Proof. If f is a smooth function, then

$$\frac{\partial}{\partial t}P_t f(x) = \int \left(-e^{-t}x + \frac{e^{-2t}}{\sqrt{1 - e^{-2t}}}y \right) \cdot \nabla f\left(e^{-t}x + \sqrt{1 - e^{-2t}}y \right) d\gamma(y).$$

By definition of the Ornstein–Uhlenbeck semigroup one gets

$$-xe^{-t} \cdot \int \nabla f\left(e^{-t}x + \sqrt{1 - e^{-2t}}y \right) d\gamma(y) = -x \cdot \nabla P_t f(x),$$

whereas the second term, after an integration by parts, gives

$$\frac{e^{-2t}}{\sqrt{1 - e^{-2t}}} \int y \cdot \nabla f\left(e^{-t}x + \sqrt{1 - e^{-2t}}y \right) d\gamma(y) = \Delta P_t f(x),$$

which finishes the proof.

Using the same computation one can prove the commutation property between P_t and the generator L. $\square$

More generally, if **L** is an infinitesimal generator associated with a linear semigroup $(\mathbf{P_t})_{t \geqslant 0}$ (not necessarily a Markov semigroup) then the commutation $\mathbf{LP_t} = \mathbf{P_t L}$ holds.

Proposition 3.3 (Some properties of the O–U semigroup). *The Ornstein–Uhlenbeck semigroup is γ-ergodic, which means for all $f \in C_b(\mathbb{R}^n)$ that*

$$\forall x \in \mathbb{R}^n, \ \lim_{t \to \infty} \mathbf{P_t} f(x) = \int f d\gamma, \tag{3.3}$$

in $L^2(d\gamma)$.

The probability measure γ is then the unique invariant probability measure, for all smooth functions $f \in C_b(\mathbb{R}^n)$:

$$\int \mathbf{P_t} f d\gamma = \int f d\gamma, \tag{3.4}$$

or equivalently for all smooth functions f,

$$\int \mathbf{L} f d\gamma = 0.$$

In fact, we have the fundamental identity

$$\int g \mathbf{L} f d\gamma = \int f \mathbf{L} g d\gamma = -\int \nabla f \cdot \nabla g d\gamma \tag{3.5}$$

*for all smooth functions f and g on $\mathbb{R}^n$. We say that the Gaussian distribution is reversible with respect to the Ornstein–Uhlenbeck semigroup; **L** is symmetric in $L^2(d\gamma)$.*

Proof. Let us give the proof of (3.5):

$$\begin{aligned}
\int f \mathbf{L} g d\gamma &= \int f \Delta g d\gamma - \int (fx \cdot \nabla g) d\gamma \\
&= -\int \nabla \cdot (f\gamma) \cdot \nabla g dx - \int fx \cdot \nabla g d\gamma \\
&= -\int \nabla f \cdot \nabla g d\gamma,
\end{aligned}$$

where $\nabla \cdot f$ stands for the divergence of f.

In fact, (3.4) is clear due to the fact if a semigroup is ergodic for some probability measure then the measure is always invariant. $\qquad\square$

As we have seen in the proof of Proposition 3.2, the Ornstein–Uhlenbeck semigroup satisfies the equality for all f and x:

$$\forall t \geqslant 0, \ \nabla \mathbf{P_t} f(x) = e^{-t} \mathbf{P_t} \nabla f(x), \tag{3.6}$$

where $\mathbf{P_t} \nabla f = (\mathbf{P_t} \partial_i f)_{1 \le i \le n}$ and for all norms $\|\cdot\|$ in $\mathbb{R}^n$, one gets easily

$$\forall t \ge 0, \quad \|\nabla \mathbf{P_t} f(x)\| \le e^{-t} \mathbf{P_t} \|\nabla f\|(x), \tag{3.7}$$

those equations are known as the commutation property of the gradient and the Ornstein–Uhlenbeck semigroup. Inequality (3.7) is the key formula to get classical inequalities.

3.2.1.1 The Poincaré and logarithmic Sobolev inequalities

Theorem 3.4. *The following Poincaré inequality for the Gaussian measure holds, for all smooth functions f on $\mathbb{R}^n$,*

$$\mathbf{Var}_\gamma(f) := \int f^2 d\gamma - \left(\int f d\gamma \right)^2 \le \int |\nabla f|^2 d\gamma. \tag{3.8}$$

The term $\mathbf{Var}_\gamma(f)$ is the variance of f under γ. Moreover, the inequality is optimal, and extremal functions are given by smooth functions satisfying $\nabla f = C$ for some constant $C \in \mathbb{R}^n$.

Proof. Let f be a smooth function on $\mathbb{R}^n$, then $\mathbf{P_0} f = f$ and $\mathbf{P_\infty} f = \int f d\gamma$ (see (3.3)); therefore, the Ornstein–Uhlenbeck semigroup gives a nice interpolation between f and $\int f d\gamma$.

$$
\begin{aligned}
\mathbf{Var}_\gamma(f) &= -\int_0^{+\infty} \frac{d}{dt} \int (\mathbf{P_t} f)^2 d\gamma \, dt \\
&= -2 \int_0^{+\infty} \int \mathbf{L} \mathbf{P_t} f \mathbf{P_t} f \, d\gamma \, dt \\
&= 2 \int_0^{+\infty} \int |\nabla \mathbf{P_t} f|^2 d\gamma \, dt \\
&\le 2 \int_0^{+\infty} \int e^{-2t} (\mathbf{P_t} |\nabla f|)^2 d\gamma \, dt \\
&\le 2 \int_0^{+\infty} \int e^{-2t} \mathbf{P_t} \left(|\nabla f|^2 \right) d\gamma \, dt \\
&= 2 \int_0^{+\infty} \int e^{-2t} |\nabla f|^2 d\gamma \, dt \\
&= \int |\nabla f|^2 d\gamma,
\end{aligned}
$$

where we use equality (3.7), the Cauchy–Schwarz inequality and the invariance property of the standard Gaussian distribution (3.4).

One can check that, in all stages of the proof, smooth functions satisfying $\nabla f = C$ are the unique function such that the two inequalities become equalities. $\qquad \square$

Theorem 3.5. *The following logarithmic Sobolev inequality for the Gaussian measure holds, for all smooth and non-negative functions f on $\mathbb{R}^n$:*

$$\mathbf{Ent}_\gamma(f) := \int f \log \frac{f}{\int f d\gamma} d\gamma \le \frac{1}{2} \int \frac{|\nabla f|^2}{f} d\gamma. \tag{3.9}$$

The term $\mathbf{Ent}_\gamma(f)$ is known as the entropy of f under γ. Moreover, inequality (3.9) is optimal, and extremal functions are given by $\nabla f = Cf$ for some constant $C \in \mathbb{R}^n$.

Proof. Let us mimic the proof of the Poincaré inequality. Let f be a smooth and non-negative function on $\mathbb{R}^n$, then

$$\begin{aligned}
\mathbf{Ent}_\gamma(f) &= -\int_0^{+\infty} \frac{d}{dt} \int \mathbf{P}_t f \log \mathbf{P}_t f d\gamma dt \\
&= -\int_0^{+\infty} \int \mathbf{L}\mathbf{P}_t f \log \mathbf{P}_t f d\gamma dt \\
&= \int_0^{+\infty} \int \nabla \mathbf{P}_t f \cdot \nabla \log \mathbf{P}_t f d\gamma dt \\
&= \int_0^{+\infty} \int \frac{|\nabla \mathbf{P}_t f|^2}{\mathbf{P}_t f} d\gamma dt, \\
&\le \int_0^{+\infty} \int e^{-2t} \frac{(\mathbf{P}_t|\nabla f|)^2}{\mathbf{P}_t f} d\gamma dt,
\end{aligned}$$

where we have used the same argument as for the Poincaré inequality. Now, to Cauchy–Schwarz inequality or the convexity of the map

$$(x, y) \mapsto x^2/y$$

for $x, y > 0$ implies

$$\frac{(\mathbf{P}_t|\nabla f|)^2}{\mathbf{P}_t f} \le \mathbf{P}_t \left(\frac{|\nabla f|^2}{f} \right).$$

Then one gets

$$\mathbf{Ent}_\gamma(f) \le \int_0^{+\infty} \int e^{-2t} \mathbf{P}_t \left(\frac{|\nabla f|^2}{f} \right) d\gamma dt = \frac{1}{2} \int \frac{|\nabla f|^2}{f} d\gamma.$$

One obtains extremal functions in the same way as for the Poincaré inequality. □

The logarithmic Sobolev inequality is often noted for f^2 instead of f, which gives for all smooth functions f

$$\mathbf{Ent}_\gamma(f^2) \le 2 \int |\nabla f|^2 d\gamma.$$

In light of Theorems 3.4 and 3.5, we say that the standard Gaussian satisfies a Poincaré and a logarithmic Sobolev inequality.

More generally, a logarithmic Sobolev inequality always implies a Poincaré inequality by a Taylor expansion (see Chapter 1 of [ABC+00]).

In Proposition 3.3, we proved that the Ornstein–Uhlenbeck semigroup is ergodic with respect to the Gaussian distribution. In fact, one of the main applications of the Poincaré and the logarithmic Sobolev inequalities is to give an estimate of the speed of convergence in two different spaces.

Theorem 3.6. *The Poincaré inequality (3.8) is equivalent to the following inequality:*

$$\mathbf{Var}_\gamma(\mathbf{P}_t f) \le e^{-2t} \mathbf{Var}_\gamma(f), \tag{3.10}$$

for all smooth functions f.

And in the same way, the logarithmic Sobolev inequality (3.9) is equivalent to

$$\mathbf{Ent}_\gamma(\mathbf{P}_t f) \le e^{-2t} \mathbf{Ent}_\gamma(f), \tag{3.11}$$

for all non-negative and smooth functions f.

Proof. For the first assertion, an elementary computation gives that

$$\frac{d}{dt}\mathbf{Var}_\gamma(\mathbf{P}_t f) = -2\int |\nabla \mathbf{P}_t f|^2 d\gamma,$$

and then the Poincaré inequality and Grönwall lemma implies (3.25). Conversely, the derivation at time $t = 0$ of (3.25) implies the Poincaré inequality.

For the second assertion, we use the same method and the derivation of the entropy

$$\frac{d}{dt}\mathbf{Ent}_\gamma(\mathbf{P}_t f) = -\int \frac{|\nabla \mathbf{P}_t f|^2}{\mathbf{P}_t f} d\gamma. \tag{3.12}$$

$\square$

One of the main difference between the two inequalities is that the initial condition is in $L^2(d\gamma)$ for the Poincaré inequality, whereas the initial condition is in $L \log L(d\gamma)$ for the logarithmic Sobolev inequality.

3.3 Poincaré and logarithmic Sobolev inequalities under curvature criterion

The main idea of this section is to obtain criteria for a probability measure μ such that the two inequalities (3.8) and (3.9) hold for the measure μ. We will study a particular case of the curvature-dimension criterion (or Γ_2-criterion) introduced by D. Bakry and M. Emery in [BÉ85]. This criterion gives conditions on an infinitesimal generator $\mathbf{L}$ such that all the computations done for the Ornstein–Uhlenbeck semigroup could be applied to $\mathbf{L}$.

Let a function $\psi \in C^2(\mathbb{R}^n)$, and define the infinitesimal generator

$$\mathbf{L}f = \Delta f - \nabla\psi \cdot \nabla f \qquad (3.13)$$

for all smooth functions f.

Assume that $\int e^{-\psi}dx < +\infty$ and define the probability measure $d\mu_\psi(x) = \frac{e^{-\psi}dx}{Z_\psi}dx$, where $Z_\psi = \int e^{-\psi}dx$. It is easy to see that the operator $\mathbf{L}$ satisfies for all smooth functions f and g on $\mathbb{R}^n$

$$\int f\mathbf{L}gd\mu_\psi = \int g\mathbf{L}fd\mu_\psi = -\int \nabla f \cdot \nabla g d\mu_\psi, \qquad (3.14)$$

and $\int \mathbf{L}fd\mu_\psi = 0$. We recover the same property as for the Ornstein–Uhlenbeck semigroup; see (3.5). The generator $\mathbf{L}$ is symmetric in $L^2(d\mu_\psi)$ and the probability measure μ_ψ is also invariant with respect to $\mathbf{L}$.

Let us define the *carré du champ*, for all smooth functions f:

$$\Gamma(f, f) = \frac{1}{2}\left(\mathbf{L}(f^2) - 2f\mathbf{L}f\right); \qquad (3.15)$$

we note usually $\Gamma(f)$ instead of $\Gamma(f, f)$. The carré du champ is a quadratic form, and the bilinear form associated is given by

$$\Gamma(f, g) = \frac{1}{2}(\mathbf{L}(fg) - f\mathbf{L}g - g\mathbf{L}f).$$

If we iterate the process one obtains the Γ_2-operator, for all smooth functions f:

$$\Gamma_2(f, f) = \frac{1}{2}(\mathbf{L}(\Gamma(f)) - 2\Gamma(f, \mathbf{L}f)). \qquad (3.16)$$

We assume in this section that there exists a set of function $\mathcal{A}$, dense in $L^2(d\mu)$, such that all computations can be done in this class of function. In the previous section, the set $\mathcal{A}$ was $C_c^\infty(\mathbb{R}^n)$ and one of the main problems is to describe this class of functions. It can be done under the Γ_2-criterion $CD(\rho, +\infty)$ (see the definition below); we refer to [ABC+00, Bak06] and references therein to get more information.

Definition 2. We say that the linear operator $\mathbf{L}$, satisfies the Γ_2-criterion $CD(\rho, +\infty)$ with some $\rho \in \mathbb{R}$, if for all functions $f \in \mathcal{A}$

$$\Gamma_2(f) \geqslant \rho\Gamma(f). \qquad (3.17)$$

Remark 2. Since for all smooth functions f, $\mathbf{L}f = \Delta f - \nabla\psi \cdot \nabla f$, a straightforward computation gives

$$\Gamma(f) = |\nabla f|^2,$$

and

$$\Gamma_2(f) = \|\text{Hess}(f)\|_{H.S.}^2 + \langle \nabla f, \text{Hess}(\psi)\nabla f\rangle,$$

where the Hilbert–Schmidt norm is given by $\|\text{Hess}(f)\|_{H.S.}^2 = \sum_{i,j}\left(\frac{\partial^2}{\partial x_i \partial x_j}f\right)^2$.

Then the linear operator $\mathbf{L}$ defined in (3.13) satisfies the Γ_2-criterion $CD(\rho, +\infty)$ with some $\rho \in \mathbb{R}$ if for all $x \in \mathbb{R}^n$

$$\text{Hess}(\psi)(x) \geqslant \rho Id, \tag{3.18}$$

in the sense of the symmetric matrix. That is, for all $Y \in \mathbb{R}^n$,

$$\langle Y, \text{Hess}(\psi)(x)Y\rangle \geqslant \rho|Y|^2,$$

where $\langle \cdot, \cdot \rangle$ is the Euclidean scalar product.

Theorem 3.7. *Let $\psi \in C^2(\mathbb{R}^n)$ and assume that there exists $\rho > 0$ such that the linear operator (3.13) satisfies a Γ_2-criterion $CD(\rho, +\infty)$, then the probability measure μ_ψ satisfies a Poincaré inequality*

$$\mathbf{Var}_{\mu_\psi}(f) \leq \frac{1}{\rho}\int |\nabla f|^2 d\mu_\psi \tag{3.19}$$

for all $f \in \mathcal{A}$ and a logarithmic Sobolev inequality

$$\mathbf{Ent}_\gamma(f) \leq \frac{1}{2\rho}\int \frac{|\nabla f|^2}{f}d\mu_\psi \tag{3.20}$$

for all smooth and non-negative functions $f \in \mathcal{A}$.

Lemma 3.8. *Let $(\mathbf{P_t})_{t\geqslant 0}$ be the Markov semigroup associated with the infinitesimal generator $\mathbf{L}$. Assume that $\rho > 0$; then $(\mathbf{P_t})_{t\geqslant 0}$ is μ_ψ-ergodic, which means for all functions $f \in \mathcal{A}$*

$$\lim_{t\to+\infty}\mathbf{P_t}f(x) = \int f d\mu_\psi,$$

in $f \in L^2(d\mu_\psi)$ and μ_ψ almost surely.

Lemma 3.9. *Let φ be a C^2 function; then, for all functions $f \in \mathcal{A}$,*

$$\mathbf{L}\varphi(f) = \varphi'(f)\mathbf{L}f + \varphi''(f)\Gamma(f) \text{ and } \Gamma(\log f) = \frac{1}{f^2}\Gamma(f). \tag{3.21}$$

Moreover, one has

$$\Gamma_2(\log f) = \frac{1}{f^2}\Gamma_2(f) - \frac{1}{f^3}\Gamma(f, \Gamma(f)) + \frac{1}{f^4}(\Gamma(f))^2. \tag{3.22}$$

Proof of Theorem 3.7. First we prove the inequality (3.19). As for the Ornsten–Uhlenbeck semigroup, one gets if $(\mathbf{P_t})_{t \geq 0}$ is the Markov semigroup associated with the infinitesimal generator $\mathbf{L}$, for all functions $f \in \mathcal{A}$,

$$\mathbf{Var}_{\mu_\psi}(f) = -\int_0^{+\infty} \frac{d}{dt} \int (\mathbf{P_t}f)^2 d\mu_\psi \, dt$$
$$= -2\int_0^{+\infty} \int \mathbf{LP_t}f\mathbf{P_t}f \, d\mu_\psi \, dt.$$

Since μ_ψ is invariant,

$$\int 2\mathbf{P_t}f\mathbf{LP_t}f \, d\mu_\psi = \int (2\mathbf{P_t}f\mathbf{LP_t}f - \mathbf{L}(\mathbf{P_t}f)^2)d\mu_\psi = -2\int \Gamma(\mathbf{P_t}f)d\mu_\psi,$$

which gives

$$\mathbf{Var}_{\mu_\psi}(f) = \int_0^{+\infty} 2\int \Gamma(\mathbf{P_t}f)d\mu_\psi \, dt. \tag{3.23}$$

Let now consider, for all $t > 0$,

$$\Phi(t) = 2\int \Gamma(\mathbf{P_t}f)d\mu_\psi.$$

The time derivative of Φ is equal to

$$\Phi'(t) = 4\int \Gamma(\mathbf{P_t}f, \mathbf{LP_t}f)d\mu_\psi$$
$$= 2\int (2\Gamma(\mathbf{P_t}f, \mathbf{LP_t}f) - \mathbf{L}(\Gamma(\mathbf{P_t}f)))d\mu_\psi = -4\int \Gamma_2(\mathbf{P_t}f)d\mu_\psi.$$

The Γ_2-criterion implies that $\Phi'(t) \leq -2\rho\Phi(t)$, which gives $\Phi(t) \leq e^{-t2\rho}\Phi(0)$. The last inequality with (3.23) implies

$$\mathbf{Var}_{\mu_\psi}(f) \leq \int_0^{+\infty} e^{-t2\rho}dt \int 2\Gamma(f)d\mu_\psi = \frac{1}{\rho}\int \Gamma(f)d\mu_\psi \, dt.$$

Let now prove the logarithmic Sobolev inequality for the measure μ_ψ. Let f be a non-negative and smooth function on $\mathbb{R}^n$:

$$\mathbf{Ent}_{\mu_\psi}(f) = -\int_0^{+\infty} \frac{d}{dt} \int \mathbf{P_t}f \log \mathbf{P_t}f \, d\mu_\psi \, dt$$
$$= -\int_0^{+\infty} \int \mathbf{LP_t}f \log \mathbf{P_t}f \, d\mu_\psi \, dt.$$

Since **L** is symmetric and by Lemma 3.9 one gets

$$\int \mathbf{L}\mathbf{P}_t f \log \mathbf{P}_t f \, d\mu_\psi = \int \mathbf{P}_t f \mathbf{L} \log \mathbf{P}_t f \, d\mu_\psi = -\int \frac{\Gamma(\mathbf{P}_t f)}{\mathbf{P}_t f} d\mu_\psi$$

$$= -\int \Gamma(\log \mathbf{P}_t f) \mathbf{P}_t f \, d\mu_\psi,$$

which gives

$$\mathbf{Ent}_{\mu_\psi}(f) = \int_0^{+\infty} \int \Gamma(\log \mathbf{P}_t f) \mathbf{P}_t f \, d\mu_\psi dt. \qquad (3.24)$$

As for the Poincaré inequality, let us consider, for all $t > 0$,

$$\Phi(t) = \int \frac{\Gamma(\mathbf{P}_t f)}{\mathbf{P}_t f} d\mu_\psi,$$

where $\mathbf{P}_t f = g$. The time derivative of Φ is equal to

$$\Phi'(t) = \int \left(2\frac{\Gamma(\mathbf{L}g, g)}{g} - \frac{\mathbf{L}g\Gamma(g)}{g^2} \right) \mu_\psi$$

$$= \int \left(2\frac{\Gamma(\mathbf{L}g, g)}{g} - \frac{\mathbf{L}g\Gamma(g)}{g^2} - \mathbf{L}\left(\frac{\Gamma(g)}{g} \right) \right) \mu_\psi.$$

Since

$$\mathbf{L}\left(\frac{\Gamma(g)}{g} \right) = 2\Gamma\left(\Gamma(g), \frac{1}{g} \right) + \frac{1}{g}\mathbf{L}\Gamma(g) + \mathbf{L}\left(\frac{1}{g} \right)\Gamma(g),$$

by Lemma 3.9 one has

$$\Phi'(t) = -2\int \Gamma_2(\log \mathbf{P}_t f)\mathbf{P}_t f \, d\mu_\psi.$$

The Γ_2-criterion implies that $\Phi'(t) \leq -2\rho\Phi(t)$, which gives $\Phi(t) \leq e^{-2\rho t}\Phi(0)$. This inequality with (3.24) implies that

$$\mathbf{Ent}_{\mu_\psi}(f) \leq \int_0^{+\infty} e^{-2\rho t} dt \int \Gamma(\log f) f \, d\mu_\psi = \frac{1}{2\rho}\int \Gamma(\log f) f \, d\mu_\psi$$

$$= \frac{1}{2\rho}\int \frac{|\nabla f|}{f} d\mu_\psi. \qquad \square$$

The meaning of this result is that if μ_ψ is more log-concave than the Gaussian distribution then μ_ψ satisfies both inequalities.

Remark 3. The Γ_2-criterion is in fact a more general criterion. The definition of a diffusion semigroup could be a Markov semigroup such that, for all smooth functions φ, equations (3.21) and (3.22) hold for the generator associated with the semigroup.

In fact on $\mathbb{R}^n$ (or on a manifold on a local chart) that means that the infinitesimal generator $\mathbf{L}$ of the Markov semigroup is given by

$$\forall x \in \mathbb{R}^n, \ \mathbf{L}f(x) = \sum_{i,j} D_{i,j}(x)\partial_{i,j}f(x) - \sum_{i} a_i(x)\partial_i f(x),$$

where $D(x) = (D_{i,j}(x))_{i,j}$ is a symmetric and non-negative matrix and $a(x) = (a_i(x))_i$ is a vector.

Then the conditions $\Gamma_2(f) \geqslant \rho\Gamma(f)$ for some $\rho > 0$ implies that there exists an invariant measure μ of the semigroup and μ satisfies the Poincaré and a logarithmic Sobolev inequality with the same constant as before. One of the difficulties of this general case is to find tractable conditions on functions D and such that the Γ_2-criterion holds. Some other examples can be found in [BG10].

Let us also note that the Γ_2-criterion $CD(\rho, \infty)$ is a particular case of the $CD(\rho, n)$ criterion where $n \in \mathbb{N}^*$:

$$\Gamma_2(f) \geqslant \rho\Gamma(f) + \frac{1}{n}(\mathbf{L}f)^2,$$

for all smooth functions f. For example, the Ornstein–Uhlenbeck semigroup satisfies the $CD(1, \infty)$ criterion and the heat equation $\mathbf{L} = \Delta$ satisfies the $CD(0, n)$. One can observe that the Ornstein–Uhlenbeck semigroup does not satisfies a $CD(r, m)$ criterion for any $r, m > 0$.

Theorem 3.10. *As for the Ornstein–Uhlenbeck semigroup, the Poincaré inequality (3.19) is equivalent to the following inequality:*

$$\mathbf{Var}_{\mu_\psi}(\mathbf{P_t}f) \leq e^{-\frac{2}{\rho}t}\mathbf{Var}_{\mu_\psi}(f), \qquad (3.25)$$

for all functions $f \in \mathcal{A}$.

And in the same way, the logarithmic Sobolev inequality (3.20) is equivalent to

$$\mathbf{Ent}_{\mu_\psi}(\mathbf{P_t}f) \leq e^{-2t}\mathbf{Ent}_{\mu_\psi}(f), \qquad (3.26)$$

for all non-negative functions $f \in \mathcal{A}$.

The logarithmic Sobolev inequality has two main applications. The first one, the asymptotic behaviour in terms of entropy, is the result of Theorem 3.10. The second application is about concentration inequality, a probability measure μ satisfying a logarithmic Sobolev inequality has the same tail as the Gaussian distribution.

These properties are also a consequence of the Talagrand inequality described in Section 3.4.

3.4 The logarithmic Sobolev and transport inequalities by transportation method

We will see how Brenier's theorem can be used in this context to give a new proof of the logarithmic Sobolev inequality; the method is called the mass transportation method.

We will illustrate this method for the Gaussian measure, but it could be generalized for a large class of measures; this will be discussed later. The method comes from [OV00, CE02] and has been generalized for many Euclidean inequalities, such as Sobolev and Gagliardo–Nirenberg inequalities; see [AGK04, CENV04, Naz06].

The Wasserstein distance between two probability measures μ and ν is defined by

$$W_2(\mu, \nu) = \left(\inf \int |x - y|^2 d\pi(x, y) \right)^{1/2}, \qquad (3.27)$$

where the infimum is running over all probability measures π on $\mathbb{R}^n \times \mathbb{R}^n$ with respective marginals μ and ν, i.e. for all bounded functions g and h,

$$\int (g(x) + h(y)) d\pi(x, y) = \int g d\mu + \int h d\nu.$$

Such a probability is called a coupling of (μ, ν).

Brenier's theorem says that that there exists an optimal deterministic coupling of (μ, ν), i.e. there exists a convex map Φ satisfying

$$\int h(\nabla \Phi) d\nu = \int h d\mu,$$

for all bounded functions h. Moreover,

$$W_2^2(d\nu, d\mu) = \int |\nabla \theta|^2 d\nu,$$

where $\theta(x) = \Phi(x) - \frac{1}{2}|x|^2$. This result has been proved by Brenier; $\nabla \Phi$ is called the Brenier map between ν and μ; see [Vil09].

We apply this result in the Gaussian case. Letting f be a smooth and positive function such that $\int f d\gamma = 1$, Brenier's theorem implies that there exists a convex map Φ satisfying

$$\int h(\nabla \Phi) f d\gamma = \int h d\gamma, \qquad (3.28)$$

for all bounded and measurable functions h. Moreover,

$$W_2^2(f d\gamma, d\gamma) = \int |\nabla \theta|^2 f d\gamma,$$

where $\theta(x) = \Phi(x) - \frac{1}{2}|x|^2$.

If now Φ is a $C^2(\mathbb{R}^n)$ function, then coming from (3.28), the Monge–Ampère equation holds, i.e. $f\,d\gamma$-a.e.:

$$f(x)e^{-|x|^2/2} = \det(Id + \text{Hess}(\theta))e^{-|x+\nabla\theta(x)|^2/2}. \qquad (3.29)$$

After taking the logarithm, we get

$$\begin{aligned}
\log f(x) &= -\frac{1}{2}|x + \nabla\theta(x)|^2 + \frac{1}{2}|x|^2 + \log\det(Id + \text{Hess}(\theta)) \\
&= -x \cdot \nabla\theta(x) - \frac{1}{2}|\nabla\theta(x)|^2 + \log\det(Id + \text{Hess}(\theta)) \\
&\leq -x \cdot \nabla\theta(x) - \frac{1}{2}|\nabla\theta(x)|^2 + \Delta\theta(x),
\end{aligned}$$

where we used inequality $\log(1+t) \leq t$ whenever $1+t > 0$. We integrate with respect to $f\,d\gamma$:

$$\mathbf{Ent}_\gamma(f) \leq \int f(\Delta\theta - x \cdot \nabla\theta)d\gamma - \int \frac{1}{2}|\nabla\theta(x)|^2 f\,d\gamma.$$

The integration by parts implies

$$\begin{aligned}
\mathbf{Ent}_\gamma(f) &\leq -\int \nabla\theta \cdot \nabla f\,d\gamma - \int \frac{1}{2}|\nabla\theta(x)|^2 f\,d\gamma \\
&\leq -\frac{1}{2}\int \left|\sqrt{f}\nabla\theta + \frac{\nabla f}{\sqrt{f}}\right|^2 d\gamma + \frac{1}{2}\int \frac{|\nabla f|^2}{f}d\gamma \\
&\leq \frac{1}{2}\int \frac{|\nabla f|^2}{f}d\gamma,
\end{aligned}$$

which is the optimal logarithmic Sobolev inequality (3.9).

Hence, we have proved, using Brenier's map, the logarithmic Sobolev inequality for the Gaussian measure with the optimal constant. As we can see in the proof, one has assumed that Φ is a C^2 function. It can be obtained using Caffarelli's regularity theory : it needs another assumptions, f has to be smooth with a compact and convex support. We skip it for simplicity of the description of the method; many information can be bound in [Vil09].

Let us see what can be done if now $\nabla\Phi$ is the Brenier map between $d\gamma$ and $f\,d\gamma$ instead $f\,d\gamma$ and $d\gamma$; that is, for all bounded and measurable functions h:

$$\int hf\,d\gamma = \int h(\nabla\Phi)d\gamma,$$

and if $x + \nabla\theta(x) = \nabla\Phi$ then

$$W_2^2(f\,d\gamma, d\gamma) = \int |\nabla\theta|^2 d\gamma.$$

In that case the Monge–Ampère equation gives

$$\det(Id + \text{Hess}(\theta))f(x + \nabla\theta(x))e^{-|x+\nabla\theta(x)|^2/2} = e^{-|x|^2/2}, \qquad (3.30)$$

which implies

$$\log f(x + \nabla\theta(x)) = \frac{1}{2}|x + \nabla\theta(x)|^2 - \frac{1}{2}|x|^2 - \log\det(Id + \text{Hess}(\theta))$$
$$= x \cdot \nabla\theta(x) + \frac{1}{2}|\nabla\theta(x)|^2 - \log\det(Id + \text{Hess}(\theta))$$
$$\geq x \cdot \nabla\theta(x) + \frac{1}{2}|\nabla\theta(x)|^2 - \Delta\theta(x)$$
$$= -\mathbf{L}\theta + \frac{1}{2}|\nabla\theta(x)|^2,$$

where $\mathbf{L}$ is the Ornstein–Uhlenbeck generator. Then

$$\mathbf{Ent}_\gamma(f) = \int f \log f d\gamma$$
$$= \int \log f(\nabla\Phi)d\gamma$$
$$\geq \int -\mathbf{L}\theta d\gamma + \int \frac{1}{2}|\nabla\theta(x)|^2 d\gamma$$
$$= \int \frac{1}{2}|\nabla\theta(x)|^2 d\gamma = \frac{1}{2}W_2^2(f d\gamma, d\gamma).$$

We have proved that for all functions f such that $f d\gamma$ is a probability measure, one has

$$W_2(f d\gamma, d\gamma) \leq \sqrt{2\mathbf{Ent}_\gamma(f)}. \tag{3.31}$$

This inequality, called the *Talagrand inequality for the Gaussian distribution* (or $\mathcal{T}_2$ inequality), has been proved by Talagrand in [Tal96].

As for the Poincaré and logarithmic Sobolev inequalities, we say that a probability measure μ satisfies a Talagrand inequality if there exists $C \geq 0$ such that

$$T_2(f d\mu, d\mu) \leq \sqrt{C\mathbf{Ent}_\mu(f)} \tag{3.32}$$

for all functions f such that $f d\mu$ is a probability measure,

3.4.1 Remarks and extensions

This method can also be used in the context of Section 3.3. Assume that ψ is uniformly convex and satisfying

$$\text{Hess}(\psi) \geq \rho I,$$

with some $\rho > 0$. The mass transportation method implies that the measure

$$d\mu_\psi(x) = \frac{e^{-\psi}dx}{Z_\psi}dx$$

satisfies the logarithmic Sobolev inequality (3.20) with the constant $1/(2\rho)$. This is an alternative proof of Theorem 3.7. Actually, this method is not useful to obtain a Poincaré inequality directly.

Of course, as for Ornstein–Uhlenbeck semigroup, the mass transportation method also gives a Talagrand inequality (3.32) for the measure μ_ψ:

$$T_2(f d\mu_\psi, d\mu_\psi) \leq \sqrt{\frac{1}{\rho}\mathbf{Ent}_{\mu_\psi}(f)},$$

for all probability measure $f d\mu_\psi$.

In fact, one gets the following general result.

Theorem 3.11 (Otto–Villani). *Let μ be a probability measure on $\mathbb{R}^n$ satisfying a logarithmic Sobolev inequality*

$$\mathbf{Ent}_\mu\left(f^2\right) \leq C \int |\nabla f|^2 d\mu,$$

for all smooth functions f and for some constant $C \geqslant 0$. Then μ satisfies a Talagrand inequality

$$T_2(f d\mu, d\mu) \leq \sqrt{2C\mathbf{Ent}_\mu(f)},$$

for all probability measures $f d\mu$.

The original proof comes from [OV00] and an easier one, using the Hamilton–Jacobi equation, has been given in [BGL01]. These two inequalities are quite similar, but it has been proved in [CG06, Goz07] that they are not equivalent.

References

[ABC+00] C. Ané, S. Blachère, D. Chafaï, P. Fougères, I. Gentil, F. Malrieu, C. Roberto and G. Scheffer. *Sur les inégalités de Sobolev logarithmiques*, volume 10 of *Panoramas et Synthèses*. Société Mathématique de France, Paris, 2000.

[AGK04] M. Agueh, N. Ghoussoub and X. Kang. Geometric inequalities via a general comparison principle for interacting gases. *Geom. Funct. Anal.*, 14(1):215–244, 2004.

[Bak06] D. Bakry. Functional inequalities for Markov semigroups. In *Probability Measures on Groups: Recent Directions and Trends*, pp. 91–147. Tata Inst. Fund. Res., Mumbai, 2006.

[BÉ85] D. Bakry and M. Émery. Diffusions hypercontractives. In *Séminaire de Probabilités, XIX, 1983/84*, Lecture Notes in Math. 1123, pp. 177–206. Springer, Berlin, 1985.

[BG10] F. Bolley and I. Gentil. Phi-entropy inequalities for diffusion semigroups. *J. Math. Pures Appl.*, 93(5):449–473, 2010.

[BGL01] S. G. Bobkov, I. Gentil and M. Ledoux. Hypercontractivity of Hamilton–Jacobi equations. *J. Math. Pures Appl.*, 80(7):669–696, 2001.

[CE02] D. Cordero-Erausquin. Some applications of mass transport to Gaussian-type inequalities. *Arch. Ration. Mech. Anal.*, 161(3):257–269, 2002.

[CENV04] D. Cordero-Erausquin, B. Nazaret and C. Villani. A mass-transportation approach to sharp Sobolev and Gagliardo–Nirenberg inequalities. *Adv. Math.*, 182(2):307–332, 2004.

[CG06] P. Cattiaux and A. Guillin. On quadratic transportation cost inequalities. *J. Math. Pures Appl.*, 86(4):341–361, 2006.

[Goz07] N. Gozlan. Characterization of Talagrand's like transportation-cost inequalities on the real line. *J. Funct. Anal.*, 250(2):400–425, 2007.

[Naz06] B. Nazaret. Best constant in Sobolev trace inequalities on the half-space. *Nonlinear Anal.*, 65(10):1977–1985, 2006.

[OV00] F. Otto and C. Villani. Generalization of an inequality by Talagrand, and links with the logarithmic Sobolev inequality. *J. Funct. Anal.*, 173(2):361–400, 2000.

[Tal96] M. Talagrand. Transportation cost for Gaussian and other product measures. *Geom. Funct. Anal.*, 6(3):587–600, 1996.

[Vil09] C. Villani. *Optimal Transport*, volume 338. Springer-Verlag, Berlin, 2009.

4

Lecture notes on variational models for incompressible Euler equations

LUIGI AMBROSIO AND ALESSIO FIGALLI

Abstract

These notes briefly summarise the lectures for the Summer School "Optimal Transportation: Theory and Applications" held by the second author in Grenoble during the week of 22–26 June 2009. Their goal is to describe some recent results on Brenier's variational models for incompressible Euler equations [Ambrosio and Figalli, *Arch. Ration. Math. Anal.*, **194** (2009) 421–462; Ambrosio and Figalli, Calc. Var. Partial Dif. Equations, **31** (2008) 497–509; Bernot *et al.*, *J. Math. Pures Appl.*, **91** (2008) 137–155].

4.1 Euler incompressible equations and Arnold geodesics

Let D denote either a bounded domain of $\mathbb{R}^d$ or the d-dimensional torus $\mathbb{T}^d := \mathbb{R}^d/\mathbb{Z}^d$. We consider an incompressible fluid moving inside D with velocity $\boldsymbol{u}$. The Euler equations for $\boldsymbol{u}$ describe the evolution in time of the velocity field, and are given by

$$\begin{cases} \partial_t \boldsymbol{u} + (\boldsymbol{u} \cdot \nabla)\boldsymbol{u} = -\nabla p & \text{in } [0, T] \times D, \\ \operatorname{div} \boldsymbol{u} = 0 & \text{in } [0, T] \times D, \end{cases}$$

coupled with the boundary condition

$$\boldsymbol{u} \cdot \nu = 0 \qquad \text{on } [0, T] \times \partial D$$

when $D \neq \mathbb{T}^d$. Here, p is the pressure field, and arises as a Lagrange multiplier for the divergence-free constraint on the velocity $\boldsymbol{u}$.

If $\boldsymbol{u}$ is smooth we can write the above equations in Lagrangian coordinates: let g denote the flow map of $\boldsymbol{u}$; that is,

$$\begin{cases} \dot{g}(t, a) = \boldsymbol{u}(t, g(t, a)), \\ g(0, a) = a. \end{cases}$$

By the incompressibility condition, and the classical differential identity

$$\frac{d}{dt}\det\nabla_a g(t, a) = \operatorname{div} \boldsymbol{u}(t, g(t, a))\det\nabla_a g(t, a),$$

(here and in the following div denotes the spatial divergence of a possibly time-dependent vector field) we get $\det\nabla_a g(t, a) \equiv 1$. This means that $g(t, \cdot) : D \to D$ is a measure-preserving diffeomorphism of D:

$$g(t, \cdot)_{\#}\mu_D = \mu_D \qquad \left(\text{i.e. } \mu_D(g(t, \cdot)^{-1}(E)) = \mu_D(E) \;\; \forall E\right).$$

Here and in the following $f_{\#}\mu$ is the push-forward of a Borel measure μ through a map $f : X \to Y$ (i.e. $\int_Y \phi \, df_{\#}\mu = \int_X \phi \circ f \, d\mu$ for all Borel bounded functions $\phi : Y \to \mathbb{R}$), and μ_D is the volume measure of D, renormalised by a constant so that $\mu_D(D) = 1$.

Writing Euler's equations in terms of g we obtain an ordinary differential equation (ODE) for $t \mapsto g(t)$ in the space SDiff(D) of measure-preserving smooth diffeomorphisms of D:

$$\begin{cases} \ddot{g}(t, a) = -\nabla p \, (t, g(t, a)) & (t, a) \in [0, T] \times D, \\ g(0, a) = a & a \in D, \\ g(t, \cdot) \in \text{SDiff}(D) & t \in [0, T]. \end{cases} \qquad (4.1)$$

4.1.1 Weak solutions to Euler's equations

In the case $d = 2$, existence of distributional solutions can be proved through the *vorticity* formulation; setting $\omega_t(\cdot) = \operatorname{curl} \boldsymbol{u}(t, \cdot)$, so that $\boldsymbol{u}(t, \cdot) = \nabla^{\perp}\Delta^{-1}\omega_t$, the Euler equations can be read as follows:

$$\frac{d}{dt}\omega_t(x) + \operatorname{div}\left(\omega_t(x)\boldsymbol{u}(t, x)\right) = 0.$$

Formally, this equation preserves all L^p norms of solutions, and indeed existence is not hard to obtain if $\omega_0 \in L^p$ for $1 < p \le \infty$. Delort improved the existence theory up to L^1 or measure initial conditions ω_0 whose positive (or negative) part is absolutely continuous, and the problem of getting a solution for all measure initial data is still open. As shown by Yudovitch [15,16], uniqueness holds for $p = \infty$, while it is still open in all the other cases.

In the case $d > 2$ much less is known: *no* general global existence results of distributional solutions are presently available.

4.1.2 Arnold's geodesic interpretation

At least formally, one can view the space SDiff(D) of measure-preserving diffeomorphisms of D as an infinite-dimensional manifold with the metric inherited from the embedding in $L^2(D; \mathbb{R}^d)$, and with tangent space made by the divergence-free vector fields. Using this viewpoint, Arnold interpreted the ODE (4.1), and therefore Euler's equations, as a *geodesic* equation on SDiff(D) [3]. Therefore, one can look for solutions of Euler's equations on $[0, 1] \times D$ by minimizing the action functional

$$\mathscr{A}(g) := \int_0^1 \int_D \frac{1}{2} |\dot{g}(t, x)|^2 \, d\mu_D(x) \, dt$$

among all paths $g(t, \cdot) : [0, 1] \to \text{SDiff}(D)$ with $g(0, \cdot) = f$ and $g(1, \cdot) = h$ prescribed (typically, by right invariance, f is taken as the identity map i). Ebin and Marsden proved in [10] that this problem has indeed a unique solution when $h \circ f^{-1}$ is sufficiently close, in a strong Sobolev norm, to i. We shall denote by $\delta(f, h)$ the Arnold distance in SDiff(D) induced by this minimisation problem.

Of course, this variational problem differs from Euler's problem, because the initial and final diffeomorphisms, and not the initial velocity, are prescribed. Nevertheless, the investigation of this problem leads to difficult and still not completely understood questions (typical of calculus of variations); namely:

(a) necessary and sufficient optimality conditions;
(b) regularity of the pressure field;
(c) regularity of (relaxed) curves with minimal length.

Before describing some of the main contributions in this field, let us recall some 'negative' results that motivate somehow the necessity of relaxed formulations of this minimisation problem.

4.1.3 Non-attainment and non-existence results

Shnirelman [12, 13] found the example of a map $\bar{g} \in \text{SDiff}([0, 1]^2)$ which cannot be connected to i by a path with finite action, i.e. $\delta(i, \bar{g}) = +\infty$. Furthermore, he proved that for $h \in \text{SDiff}([0, 1]^3)$ of the form

$$h(x_1, x_2, x_3) = (\bar{g}_1(x_1, x_2), \bar{g}_2(x_1, x_2), x_3), \quad \text{with} \quad (\bar{g}_1, \bar{g}_2) = \bar{g} \text{ as above,}$$

$\delta(i, h)$ is not attained, i.e. no minimizing path between i and h exists (although there exist paths with a finite action). This fact can be easily explained as follows (see also [8, Paragraph 1.3]): since there is no two-dimensional path with finite action connecting i to $\bar{g}$ while in three dimensions it is known that the minimal action is finite [12], if a minimising path $t \mapsto g(t)$ exists then it has a non-trivial third component, i.e. $g_3(t, x) \neq x_3$. Set $\eta(x_3) := \min\{2x_3, 2 - 2x_3\}$, and let $\boldsymbol{u}$ denote the velocity field associated with g, i.e. $\boldsymbol{u} = \dot{g} \circ g^{-1}$. Then it is easily seen that the velocity field

$$\tilde{u}(x_1, x_2, x_3) := \begin{cases} u_1(x_1, x_2, \eta(x_3)) \\ u_2(x_1, x_2, \eta(x_3)) \\ \frac{1}{2}u_3(x_1, x_2, \eta(x_3)) \end{cases}$$

induces a path $\tilde{g}$ which still joins i to h, but with strictly less action (since u_3 is not identically zero). This contradicts the minimality of g, and proves that there is no minimizing path between i and h. (See also [8, Paragraph 1.3].)

Let us point out that the above argument shows that minimising sequences exhibit oscillations on small scales, and strongly suggest the analysis of weak solutions.

4.1.4 Time discretisation, minimal projection and optimal transport

Before describing the concept of relaxed solutions to the Euler equations introduced by Brenier, let us first see what happens when one tries to attack the above variational problem by time-discretisation: assume $D \subset \mathbb{R}^d$, and fix $g_0, g_1 \in \mathrm{SDiff}(D)$. We want to find the 'midpoint' $g_{1/2}$ between g_0 and g_1; that is, we consider

$$\min_{g \in \mathrm{SDiff}(D)} \left\{ \frac{1}{2}\|g - g_0\|^2_{L^2(D;\mathbb{R}^d)} + \frac{1}{2}\|g_1 - g\|^2_{L^2(D;\mathbb{R}^d)} \right\}.$$

Up to rearranging the terms and removing all the quantities independent on g, the above problem is equivalent to minimising

$$\min_{g \in \mathrm{SDiff}(D)} \left\| g - \frac{g_0 + g_1}{2} \right\|^2_{L^2(D;\mathbb{R}^d)},$$

i.e. we have to find the L^2-projection on $\mathrm{SDiff}(D)$ of the function $\frac{g_0 + g_1}{2} \in L^2(D; \mathbb{R}^d)$. Since the set $\mathrm{SDiff}(D)$ is neither closed nor convex, no classical theory is available to ensure the existence of such projection.

In order to make the problem more treatable, let us close $\mathrm{SDiff}(D)$: as shown for instance in [9], if $D = [0, 1]^d$ or $D = \mathbb{T}^d$ then the L^2-closure of $\mathrm{SDiff}(D)$ in $L^2(D; \mathbb{R}^d)$ coincides with the space $S(D)$ of measure-preserving maps:

$$S(D) := \left\{ g : D \to D : \mu_D(g^{-1}(A)) = \mu_D(A) \ \forall A \in \mathcal{B}(D) \right\}.$$

Then the general problem we want to study becomes the following: given $h \in L^2(D; \mathbb{R}^d)$, solve

$$\min_{s \in S(D)} \int_D |h - s|^2 \, d\mu_D. \tag{4.2}$$

As in the classical optimal transport problem, one can consider the following
Kantorovich relaxation: denoting by $\Pi(\mathbb{R}^d)$ the set of probability measures on
$\mathbb{R}^d \times \mathbb{R}^d$ with first marginal μ_D and second marginal $\nu := h_\# \mu_D$, we minimise

$$\min_{\gamma \in \Pi(\mathbb{R}^d)} \int_{\mathbb{R}^d \times \mathbb{R}^d} |x - y|^2 \, d\gamma(x, y). \tag{4.3}$$

Assume the *non-degeneracy condition* $\nu \ll dx$. Then we can apply the classical
theory of optimal transport with quadratic cost for the problem of sending ν
onto μ_D [6]: there exists a unique optimal transport map $\nabla \phi : \mathbb{R}^d \to \mathbb{R}^d$ such
that $(\nabla \phi)_\# \nu = \mu_D$. Moreover, the unique optimal measure $\bar{\gamma}$ which solves (4.3)
is given by

$$\bar{\gamma} = (\nabla \phi \times \mathrm{Id})_\# \nu.$$

Then it is easily seen that the map

$$\bar{s} := \nabla \phi \circ h$$

belongs to $S(D)$ and uniquely solves (4.2) (see [6] or [14, Chapter 3] for more
details).

4.2 Relaxed solutions

In Section 4.1 we have seen how the attempt of attacking Arnold's geodesics
problem by time discretisation leads to study the existence of the L^2-projection
onto SDiff(D), and that the projection of a function h onto its closure $S(D)$
exists and is unique whenever h satisfies a non-degeneracy condition. Instead of
going on with this strategy, we now want to change the point of view, attacking
the problem by a relaxation in 'space'.

Two levels of relaxation can be imagined: the first one is to relax the smoothness and injectivity constraints, and this leads to the definition of the space
$S(D)$ of *measure-preserving maps*. However, we will see that a second level
is necessary, giving up the idea that $g(t, \cdot)$ is a map, but allowing it to be a
measure-preserving plan (roughly speaking, a multivalued map). This leads to
the space

$$\Gamma(D) := \{\eta \in \mathscr{P}(D \times D) : \ \eta(A \times D) = \mu_D(A) = \eta(D \times A) \ \forall A \in \mathcal{B}(D)\}.$$

The space $S(D)$ 'embeds' into $\Gamma(D)$ considering

$$S(D) \ni g \mapsto (i \times g)_\# \mu_D \in \Gamma(D).$$

Conversely, any $\eta \in \Gamma(D)$ concentrated on a graph is induced by a map $g \in S(D)$.

Even from the Lagrangian viewpoint, it is natural to follow the path of each particle, and to relax the smoothness and injectivity constraints, allowing fluid paths to split, forward or backward in time. These remarks led Brenier in 1989 to the following model [5]: let

$$\Omega(D) := C([0, 1]; D), \qquad e_t(\omega) := \omega(t), \quad t \in [0, 1].$$

Then, denoting by $\mathscr{P}(\Omega(D))$ the family of probability measures in $\Omega(D)$, we minimise the action functional

$$\mathscr{A}(\eta) := \int_{\Omega(D)} \frac{1}{2} \int_0^1 |\dot{\omega}|^2 \, dt \, d\eta(\omega), \qquad \eta \in \mathscr{P}(\Omega(D))$$

with the endpoint and incompressibility constraints

$$(e_0, e_1)_\# \eta = (i \times h)_\# \mu_D, \qquad (e_t)_\# \eta = \mu_D \ \forall t \in [0, T].$$

In Brenier's model, a flow is modelled by a random path with some constraints on the expectations of this path. As we will see below, this problem can be recast in the optimal transportation framework, dealing properly with the incompressibility constraint.

Classical flows $g(t, a)$ induce generalised ones, with the same kinetic action, via the relation $\eta = (\Phi_g)_\# \mu_D$, with

$$\Phi_g : D \to \Omega(D), \qquad \Phi_g(a) := g(\cdot, a).$$

In this relaxed model, some obstructions of the original one disappear; for instance, in the case $D = [0, 1]^d$ or $D = \mathbb{T}^d$ it is always possible to connect any couple of measure-preserving diffeomorphism by a path with action less than $\sqrt{d}$. Actually, this allows are to prove that finite-action paths exist in many situation: as shown in [1, Theorem 3.3], given a domain D for which there exists a bi-Lipschitz measure-preserving diffeomorphism $\Phi : D \to [0, 1]^d$, by considering composition of generalised flows with Φ one can easily constructs a generalised flow with finite action between any $h_0, h_1 \in \mathrm{SDiff}(D)$. Moreover, standard compactness/lower semicontinuity arguments in the space $\mathscr{P}(\Omega(D))$ provide existence of generalised flows with minimal action.

4.2.1 Eulerian–Lagrangian model

Coming back to the relaxed model described above, we observe that the endpoint constraint $(e_0, e_1)_\# \eta = (i \times h)_\# \mu_D$ cannot be modified to deal with the more general problem of connecting $f \in S(D)$ to $h \in S(D)$; indeed, by right invariance, this is clear only if f is invertible (in this case, one looks for the optimal connection between i and $h \circ f^{-1}$). These remarks led to a more general model, which allows are to connect $\eta = \eta_a \otimes \mu_D$ to $\gamma = \gamma_a \otimes \mu_D$ [2]. (Here, we are disintegrating both the initial and final plans with respect to the first variable.) The idea, which appears first in Brenier's *Eulerian–Lagrangian* model [8] is to 'double' the state space, adding to the Eulerian state space D a Lagrangian state space A. Even though A could be thought of as an identical copy of D, it is convenient to denote it by a different symbol.

Let

$$\Omega^*(D) := \Omega(D) \times A.$$

Then, consider probability measures $\eta = \eta_a \otimes \mu_D$ in $\Omega^*(D)$; this means that η has μ_D as second marginal, and that

$$\int \phi(\omega, a)\, d\eta(\omega, a) = \int_A \left(\int_{\Omega(D)} \phi(\omega, a)\, d\eta_a \right) d\mu_D(a)$$

for all bounded Borel functions ϕ on $\Omega^*(D)$.

Again, one minimises the action

$$\mathscr{A}(\eta) := \int_{\Omega^*(D)} \frac{1}{2} \int_0^1 |\dot\omega|^2\, dt\, d\eta(\omega, a)$$

with the incompressibility constraint $(e_t)_\#\eta = \mu_D$ for all t (here, $e_t(\omega, a) = \omega(t)$) and the family of endpoint constraints

$$(e_0)_\#\eta_a = \gamma_a, \quad (e_1)_\#\eta_a = \eta_a \qquad \text{for } \mu_D\text{-a.e. } a \in D.$$

As in Section 4.2, we are using $\eta_a \otimes \mu_D$ and $\gamma_a \otimes \mu_D$ to denote the disintegrations of η and γ respectively.

Denoting by $\bar\delta(\eta, \gamma)^2$ the minimal action, it turns out that one can define natural operations of *reparameterisation*, *restriction* and *concatenation* in this class of flows. These imply that $(\bar\delta, \Gamma(D))$ is a metric space.

Indeed, it is proved in [1] that it is *complete* and a *length* space, whose convergence is stronger than weak convergence in $\mathscr{P}(D \times D)$.

4.2.2 Motivation for the extension to $\Gamma(D)$

Even for deterministic initial and final data, there exist examples of minimising geodesics η that are *not deterministic* in between; this means that $(e_0, e_t)_\#\eta \in \Gamma(D) \setminus S(D), t \in (0, 1)$.

To show this phenomenon, consider the problem of connecting up to additive constants in $D = B_1(0) \subset \mathbb{R}^2$ the identity map i to $-i$. For convenience, up to a reparameterisation, we can choose the time interval as $[0, \pi]$. Two classical solutions are

$$[0, \pi] \ni t \mapsto (x_1 \cos \pm t + x_2 \sin \pm t, x_1 \sin \pm t + x_2 \cos \pm t),$$

corresponding to a clockwise and an anti-clockwise rotation.

On the other hand, one can consider the family of maps $\omega_{x,\theta}$ connecting x to $-x$:

$$\omega_{x,\theta}(t) := x \cos t + \sqrt{1 - |x|^2}(\cos \theta, \sin \theta) \sin t \qquad \theta \in (0, \pi) \qquad (4.4)$$

and define $\eta := (\omega_{x,\theta})_\sharp(\frac{1}{2\pi^2}\mathscr{L}^2 \lfloor D \times \mathscr{L}^1 \lfloor (0, 2\pi))$.

It turns out that η is optimal as well, and non-deterministic in between. Moreover, as shown in [4], it is possible to construct infinitely many other solutions to the above minimisation problem which are not induced by maps. For instance, one can split the measure η above as $\frac{1}{2}(\eta_+ + \eta_-)$, where η_+ consists of the curves such that $(\cos \theta, \sin \theta) \cdot x^\perp \geq 0$, and η_- consists of the curves such that $(\cos \theta, \sin \theta) \cdot x^\perp \leq 0$, where $x^\perp = (x_2, -x_1)$, and the two flows η_+ and η_- can be shown to be still incompressible (see [4, Paragraph 4.1]). We will say more about these important examples later on, as more results on the theory will be available.

4.3 The pressure field

Brenier proved in [7] a surprising result: even though geodesics are not unique in general, given the initial and final conditions, there is a *unique*, up to an additive time-dependent constant, pressure field. The pressure field arises if one relaxes the incompressibility constraint, considering *almost incompressible* flows v. Denoting by ρ^v the density produced by the flow, defined by

$$(e_t)_\sharp v = \rho^v(t, \cdot)\mu_D \quad \left(\text{i.e. } \int \phi(\omega(t))\, d\eta(\omega) = \int_D \phi\rho^v(t, \cdot)\, d\mu_D \text{ for all } \phi\right),$$

we say that v is almost incompressible if $\|\rho^v - 1\|_{C^1} \leq 1/2$.

Theorem 4.1 (Pressure as a Lagrange multiplier, [1,7]). *Let η be optimal between η and γ. There exists a distribution $p \in (C^1)^*$ such that*

$$\mathscr{A}(v) + \langle p, \rho^v - 1 \rangle \geq \mathscr{A}(\eta) \qquad (4.5)$$

for all almost incompressible flows v between η and γ with $\rho^v(t, \cdot) = 1$ for t sufficiently close to 0 and to 1.

Using this result one can make first variations as follows: given a smooth field $w(t, x)$, vanishing for t close to 0 and 1, one can consider the family (X^t)

of flow maps

$$\frac{d}{d\varepsilon} X^t(\varepsilon, x) = \boldsymbol{w}(t, X^t(\varepsilon, x)), \qquad X^t(0, x) = x$$

and perturb (smoothly) the paths ω by $\omega(t) \mapsto X^t(\varepsilon, \omega(t)) \sim \omega(t) + \varepsilon \, \boldsymbol{w}(t, \omega(t))$. Denoting by

$$\Phi_\varepsilon : \Omega^*(D) \to \Omega^*(D), \qquad \Phi_\varepsilon(\omega, a)(t) := \big(X^t(\varepsilon, \omega(t)), a\big),$$

the induced perturbations in $\Omega^*(D)$, these in turn induce perturbations $\boldsymbol{\eta}_\varepsilon := (\Phi_\varepsilon)_\# \boldsymbol{\eta}$ of $\boldsymbol{\eta}$ which are almost incompressible. Then, the first variation gives

$$\int_{\Omega^*(D)} \int_0^1 \dot\omega(t) \cdot \frac{d}{dt} \boldsymbol{w}(t, \omega(t)) \, dt \, d\boldsymbol{\eta}(\omega, a) + \langle p, \operatorname{div} \boldsymbol{w} \rangle = 0.$$

This equation uniquely determines ∇p as a distribution, independently of the chosen minimizer $\boldsymbol{\eta}$; indeed, $\boldsymbol{\eta}$ enters in (4.5) only through $\mathscr{A}(\boldsymbol{\eta})$, which obviously is independent of the chosen minimiser, and so the above equation holds true for *every* minimiser $\boldsymbol{\eta}$. Since $\boldsymbol{w}$ is arbitrary, the first variation also leads to a weak formulation of Euler's equations

$$\partial_t \overline{\boldsymbol{v}}_t(x) + \operatorname{div} \big(\overline{\boldsymbol{v} \otimes \boldsymbol{v}}_t(x) \big) + \nabla_x p(t, x) = 0,$$

where $\overline{\boldsymbol{v}}_t$ and $\overline{\boldsymbol{v} \otimes \boldsymbol{v}}_t$ are implicitly defined by

$$\overline{\boldsymbol{v}}_t \mu_D = (e_t)_\#(\dot\omega(t)\boldsymbol{\eta}), \qquad \overline{\boldsymbol{v} \otimes \boldsymbol{v}}_t \mu_D = (e_t)_\#(\dot\omega(t) \otimes \dot\omega(t)\boldsymbol{\eta}).$$

Observe that in general $\overline{\boldsymbol{v} \otimes \boldsymbol{v}}_t \neq \overline{\boldsymbol{v}}_t \otimes \overline{\boldsymbol{v}}_t$. Indeed, since these models allow the passage of many fluid paths at the same point at the same time (i.e. branching and multiple velocities are possible), the product $\overline{\boldsymbol{v}}_t(x) \otimes \overline{\boldsymbol{v}}_t(x)$ of the mean velocity $\overline{\boldsymbol{v}}_t(x)$ with itself might be quite different from the mean value $\overline{\boldsymbol{v} \otimes \boldsymbol{v}}_t(x)$ of the product. This gap precisely marks the difference between genuine distributional solutions to Euler's equation and 'generalised' ones (see also [4, Section 2 and Paragraph 4.4] for more comments on this fact).

4.4 Necessary and sufficient optimality conditions

In this section we study necessary and sufficient optimality conditions for Brenier's variational problem and its extensions.

The basic remark is that *any* Borel integrable function $q : [0, 1] \times D \to \mathbb{R}$ with $\int_D q(t, \cdot) d\mu_D = 0$ for every $t \in [0, 1]$ induces a null-Lagrangian for the minimisation problem, with the incompressibility constraint; indeed,

$$\int_{\Omega^*(D)} \int_0^1 q(t, \omega(t)) \, dt \, d\eta(\omega, a) = \int_0^1 \int_D q(t, x) \, d\mu_D(x) \, dt = 0$$

for any generalized incompressible flow η. If we denote by

$$c_q^{0,1}(x, y) := \inf \left\{ \int_0^1 \frac{1}{2} |\dot{\omega}(t)|^2 - q(t, \omega(t)) \, dt : \omega(0) = x, \ \omega(1) = y \right\}$$

the value function for the Lagrangian $\mathcal{L}_q(\omega) := \int \frac{1}{2} |\dot{\omega}(t)|^2 - q(t, \omega(t)) \, dt$, we also have

$$\int_{\Omega^*(D)} \int_0^1 \frac{1}{2} |\dot{\omega}(t)|^2 - q(t, \omega(t)) \, dt \, d\eta(\omega, a) \geq \int_D c_q^{0,1}(a, h(a)) \, d\mu_D(a)$$

for any incompressible flow η between i and h. Moreover, equality holds if and only if η-almost every (ω, a) is a $c_q^{0,1}$-minimising path.

The following result, proved in [8, Section 3.6], shows that this lower bound is sharp with $q = p$, if p is sufficiently smooth.

Theorem 4.2. *Let u be a C^1 solution to the Euler equations in $[0, T] \times D$, whose pressure field p satisfies*

$$(*) \qquad T^2 \sup_{t \in [0,T]} \sup_{x \in D, |\xi| \leq 1} \langle \nabla_x^2 p(t, x) \xi, \xi \rangle \leq \pi^2.$$

Then the measure η induced by u via the flow map is optimal on $[0, T]$.

This follows by the fact that the integral paths of u satisfy $\ddot{\omega}(t) = -\nabla p(t, \omega(t))$, and $(*)$ implies that stationary paths for the action are also minimal for $\mathcal{L}_p$. (This is a consequence of the one-dimensional Poincaré inequality $\int_0^T |\dot{u}(t)|^2 \, dt \geq \frac{\pi^2}{T^2} \int_0^T |u(t)|^2 \, dt$ for all $u : [0, T] \to \mathbb{R}$ such that $\int_0^T u \, dt = 0$; see [5, Section 5] or [8, Proposition 3.2] for more details.)

The question investigated in [1] is: How far are these conditions from being *necessary*? C^1 regularity or even one-sided bounds on $\nabla^2 p$ are not realistic, so one has to look for necessary (and sufficient) conditions under much weaker regularity assumptions on p.

From now on, we restrict for simplicity to the case $D = \mathbb{T}^d$. The following regularity result for the pressure field has been obtained in [2], improving the regularity $\nabla p \in \mathcal{M}_{\text{loc}}((0, 1) \times \mathbb{T}^d)$ obtained in [8].

Theorem 4.3. *For any $\gamma, \eta \in \Gamma(\mathbb{T}^d)$ the unique pressure field given by Theorem 4.1 belongs to $L_{\text{loc}}^2((0, 1); BV(\mathbb{T}^d))$.*

The above result says in particular that p is a function, and not just a distribution. This allows one to define the value of p pointwise, which as we will see below will play a key role.

In order to guess the right optimality conditions, we recall that the two main degrees of freedom in optimal transport problems are:

- in moving mass from x to y, the path, or the family of paths, that should be followed;
- the amount of mass that should be moved, on each such path, from x to y.

The second degree of freedom is even more important in situations when more than one optimal path between x and y is available. As we will see, both things will depend on $\mathcal{L}_p$. But, since p is defined only up to negligible sets, the value of the Lagrangian $\mathcal{L}_p$ on a path ω is *not* invariant in the Lebesgue equivalence class; furthermore, no local pointwise bounds on p are available (remember that $p(t, \cdot)$ is only a BV function, with BV norm in $L^2_{\text{loc}}(0, 1)$). Therefore, as done in [1], one has to:

- Define a *precise* representative $\bar{p}$ in the Lebesgue equivalence class of p; it turns out that the 'correct' definition is

$$\bar{p}(t, x) := \liminf_{\varepsilon \downarrow 0} p(t, \cdot) * \phi_\varepsilon(x),$$

where $p(t, \cdot) * \phi_\varepsilon$ are suitable mollifications of $p(t, \cdot)$. Of course, this definition depends on the choice of the mollifiers, but we prove that a suitable choice of them provides a well-behaved (in the sense stated in Theorem 4.4) function $\bar{p}$.
- Consider, in the minimisation problem, only paths ω satisfying

$$Mp(t, \omega(t)) \in L^1_{\text{loc}}(0, 1),$$

where $Mp(t, \cdot)$ is a suitable maximal function of $p(t, \cdot)$ (see [1] for a more precise definition of the maximal operator).

With these constraints one can talk of *locally minimising path* ω for the Lagrangian $\mathcal{L}_{\bar{p}}$ and, correspondingly, define a family of value functions

$$c_{\bar{p}}^{s,t} : \mathbb{T}^d \times \mathbb{T}^d \to [-\infty, +\infty], \qquad [s, t] \subset (0, 1),$$

representing the cost of the minimal connection between x and y in the time interval (s, t):

$$c_{\bar{p}}^{s,t}(x, y) := \inf\left\{\int_s^t \frac{1}{2}|\dot{\omega}(\tau)|^2 - \bar{p}(\tau, \omega)\,d\tau : \omega(s) = x,\right.$$

$$\left. \omega(t) = y, Mp(\tau, \omega(\tau)) \in L^1(s, t)\right\}.$$

With this notation, the following result proved in [1] provides necessary and sufficient optimality conditions.

Theorem 4.4. *Let* $\eta = \eta_a \otimes \mu_{\mathrm{T}}$ *be an optimal incompressible flow between* $\eta = \eta_a \otimes \mu_{\mathrm{T}}$ *and* $\gamma = \gamma_a \otimes \mu_{\mathrm{T}}$. *Then*

(i) η *is concentrated on locally minimising paths for* $\mathcal{L}_{\bar{p}}$;
(ii) for all intervals $[s, t] \subset (0, T)$, *for* μ_{T}-*a.e.* a, *the plan* $(e_s, e_t)_{\#}\eta_a$ *is* $c_{\bar{p}}^{s,t}$-*optimal, i.e.*

$$\int_{\mathbb{T}^d \times \mathbb{T}^d} c_{\bar{p}}^{s,t}(x, y)\,d(e_s, e_t)_{\#}\eta_a \leq \int_{\mathbb{T}^d \times \mathbb{T}^d} c_{\bar{p}}^{s,t}(x, y)\,d\lambda$$

for any $\lambda \in \mathscr{P}(\mathbb{T}^d \times \mathbb{T}^d)$ *having the same marginals of* $(e_s, e_t)_{\#}\eta_a$.

Conversely, if (i), (ii) hold with $\bar{p}$ *replaced by some function* q *satisfying* $Mq \in L^1_{\mathrm{loc}}\left((0, 1); L^1(\mathbb{T}^d)\right)$, *then* η *is optimal, and* q *is the pressure field.*

Notice that an optimal transport problem is trivial if either the initial or the final measure is a Dirac mass; therefore, the second condition becomes meaningful when either $(e_s)_{\#}\eta_a$ or $(e_t)_{\#}\eta_a$ is not a Dirac mass. This corresponds to the case when $(e_s, \pi_a)_{\#}\eta$ is *not* induced by a map, a phenomenon that cannot be ruled out, as we discussed in Section 4.2.2. In the example presented in Section 4.2.2 the pressure field $p(x) = |x|^2/2$ is smooth and time independent, but the initial and final conditions are chosen in such a way that a continuum of action-minimising paths (4.4) between x and $-x$ exists. As shown in [4], there are infinitely many incompressible flows connecting the identity map i to $-i$, which moreover induce infinitely many distributional solutions to the Euler equations [4, Paragraph 4.4].

The results in [1] show a connection with the theory of action-minimising measures, though in this case the Lagrangian $\int_0^1 \frac{1}{2}|\dot{\omega}(t)|^2 - \bar{p}(t, \omega(t))\,dt$ is possibly non-smooth and not given a priori, but *generated* by the variational problem itself.

Here, we see a nice variation on a classical theme of calculus of variations: a field of (smooth, non-intersecting) *extremals* gives rise both to *minimisers* and to an incompressible flow *in phase space*. Here, instead, we have a field of (possibly non-smooth, or intersecting) *minimisers* which has to produce an incompressible flow *in the state space*. This structure seems to be rigid, and it might lead to new regularity results for the pressure field.

Let us also recall that, as recently shown in [11], under a $W^{1,p}$-regularity of the pressure p one can show that η-a.e. ω solves the Euler–Lagrange equations and belongs to $W^{2,p}([0, 1]) \subset C^1([0, 1])$. This result is a first step towards the BV case, where one can still expect that the minimality of η may allow one to prove higher regularity on the minimising curves ω (like $\dot{\omega} \in BV$).

References

[1] L. Ambrosio and A. Figalli, Geodesics in the space of measure-preserving maps and plans, *Arch. Ration. Math. Anal.*, **194** (2009), no. 2, 421–462.

[2] L. Ambrosio and A. Figalli, On the regularity of the pressure field of Brenier's weak solutions to incompressible Euler equations, *Calc. Var. Partial Dif. Equations*, **31** (2008), 497–509.

[3] V. Arnold, Sur la géométrie différentielle des groupes de Lie de dimension infinie et ses applications à l'hydrodynamique des fluides parfaits, *Ann. Inst. Fourier (Grenoble)*, **16** (1966), fasc. 1, 319–361.

[4] M. Bernot, A. Figalli and F. Santambrogio, Generalized solutions for the Euler equations in one and two dimensions, *J. Math. Pures Appl.*, **91** (2008), no. 2, 137–155.

[5] Y. Brenier, The least action principle and the related concept of generalized flows for incompressible perfect fluids, *J. Am. Math. Soc.*, **2** (1989), 225–255.

[6] Y. Brenier, Polar factorization and monotone rearrangement of vector-valued functions, *Commun. Pure Appl. Math.*, **44** (1991), 375–417.

[7] Y. Brenier, The dual least action problem for an ideal, incompressible fluid, *Arch. Rational Mech. Anal.*, **122** (1993), 323–351.

[8] Y. Brenier, Minimal geodesics on groups of volume-preserving maps and generalized solutions of the Euler equations, *Commun. Pure Appl. Math.*, **52** (1999), 411–452.

[9] Y. Brenier and W. Gangbo, L^p approximation of maps by diffeomorphisms, *Calc. Var. Partial Dif. Equations*, **16** (2003), no. 2, 147–164.

[10] D.G. Ebin and J. Marsden, Groups of diffeomorphisms and the motion of an ideal incompressible fluid, *Ann. Math.*, **2** (1970), 102–163.

[11] A. Figalli and V. Mandorino Fine properties of minimizers of mechanical Lagrangians with Sobolev potentials, *Discrete Contin. Dyn. Syst.*, **31** (2011), no. 4, 1325–1346.

[12] A.I. Shnirelman, The geometry of the group of diffeomorphisms and the dynamics of an ideal incompressible fluid, *Mat. Sb. (N.S.)*, **128 (170)** (1985), no. 1, 82–109 (in Russian).

[13] A.I. Shnirelman, Generalized fluid flows, their approximation and applications, *Geom. Funct. Anal.*, **4** (1994), no. 5, 586–620.

[14] C. Villani, *Topics in Optimal Transportation,* Graduate Studies in Mathematics, vol. 58, American Mathematical Society, Providence, RI, 2003.

[15] V. Yudovich, Nonstationary flow of an ideal incompressible liquid, *Zhurn. Vych. Mat.,* **3** (1963), 1032–1066.

[16] V. Yudovich, Some bounds for solutions of elliptic equations, *Am. Math. Soc. Transl.* (2) **56** (1962).

5

Ricci flow: the foundations via optimal transportation

PETER TOPPING

Contents

5.1 Overview

Since the creation of Ricci flow by Hamilton in 1982, a rich theory has been developed in order to understand the behaviour of the flow, and to analyse the singularities that may occur, and these developments have had profound applications, most famously to the Poincaré conjecture. At the heart of the theory lie a large number of *a priori* estimates and geometric constructions, which include most notably the Harnack estimates of Hamilton, the $\mathcal{L}$-length of Perelman (in the spirit of Li-Yau), the logarithmic Sobolev inequality arising from Perelman's $\mathcal{W}$-entropy, and the reduced volume of Perelman, amongst others.

The objective of these lectures is to explain this theory from the point of view of optimal transportation. As I explain in Section 5.4, Ricci flow and optimal transportation combine rather well, and we will see fundamental but elementary aspects of this when we see in Theorem 5.2 how diffusions contract under reverse-time Ricci flow. However, the key to the whole theory is to realise to which object one should apply this result: not the original Ricci flow, but a new Ricci flow derived from the original one, on a base manifold of one higher dimension, that we call the canonical soliton. In this way, essentially the entire foundational theory of Ricci flow mentioned above drops out naturally.

Throughout the lectures I emphasise the intuition; the objective is to demonstrate how one can discover the theory rather than treat it as a black box that just happens to work.

5.2 Prerequisites

It would be helpful, but not essential, to have some basic idea of what Ricci flow is before proceeding. The relevant reference is [23], and I will refer back to that monograph during these lectures.

I will assume the basic notions of manifolds and Riemannian geometry, including Riemannian metrics, Riemannian distance, geodesics and geodesic balls. To understand the details of the lectures, one would need also to understand a little about curvature, including the curvature tensor Rm, Ricci tensor Ric, scalar curvature R, sectional curvature and hopefully also the curvature operator $\mathcal{R}$. Occasionally we will pull back tensors by diffeomorphisms and consider Lie derivatives, etc. We will need the most basic Bochner formula (which follows from a short computation in Riemannian geometry). I will assume you know how to integrate vector fields on manifolds (i.e. follow flow-lines) and understand the corresponding notion of 'complete vector field'.

From partial differential equation theory I will assume very little, except basic facts and intuition concerning the heat equation. In particular, we need the parabolic maximum principle.

I will describe the theory of optimal transportation that we require, such as the definition of Wasserstein distance, but it may help to have seen these basic notions already.

5.3 Introduction to Ricci flow and diffusions

5.3.1 The equation

The Ricci flow is a way of deforming a Riemannian manifold under a nonlinear evolution equation in order to process or improve it. To be a Ricci flow, we ask a smooth one-parameter family $g(t)$ of Riemannian metrics on a manifold $\mathcal{M}$ to satisfy the nonlinear partial differential equation

$$\frac{\partial g}{\partial t} = -2\text{Ric}(g), \tag{5.1}$$

where $\text{Ric}(g)$ is the Ricci curvature of the metric $g(t)$, a section of $\text{Sym}^2 T^*\mathcal{M}$; see [9].

There are a number of points to recall about Ricci curvature that may help here and later in these notes. First, when $\mathcal{M}$ is two-dimensional, $\text{Ric}(g)$ is just the Gaussian curvature times the metric g. More generally, one should view $-\text{Ric}(g)$ as some sort of Laplacian of g, as discussed in [23, Section 1.1], so the Ricci flow is like a heat equation for g (albeit nonlinear, and liable to develop singularities). Generally in the subject of these notes, it is useful to view $\text{Ric}(g)$ as a quantity which controls *volume growth*. In particular, it governs the change in the volume element along a geodesic.

The interpretation of Ricci flow as some sort of diffusion equation for a Riemannian metric is brought out when one computes the evolution equation governing the curvatures of a Ricci flow. For example, the scalar curvature R (which is the trace of the Ricci tensor) satisfies

$$\frac{\partial R}{\partial t} = \Delta_{g(t)} R + 2|\text{Ric}|^2, \tag{5.2}$$

while the full curvature tensor Rm and the Ricci tensor Ric satisfy related equations [23, Section 2.5]. Note that $|\text{Ric}|$ is the norm of Ric using the norm induced by g (i.e. $|\text{Ric}|^2 = R_{ij} R^{ij}$). For further explanation, see [23, Chapter 2].

One useful consequence of (5.2), thanks to the maximum principle (whenever it applies), is that positive (or weakly positive) scalar curvature is preserved. That is, if $R(g(0)) \geq 0$ then $R(g(t)) \geq 0$ for $t > 0$. This is just one instance of a recurring theme; many positive curvature conditions are preserved; for

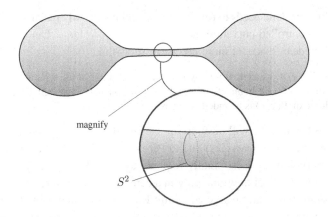

Figure 5.1 Blowing up.

example, positive curvature operator $\mathcal{R}(g) \geq 0$. (See [23] for the definition of curvature operator.) Other recent breakthoughs [3, 16] have involved showing that the notion of *positive isotropic curvature* is preserved, and we will briefly return to this notion in Section 5.6.2.

Before proceeding, we note some simple examples of Ricci flows relevant to Section 5.3.2. First, if we deform a round sphere under the Ricci flow then it will shrink, owing to its positive curvature, and disappear to nothing within a finite time. (See Section 5.3.3 for more details.) Also, the Cartesian product of two Ricci flows is another Ricci flow; in particular, a cylinder $S^n \times \mathbb{R}$ can be evolved under Ricci flow simply by shrinking the S^n component.

5.3.2 Rapid overview of Ricci flow theory, and motivations

Ricci flow has become a powerful tool for understanding the topology of manifolds, particularly three-dimensional manifolds because it is in that dimension that the singularities of Ricci flow can be analysed to a sufficiently fine degree. A typical singularity occurs because a part of the manifold 'pinches a neck'.[1]

At a singularity, the curvature blows up. More precisely, the maximum sectional curvature of $g(t)$ becomes unbounded as t increases to some singular time T. One would like to look at this flow through a microscope (i.e. rescale parabolically, thus reducing the curvature) and try to extract a limiting flow. For example, in the case of a neck pinch, the limit flow should be a cylinder $S^2 \times \mathbb{R}$ with the S^2 part shrinking as time increases (see Figure 5.1).

The technical details of this process need not concern us here, except that it turns out to be possible to find this limit only if the flow does not degenerate in

---

[1] Thanks to Neil Course for drawing Figure 5.1.

a particular way. One needs to rule out 'local collapsing' where balls of a given radius, and controlled curvature, have much smaller volume than they would have in Euclidean space, or more precisely, show that:

> There exists $\kappa > 0$ such that for any $t < T$ and $r \in (0, 1]$, if the sectional curvatures of $g(t)$ are bounded in norm by r^{-2} in some geodesic ball $B(x, r)$ then the volume of $B(x, r)$ is bounded below by κr^n.

In these lectures we will develop the elegant theory which guarantees this (and many other things besides) due to Perelman [18], via the theory of optimal transportation. This essentially requires us to find a certain log-Sobolev inequality on our Ricci flow, and plug in a carefully chosen function.

Now suppose we have extracted a limit Ricci flow (which we will also call $g(t)$), for example the shrinking cylinder mentioned above. This turns out to be defined for all $t < 0$, and when $\mathcal{M}$ is three-dimensional, a result of Hamilton and Ivey [11] tells us that it has positive curvature. It is a crucial step to analyse the geometry of limits such as $g(t)$, and in these lectures we will see a number of important techniques and tools required in this analysis springing out of the optimal transportation viewpoint, including Perelman's $\mathcal{L}$-length and the theory thereof. A further big step is to be able to relate the (scalar) curvatures at different points in space and time on such a flow, and in this direction we will be led very naturally to the following *Harnack* inequality which first arises in the work of Hamilton [10]:

> For $x_1, x_2 \in \mathcal{M}$ and times $t_1 < t_2 < 0$ on a three-dimensional limit Ricci flow, we have
> $$\frac{R(x_2, t_2)}{R(x_1, t_1)} \geq \exp\left(-\frac{d_{g(t_1)}^2(x_1, x_2)}{2(t_2 - t_1)}\right).$$

One further issue in the analysis of these limit flows is to understand what they look like at very large scales, very far back in time. The theory developed in these lectures can also be used to show that they look like solitons. (The original reference for this is [18, Section 11.2].)

5.3.3 Ricci solitons

The most elementary examples of Ricci flows are induced by Einstein metrics – i.e. metrics g_0 on $\mathcal{M}$ that satisfy $\mathrm{Ric}(g_0) = \lambda g_0$, for $\lambda \in \mathbb{R}$. An obvious Ricci flow starting at g_0 is given by[2]

$$g(t) = (1 - 2\lambda t)g_0. \tag{5.3}$$

---

[2] Recall that scaling a metric does not change the Ricci tensor: $\mathrm{Ric}(\lambda g) = \mathrm{Ric}(g)$ for any $\lambda > 0$.

For example, a round sphere would shrink homothetically to nothing in a finite time, and a hyperbolic manifold would expand homothetically indefinitely.

A more general notion of self-similar solution is that of a *Ricci soliton flow*, for which the manifold $(\mathcal{M}, g(t))$ is again isometric to a homothetic scaling of $(\mathcal{M}, g(0))$, although this time not necessarily via the identity map.

Definition 1. We call the combination $(\mathcal{M}, g, f)$ of a Riemannian manifold $(\mathcal{M}, g)$ together with a smooth function $f : \mathcal{M} \to \mathbb{R}$ a *(gradient) Ricci soliton* if it satisfies the equation

$$\mathrm{Ric}(g) + \mathrm{Hess}_g(f) = \lambda g. \tag{5.4}$$

Moreover, we call $(\mathcal{M}, g, f)$ either *steady*, *expanding* or *shrinking* depending on whether $\lambda = 0$, $\lambda < 0$ or $\lambda > 0$ respectively.

The significance of such a combination of manifold and function is that it induces a Ricci flow generalising the Einstein case. It turns out (clarified only recently [26]) that if the manifold is complete, then simply by virtue of f satisfying Equation (5.4), the vector field ∇f is *complete*. (That is, one can follow the flow-lines of this vector field without falling off the manifold.) In particular, the vector field generates a family of diffeomorphisms $\psi_t : \mathcal{M} \to \mathcal{M}$ with ψ_0 the identity, according to

$$\frac{\partial}{\partial t}(\psi_t(y)) = \frac{\nabla f(\psi_t(y))}{1 - 2\lambda t} \tag{5.5}$$

for $t < \frac{1}{2\lambda}$ if $\lambda > 0$ and $t > \frac{1}{2\lambda}$ if $\lambda < 0$. We may then define

$$G(t) := (1 - 2\lambda t)\psi_t^*(g), \tag{5.6}$$

and compute (see also [23, Section 1.2])

$$\begin{aligned}
\frac{\partial G(t)}{\partial t} &= -2\lambda \psi_t^*(g) + (1 - 2\lambda t)\psi_t^*(\mathcal{L}_{\frac{\nabla f(\psi_t(y))}{1 - 2\lambda t}} g) \\
&= \psi_t^*(-2\lambda g + \mathcal{L}_{\nabla f(\psi_t(y))} g) \\
&= \psi_t^*(-2\lambda g + 2\mathrm{Hess}_g(f)) \\
&= \psi_t^*(-2\mathrm{Ric}(g)) \\
&= -2\mathrm{Ric}(\psi_t^* g) \\
&= -2\mathrm{Ric}(G(t))
\end{aligned}$$

to see that $G(t)$ is in fact a Ricci flow generalising (5.3); this motivates:

Definition 2. We call a flow $G(t)$ constructed in this way a *Ricci soliton flow*.

Ricci solitons occur throughout the theory of Ricci flow. Our canonical solitons alluded to earlier will be exotic examples.

5.3.4 Diffusion and the heat equation on manifolds and flows

Consider a solution $u : \mathcal{M} \times [0, T] \to [0, \infty)$ to the heat equation

$$\frac{\partial u}{\partial t} = \Delta u,$$

on a closed (compact, no boundary) Riemannian manifold $(\mathcal{M}, g)$, with $\int_{\mathcal{M}} u(\cdot, 0) d\mu_g = 1$, where μ_g is the Riemannian volume measure. Since

$$\frac{d}{dt} \int_{\mathcal{M}} u(\cdot, t) d\mu_g = \int_{\mathcal{M}} \Delta u \, d\mu_g = 0,$$

the integral of u over $\mathcal{M}$ remains one for each $t \in [0, T]$, and in particular, u can be viewed as a probability density of a measure $\nu(t)$ defined by $d\nu(t) := u(\cdot, t) d\mu_g$. This one-parameter family of measures $\nu(t)$ represents the probability distribution of a particle moving on the manifold under Brownian motion, with initial probability distribution $\nu(0)$, and will be an example of what we will call a 'diffusion' in these lectures.

In practice, we need to generalise these familiar notions to the case of diffusion on an *evolving* manifold.

Definition 3. Suppose $g(\tau)$ is a family of Riemannian metrics on $\mathcal{M}$, for $\tau \in [0, T]$. We call a flow of measures $\nu(\tau)$ a **diffusion** (representing the probability distribution of a Brownian particle as above) if $d\nu(\tau) = u(\cdot, \tau) d\mu_{g(\tau)}$ with u satisfying the *conjugate* heat equation

$$\frac{\partial u}{\partial \tau} = \Delta_{g(\tau)} u - \left(\frac{1}{2} \mathrm{tr} \frac{\partial g}{\partial \tau} \right) u. \tag{5.7}$$

The single most important situation is that where $g(\tau)$ solves the equation $\frac{\partial g}{\partial \tau} = 2\mathrm{Ric}(g(\tau))$, which makes $g(\cdot)$ a Ricci flow with respect to a 'reverse' time coordinate τ (that is, $\tau = C - t$ for some constant C). In that case, u would satisfy

$$\frac{\partial u}{\partial \tau} = \Delta_{g(\tau)} u - Ru, \tag{5.8}$$

where $R = \mathrm{tr}\mathrm{Ric}$ is the scalar curvature as before.

Note that the extra term in (5.7) compensates for the fact that the Riemannian volume measure $\mu_{g(\tau)}$ is no longer constant in time – in fact $\frac{\partial}{\partial \tau} d\mu_{g(\tau)} = \frac{1}{2}\mathrm{tr}\frac{\partial g}{\partial \tau} d\mu_{g(\tau)}$ – to keep $\nu(\tau)$ a *probability* measure:

$$\frac{d}{d\tau} \int_{\mathcal{M}} d\nu(\tau) = \frac{d}{d\tau} \int_{\mathcal{M}} u(\cdot, \tau) d\mu_{g(\tau)} = \int_{\mathcal{M}} \left(\frac{\partial u}{\partial \tau} d\mu_{g(\tau)} + u \frac{\partial}{\partial \tau} d\mu_{g(\tau)} \right)$$

$$= \int_{\mathcal{M}} \Delta_{g(\tau)} u \, d\mu_{g(\tau)} = 0.$$

We will also need the following more general statement which justifies the terminology *conjugate* heat equation above:

Proposition 5.1. *If $g(\tau)$ is a flow of Riemannian metrics for $\tau \in [a, b]$, and $\nu(\tau)$ is a diffusion (as defined above) then given any $f : \mathcal{M} \times [a, b] \to \mathbb{R}$ solving $-\frac{\partial f}{\partial \tau} = \Delta_{g(\tau)} f$, we have*

$$\frac{d}{d\tau} \int_{\mathcal{M}} f(\cdot, \tau) d\nu(\tau) = 0. \tag{5.9}$$

In the following, we will see that the notion of diffusion discussed in this section interacts extremely well both with Ricci flow and with optimal transportation.

5.4 Optimal transportation

In this section we give an overview of the topic of optimal transportation on evolving manifolds. For an introduction to optimal transportation on fixed manifolds, see [25].

5.4.1 Wasserstein distance

Suppose that $(\mathcal{M}, g)$ is a closed Riemannian manifold, and ν_1 and ν_2 are two Borel probability measures on $\mathcal{M}$.[3] For $p \in [1, \infty)$, the p-Wasserstein distance W_p between ν_1 and ν_2 is defined to be

$$W_p^g(\nu_1, \nu_2) := \left[\inf_{\pi \in \Gamma(\nu_1, \nu_2)} \int_{\mathcal{M} \times \mathcal{M}} d^p(x, y) d\pi(x, y) \right]^{1/p}, \tag{5.10}$$

where $d(\cdot, \cdot)$ is the Riemannian distance function induced by g and $\Gamma(\nu_1, \nu_2)$ is the space of probability measures on $\mathcal{M} \times \mathcal{M}$ with marginals ν_1 and ν_2. (In other words, $\pi \in \Gamma(\nu_1, \nu_2)$ if the push forward[4] of π under the projections $\mathcal{M} \times \mathcal{M} \mapsto \mathcal{M}$ onto the first and second components are ν_1 and ν_2 respectively. Or put another way, if $A \subset \mathcal{M}$ then $\pi(A \times \mathcal{M}) = \nu_1(A)$ and $\pi(\mathcal{M} \times A) = \nu_2(A)$.)

5.4.2 Contractivity for diffusions on evolving manifolds

A basic principle in the subject is that two probability measures evolving under an appropriate diffusion equation should get closer in the Wasserstein

---

[3] From now on, all maps, measures and sets considered will be assumed to be Borel.

[4] Given a map $F : M \to \hat{M}$ between manifolds and a probability measure ν on M, the push-forward $F_{\#}\nu$ of ν under F is the probability measure on $\hat{M}$, defined by $(F_{\#}\nu)[V] = \nu[F^{-1}(V)]$ for all $V \subset \hat{M}$.

sense provided the manifold satisfies some curvature condition, most famously positive Ricci curvature. This principle has many incarnations in the literature; see Sturm and von Renesse [21] and the references therein. The geometric side of the subject can be followed back to the work of Otto and Villani [17].

In [15], McCann and Topping showed that this type of contractivity on an *evolving* manifold $(\mathcal{M}, g(\tau))$ *characterises* super-solutions of the Ricci flow (parameterised backwards in time) by which we mean solutions to

$$\frac{\partial g}{\partial \tau} \leq 2\operatorname{Ric}(g(\tau)). \tag{5.11}$$

Here, the relevant notion of diffusion on an evolving manifold $(\mathcal{M}, g(\tau))$ was given in Definition 3.

Theorem 5.2 (cf. [15, Theorem 2]). *Suppose that $\mathcal{M}$ is a closed manifold equipped with a smooth family of metrics $g(\tau)$ for $\tau \in [\tau_1, \tau_2] \subset \mathbb{R}$. Then the following are equivalent:*

(A) $g(\tau)$ is a super Ricci flow (i.e. satisfies (5.11));

(B_1) whenever $\tau_1 < a < b < \tau_2$ and $\nu_1(\tau)$, $\nu_2(\tau)$ are diffusions (Definition 3) for $\tau \in (a, b)$, the function $\tau \mapsto W_1^{g(\tau)}(\nu_1(\tau), \nu_2(\tau))$ is weakly decreasing in $\tau \in (a, b)$;

(B_2) whenever $\tau_1 < a < b < \tau_2$ and $\nu_1(\tau)$, $\nu_2(\tau)$ are diffusions (Definition 3) for $\tau \in (a, b)$, the function $\tau \mapsto W_2^{g(\tau)}(\nu_1(\tau), \nu_2(\tau))$ is weakly decreasing in $\tau \in (a, b)$;

(C) whenever $\tau_1 < a < b < \tau_2$ and $f : \mathcal{M} \times (a, b) \to \mathbb{R}$ is a solution to $-\frac{\partial f}{\partial \tau} = \Delta_{g(\tau)} f$, the Lipschitz constant of $f(\cdot, \tau)$ with respect to $g(\tau)$ (i.e. the function $\tau \mapsto \sup_{\mathcal{M}} |\nabla f(\cdot, \tau)|$) is weakly increasing in τ.

The characterisation $(A) \Longleftrightarrow (B_2) \Longleftrightarrow (C)$ was proved in [15]. It has since been extended to handle the W_p distance for arbitrary p [1]. The only other case relevant to us is the case $p = 1$ (i.e. (B_1)); the extension to $p = 1$ was first observed by Tom Ilmanen.

By taking flows $g(\tau)$ which are independent of τ in Theorem 5.2, one recovers an existing theory of contractivity on manifolds of weakly positive Ricci curvature (see [21] and the references therein).

In some sense, the most important proof is the implication $(A) \Longrightarrow (B_2)$ [15] because the techniques therein extend to give the rigorous proof of the key result, Theorem 5.5, which we describe in the Section 5.5.4. However, here we will describe the more elementary (but elegant) implications $(A) \Longrightarrow (C)$ and $(A) \Longrightarrow (B_1)$ and use them in Section 5.5 to give a *formal* derivation of Theorem 5.5.

Proof. (A) $\implies$ (C). Define $t = b - \tau$ so that f satisfies the standard heat equation $\frac{\partial f}{\partial t} = \Delta f$ on the super Ricci flow $g(\cdot)$. We remark that it is important that the heat equation 'flows' at the same rate as the Ricci flow, by which we mean that if f satisfied $\frac{\partial f}{\partial t} = c\Delta f$ for $c \neq 1$, then the implication would not work.

We compute (taking care with signs which can be confusing because g is the notation used both for the metric on $T\mathcal{M}$ and $T^*\mathcal{M}$)

$$
\begin{aligned}
\frac{\partial |\nabla f|^2}{\partial t} &= \frac{\partial |df|^2}{\partial t} \\
&= -\frac{\partial g}{\partial t}(df, df) + 2\langle d\frac{\partial f}{\partial t}, df\rangle \\
&= \frac{\partial g}{\partial \tau}(\nabla f, \nabla f) + 2\langle d\Delta f, df\rangle \\
&= \frac{\partial g}{\partial \tau}(\nabla f, \nabla f) - 2\mathrm{Ric}(\nabla f, \nabla f) + \Delta|\nabla f|^2 - 2|\mathrm{Hess}(f)|^2,
\end{aligned}
\tag{5.12}
$$

by the Bochner formula [8, (4.15)]. By the definition (5.11) of super Ricci flow, we see that

$$
\frac{\partial |\nabla f|^2}{\partial t} \leq \Delta|\nabla f|^2,
$$

and therefore by the maximum principle (for example, see [23, Section 3.1]) we conclude that $\sup_{\mathcal{M}} |\nabla f|^2$ is a decreasing function of t. $\qquad\square$

Proof. (A) $\implies$ (B_1). Suppose that $g(\tau)$, $\nu_1(\tau)$ and $\nu_2(\tau)$ are defined for τ in a neighbourhood of $\tau_0 \in (a, b)$. By Kantorovich–Rubinstein duality (e.g. see [25, Section 1.2.1]) we have

$$
W_1^{g(\tau)}(\nu_1(\tau), \nu_2(\tau)) = \max\left\{ \int_{\mathcal{M}} \varphi d\nu_1(\tau) - \int_{\mathcal{M}} \varphi d\nu_2(\tau) \;\middle|\; \varphi : \mathcal{M} \to \mathbb{R} \right.
$$
$$
\left. \text{is Lipschitz and } \|\varphi\|_{Lip} \leq 1 \text{ with respect to } g(\tau) \right\}.
\tag{5.13}
$$

Let $\varphi_0 : \mathcal{M} \to \mathbb{R}$ be a function that achieves the maximum in this variational problem at time τ_0, and extend φ_0 to a function $\varphi : \mathcal{M} \times [\tau_0 - \varepsilon, \tau_0] \to \mathbb{R}$ for some $\varepsilon > 0$ by solving the equation

$$
-\frac{\partial \varphi}{\partial \tau} = \Delta_{g(\tau)}\varphi.
$$

By the implication (A) $\implies$ (C) of the theorem (proved above) for all $\tau \in [\tau_0 - \varepsilon, \tau_0]$ we have $\|\varphi(\cdot, \tau)\|_{Lip} \leq 1$ and therefore $\varphi(\cdot, \tau)$ can be used as a competitor in the variational problem (5.13) to see that

$$W_1^{g(\tau)}(\nu_1(\tau), \nu_2(\tau)) \geq \int_{\mathcal{M}} \varphi(\cdot, \tau) d\nu_1(\tau) - \int_{\mathcal{M}} \varphi(\cdot, \tau) d\nu_2(\tau).$$

But by Proposition 5.1 we know that the functions

$$\tau \mapsto \int_{\mathcal{M}} \varphi(\cdot, \tau) d\nu_1(\tau) \qquad \text{and} \qquad \tau \mapsto \int_{\mathcal{M}} \varphi(\cdot, \tau) d\nu_2(\tau)$$

are each independent of τ, so we deduce that

$$W_1^{g(\tau)}(\nu_1(\tau), \nu_2(\tau)) \geq W_1^{g(\tau_0)}(\nu_1(\tau_0), \nu_2(\tau_0)). \qquad \square$$

As for the implication $(A) \implies (B_2)$, one considers Wasserstein geodesics between $\nu_1(\tau)$ and $\nu_2(\tau)$ and analyses the extent to which the Boltzmann entropy is convex along these geodesics. See [15] for the full story.

5.5 The canonical solitons

5.5.1 Definition of the canonical shrinking soliton

In Section 5.4, we saw an elementary but fundamental result combining Ricci flow and optimal transportation. It turns out that there is a surprising way of applying it to obtain more exotic and powerful corollaries. The trick, when trying to understand a Ricci flow $(\mathcal{M}^n, g(t))$, $t \in I$, is to not apply the result directly to that Ricci flow, but to allow that Ricci flow to induce a more elaborate Ricci flow on the $(n + 1)$-dimensional manifold $\mathcal{M} \times I$ which we engineer to be a soliton flow, and to apply the results of the previous section (and other fundamental results) to that instead.

We call these new $(n + 1)$-dimensional Ricci flows *canonical soliton flows*, and by considering them from different viewpoints we can construct the main foundations of Ricci flow in a natural way. We get steady, shrinking and expanding canonical solitons for each Ricci flow, each of which have different applications. To be precise, we get approximate solitons which can be viewed as exact in the limit of some parameter, as we now clarify in the shrinking case.

Theorem 5.3 (Cabezas-Rivas and Topping [4]). *Suppose $g(\tau)$ is a (reverse) Ricci flow – i.e. a solution of $\frac{\partial g}{\partial \tau} = 2\text{Ric}(g(\tau))$ – defined for τ within a bounded time interval $I \subset (0, \infty)$, on a manifold $\mathcal{M}$ of dimension $n \in \mathbb{N}$, and with bounded curvature. Suppose $N > 0$ is sufficiently large to give a positive definite metric $\hat{g}$ on $\hat{\mathcal{M}} := \mathcal{M} \times I$ defined by*

$$\hat{g}_{ij} = \frac{g_{ij}}{\tau}; \qquad \hat{g}_{00} = \frac{N}{2\tau^3} + \frac{R}{\tau} - \frac{n}{2\tau^2}; \qquad \hat{g}_{0i} = 0,$$

where i, j are coordinate indices on the $\mathcal{M}$ factor, 0 represents the index of the time coordinate $\tau \in I$, and the scalar curvature of g is written as R.

Then up to errors of order $\frac{1}{N}$, the metric $\hat{g}$ is a gradient shrinking Ricci soliton on the higher dimensional space $\hat{\mathcal{M}}$:

$$\text{Ric}(\hat{g}) + \text{Hess}_{\hat{g}}\left(\frac{N}{2\tau}\right) \simeq \frac{1}{2}\hat{g}, \qquad (5.14)$$

by which we mean that the quantity

$$N\left[\text{Ric}(\hat{g}) + \text{Hess}_{\hat{g}}\left(\frac{N}{2\tau}\right) - \frac{1}{2}\hat{g}\right]$$

is locally bounded independently of N, with respect to any fixed metric on $\hat{\mathcal{M}}$.

We call $\hat{g}$ the canonical shrinking soliton associated with $g(\cdot)$.

Notice that, given a Ricci flow over the time interval I, we are turning the old time parameter τ into an additional space parameter; it then makes sense to introduce a new time parameter (called s in what follows) and consider Ricci flow using (part of) $\mathcal{M} \times I$ as the underlying manifold by taking the Ricci soliton flow $G(s)$ as described in Section 5.3.3; this is what we call the canonical shrinking soliton flow.

The essential point of this theorem is the form of the metric $\hat{g}$. The verification of the result is a calculation. Some parts of that calculation are given in Proposition 5.4; more involved calculations will be required in Section 5.6 for the canonical *expanding* soliton, and can be found in [5].

Proposition 5.4 (see [4]). *Fixing $\tau > 0$, a time at which the Ricci flow exists, and fixing local coordinates $\{x^i\}$ in a neighbourhood U of some $p \in \mathcal{M}$, then in any neighbourhood $V \subset\subset U \times I$ of (p, τ), we have*

$$\hat{R}_{ij} \simeq R_{ij}; \qquad \hat{R}_{i0} \simeq -\frac{1}{2}\nabla_i R; \qquad \hat{R}_{00} \simeq -\frac{R_\tau}{2} - \frac{R}{2\tau},$$

where $\simeq$ denotes equality of the coefficients up to an error bounded in magnitude by $\frac{C}{N}$, with $C > 0$ a constant independent of N (but depending on V and the choice of coordinates). Moreover, we have

$$\hat{\nabla}^2_{ij}\left(\tfrac{N}{2\tau}\right) \simeq \frac{g_{ij}}{2\tau} - R_{ij}; \qquad \hat{\nabla}^2_{i0}\left(\tfrac{N}{2\tau}\right) \simeq \frac{\nabla_i R}{2};$$

$$\hat{\nabla}^2_{00}\left(\tfrac{N}{2\tau}\right) \simeq \frac{N}{4\tau^3} + \frac{R}{\tau} - \frac{n}{4\tau^2} + \frac{R_\tau}{2}.$$

By combining the formulae of Proposition 5.4 and the definition of $\hat{g}$, we deduce Theorem 5.3.

The creation of the canonical solitons follows a chain of previous work [6, 7, 18], which we discuss further in Section 5.6.

5.5.2 Perelman's $\mathcal{L}$-length

It is a general principle that standard concepts and theorems in Riemannian geometry applied to a canonical soliton should yield something interesting about the original Ricci flow. The first such instance of this is that considering Riemannian length or distance on $(\hat{\mathcal{M}}, \hat{g})$ induces the fundamental concept of $\mathcal{L}$-length due to Perelman [18] based on the ideas of Li and Yau [13].

Suppose that $x, y \in \mathcal{M}$ and $[\tau_1, \tau_2] \subset (0, \infty)$ lies within the time domain on which the (reverse) Ricci flow is defined. Consider paths $\Gamma : [\tau_1, \tau_2] \to \hat{\mathcal{M}}$ connecting (x, τ_1) and (y, τ_2) in $\hat{\mathcal{M}}$ of the form $\Gamma(\tau) = (\gamma(\tau), \tau)$, where $\gamma : [\tau_1, \tau_2] \to \mathcal{M}$ satisfies $\gamma(\tau_1) = x$ and $\gamma(\tau_2) = y$. Then

$$
\begin{aligned}
\text{Length}(\Gamma) &= \int_{\tau_1}^{\tau_2} \left| \gamma'(\tau) + \frac{\partial}{\partial \tau} \right|_{\hat{g}} d\tau \\
&= \int_{\tau_1}^{\tau_2} \left[\frac{|\gamma'|^2_{g(\tau)}}{\tau} + \frac{N}{2\tau^3} + \frac{R}{\tau} - \frac{n}{2\tau^2} \right]^{1/2} d\tau \\
&= \int_{\tau_1}^{\tau_2} \left[\frac{N}{2\tau^3} \right]^{1/2} \left(1 + \frac{2\tau^2 |\gamma'|^2_{g(\tau)}}{N} + \frac{2\tau^2 R}{N} - \frac{\tau n}{N} \right)^{1/2} d\tau \quad (5.15) \\
&= \int_{\tau_1}^{\tau_2} \left[\frac{N}{2\tau^3} \right]^{1/2} \left(1 + \frac{\tau^2 |\gamma'|^2}{N} + \frac{\tau^2 R}{N} - \frac{\tau n}{2N} + O(\tfrac{1}{N^2}) \right) d\tau \\
&= \sqrt{2N}(\tau_1^{-1/2} - \tau_2^{-1/2}) + \frac{1}{\sqrt{2N}} \hat{\mathcal{L}}(\gamma) + O(\tfrac{1}{N^{3/2}}),
\end{aligned}
$$

where

$$
\hat{\mathcal{L}}(\gamma) := \int_{\tau_1}^{\tau_2} \sqrt{\tau} \left(R(\gamma(\tau), \tau) + |\gamma'(\tau)|^2 - \frac{n}{2\tau} \right) d\tau. \quad (5.16)
$$

Thus, we get an expansion for the length of Γ in terms of N. Unsurprisingly, given the extreme stretching of the $\frac{\partial}{\partial \tau}$ direction in the canonical soliton, the leading order term only depends on the times τ_1 and τ_2, not on the points x and y. However, the quantity $\hat{\mathcal{L}}$ thrown up as the next order term turns out to be one of the most important concepts in the theory of Ricci flow. Traditionally, one would write

$$
\hat{\mathcal{L}}(\gamma) = \mathcal{L}(\gamma) - n(\sqrt{\tau_2} - \sqrt{\tau_1}),
$$

where

$$\mathcal{L}(\gamma) := \int_{\tau_1}^{\tau_2} \sqrt{\tau}(R(\gamma(\tau), \tau) + |\gamma'(\tau)|^2)d\tau$$

is Perelman's $\mathcal{L}$-length [18, Section 7].

Just as in Riemannian geometry, one can use such a notion of length to generate a notion of distance:

$$\hat{Q}(x, \tau_1; y, \tau_2) = \inf_\gamma \hat{\mathcal{L}}(\gamma),$$

where the infimum is over curves γ as above, and for consistency with prior notation we write $Q(x, \tau_1; y, \tau_2) = \inf_\gamma \mathcal{L}(\gamma)$. Note, however, that these lengths and distances are liable to be negative if the Ricci flow has any negative scalar curvature. For τ_2 only marginally larger than τ_1, the scalar curvature part of the definition of $\hat{\mathcal{L}}$ becomes irrelevant and one recovers the Riemannian distance in that

$$\lim_{\tau_2 \downarrow \tau_1} 2(\sqrt{\tau_2} - \sqrt{\tau_1})\hat{Q}(x, \tau_1; y, \tau_2) = d_{g(\tau_1)}^2(x, y). \qquad (5.17)$$

The considerations above suggest that minimising paths Γ should essentially arise from $\mathcal{L}$-geodesics γ for large N, and that

$$d_{\hat{g}}((x, \tau_1), (y, \tau_2)) = \sqrt{2N}(\tau_1^{-1/2} - \tau_2^{-1/2}) + \frac{1}{\sqrt{2N}}\hat{Q}(x, \tau_1; y, \tau_2) + O(\frac{1}{N^{3/2}}).$$

$$\qquad (5.18)$$

5.5.3 $\mathcal{L}$-optimal transportation

The canonical shrinking soliton also leads naturally to the notion of $\mathcal{L}$-optimal transportation and the theory concerning it (both of which first arose in [24], before the notion of canonical soliton had been discovered). Roughly speaking, $\mathcal{L}$-optimal transportation is optimal transportation in which the cost function arises from Perelman's $\mathcal{L}$-length.

The key point is that we will try to learn something significant about a Ricci flow by exploiting Theorem 5.2, but instead of applying it to the Ricci flow directly we will apply it to the Ricci flow on space-time induced by its canonical soliton.

Consider two probability measures ν_1 and ν_2 on the underlying manifold $\mathcal{M}$ of a (reverse) Ricci flow $g(\tau)$ defined for $\tau \in I \subset\subset (0, \infty)$. We can view these as measures $\hat{\nu}_1$ and $\hat{\nu}_2$ on space-time $\mathcal{M} \times I$ supported on time slices $\mathcal{M} \times \{\tau_1\}$ and $\mathcal{M} \times \{\tau_2\}$ by defining $\hat{\nu}_i$ to be the push-forward of ν_i under F_{τ_i} where, for $\tau \in I$, the map $F_\tau : \mathcal{M} \to \mathcal{M} \times I$ is defined by $x \mapsto (x, \tau)$.

If we now equip space-time $\mathcal{M} \times I$ with the canonical shrinking soliton Riemannian metric (for some large N) then we can consider the W_1 Wasserstein distance between $\hat{v}_1$ and $\hat{v}_2$. By definition (5.10) of W_1, the expansion (5.18) suggests that

$$W_1^{\hat{g}}(\hat{v}_1, \hat{v}_2) = \sqrt{2N}(\tau_1^{-1/2} - \tau_2^{-1/2})$$
$$+ \frac{1}{\sqrt{2N}} \inf_{\pi \in \Gamma(v_1, v_2)} \int_{\mathcal{M} \times \mathcal{M}} \hat{Q}(x, \tau_1; y, \tau_2) d\pi(x, y) + O(\tfrac{1}{N^{3/2}}),$$

$$(5.19)$$

motivating the following:

Definition 4 (from [24]). The $\mathcal{L}$-Wasserstein distance between v_1 at time τ_1 and v_2 at time τ_2 is defined to be

$$\mathcal{D}(v_1, \tau_1; v_2, \tau_2) := \inf_{\pi \in \Gamma(v_1, v_2)} \int_{\mathcal{M} \times \mathcal{M}} \hat{Q}(x, \tau_1; y, \tau_2) d\pi(x, y). \qquad (5.20)$$

In Section 5.5.4 we will see that this distance has useful contractivity properties when applied to two diffusions at appropriate times.

5.5.4 Contractivity of $\mathcal{L}$-Wasserstein distance

In this section we want to apply Theorem 5.2 to the canonical soliton flow in a heuristic fashion, in order to arrive at an $\mathcal{L}$-Wasserstein contractivity result which one can then go back and prove rigorously. We will be very cavalier with the distinction between 'approximate' and 'exact', and will be very loose in making precise the time intervals being considered. (You might like to consider initially only flows defined on the whole of $(0, \infty)$.) With hindsight we will see that this is not too important.

By the theory of Ricci solitons described in Section 5.3.3, we can introduce a new time variable s (since τ is now a spatial coordinate on $\hat{\mathcal{M}}$) and consider the (approximate, reverse) canonical soliton Ricci flow $G(s)$ starting at time $s = 1$ with the metric $G(1) = \hat{g}$ on $\hat{\mathcal{M}} := \mathcal{M} \times I$, given by

$$G(s) := s \, \psi_s^*(\hat{g}), \qquad (5.21)$$

where $\psi_1 : \hat{\mathcal{M}} \to \hat{\mathcal{M}}$ is the identity and the remaining ψ_s are obtained by integrating the vector field $X_s := -\frac{1}{s} \hat{\nabla}(\frac{N}{2\tau})$ with $\hat{\nabla}$ representing the gradient with respect to $\hat{g}$. Note that for convenience we have shifted time compared with the exposition in Section 5.3.3.

We may compute

$$X_s = \frac{N}{2s\tau^2} \hat{g}_{00}^{-1} \frac{\partial}{\partial \tau} = \frac{\tau}{s} \frac{\partial}{\partial \tau} + O(\tfrac{1}{N}), \qquad (5.22)$$

and by neglecting the error of order $\frac{1}{N}$ this integrates to

$$\psi_s(x, \tau) \simeq (x, s\tau). \tag{5.23}$$

Therefore, the flow $G(s)$ arises by pulling back $\hat{g}$ by a dilation of time and scaling appropriately.

If we then view $G(s)$ as a flow which essentially satisfies part (A) of Theorem 5.2, then we can hope to exploit the equivalent part (B_1) of the same theorem. The question is then which diffusions we should consider in (B_1).

The answer is to take diffusions starting at $s = 1$ with singular measures supported on time-slices. Suppose $\bar{\tau}_1, \bar{\tau}_2 \in I$, with $\bar{\tau}_1 < \bar{\tau}_2$, and $\tilde{v}_1, \tilde{v}_2$ are two measures on $\mathcal{M}$. Then the initial measures for the diffusions will be $(F_{\bar{\tau}_k})_\#(\tilde{v}_k)$ for $k = 1, 2$, which are supported on the slices $\mathcal{M} \times \{\bar{\tau}_k\}$.

Because of the extreme stretching of the τ direction in the metric $\hat{g}$ (recall that $\hat{g}_{00}$ is of order N) there is essentially no diffusion of the measures in the τ direction, and we view them at later times $s > 1$ as remaining supported in the time slices $\mathcal{M} \times \{\bar{\tau}_k\}$ and write them $(F_{\bar{\tau}_k})_\#(v_k(s\bar{\tau}_k))$ for some flows of measures $v_k(\cdot)$ with $v_k(\bar{\tau}_k) = \tilde{v}_k$. These singular measures then evolve mainly under diffusion in the $\mathcal{M}$ factor, under the Laplacian induced by $G_{ij}(s)$. By (5.21), (5.23) and the definition of $\hat{g}$ from Theorem 5.3, on the time-slice $\mathcal{M} \times \{\bar{\tau}_k\}$ we have approximately

$$G_{ij}(s) = s[\psi_s^*(\hat{g})]_{ij} = s[\hat{g}|_{\mathcal{M} \times \{s\bar{\tau}_k\}}]_{ij} = \frac{g_{ij}(s\bar{\tau}_k)}{\bar{\tau}_k}.$$

Therefore, the measures evolve by diffusion in the $\mathcal{M}$ factor under the operator

$$\Delta_{\frac{g(s\bar{\tau}_k)}{\bar{\tau}_k}} = \bar{\tau}_k \Delta_{g(s\bar{\tau}_k)},$$

which implies that the flows of measures $v_k(\cdot)$ mentioned above must themselves be diffusions (but on the original Ricci flow $g(\tau)$) with $v_k(\bar{\tau}_k) = \tilde{v}_k$.

We are now in a position to apply part (B_1) of Theorem 5.2 to deduce that our two singular diffusions should get closer in the W_1 sense, with respect to $G(s)$, as s increases. In other words, we have (formally) deduced that

$$s \mapsto W_1^{G(s)}((F_{\bar{\tau}_1})_\# v_1(s\bar{\tau}_1), (F_{\bar{\tau}_2})_\# v_2(s\bar{\tau}_2))$$

is weakly decreasing.

If we now push forward this whole construction under the maps ψ_s, adopting the abbreviation $\hat{v}_k(\tau) := (F_\tau)_\# v_k(\tau)$, then we find that

$$s \mapsto W_1^{s\hat{g}}(\hat{v}_1(s\bar{\tau}_1), \hat{v}_2(s\bar{\tau}_2)) = s^{\frac{1}{2}} W_1^{\hat{g}}(\hat{v}_1(s\bar{\tau}_1), \hat{v}_2(s\bar{\tau}_2)) \tag{5.24}$$

is weakly decreasing.

By the expansion (5.19) and the definition of (5.20), this implies that

$$s \mapsto \sqrt{2N}(\bar{\tau}_1^{-1/2} - \bar{\tau}_2^{-1/2}) + \frac{s^{1/2}}{\sqrt{2N}}\mathcal{D}(\nu_1(s\bar{\tau}_1), s\bar{\tau}_1; \nu_2(s\bar{\tau}_2), s\bar{\tau}_2)$$

is monotonically decreasing, and hence so is

$$s \mapsto s^{1/2}\mathcal{D}(\nu_1(s\bar{\tau}_1), s\bar{\tau}_1; \nu_2(s\bar{\tau}_2), s\bar{\tau}_2).$$

These heuristic arguments suggest the following theorem, which was rigorously proved in [24]:

Theorem 5.5 (equivalent to [24, Theorem 1.1]). *Suppose that $g(\tau)$ is a Ricci flow on a closed manifold $\mathcal{M}$ over an open time interval containing $[\bar{\tau}_1, \bar{\tau}_2]$, and suppose that $\nu_1(\tau)$ and $\nu_2(\tau)$ are two diffusions (in the same sense as in Theorem 5.2) defined for τ in neighbourhoods of $\bar{\tau}_1$ and $\bar{\tau}_2$ respectively. Then the distance between the diffusions decays in the sense that for $s \geq 1$ sufficiently close to 1,*

$$\mathcal{D}(\nu_1(s\bar{\tau}_1), s\bar{\tau}_1; \nu_2(s\bar{\tau}_2), s\bar{\tau}_2) \leq s^{-1/2}\mathcal{D}(\nu_1(\bar{\tau}_1), \bar{\tau}_1; \nu_2(\bar{\tau}_2), \bar{\tau}_2).$$

The rigorous proof of this result involves introducing a notion of $\mathcal{L}$-Wasserstein geodesic through a Ricci flow, and considering the convexity of the Boltzmann entropy along this geodesic (see [24]) in a manner related to the arguments in [15].

5.5.5 Applications of $\mathcal{L}$-Wasserstein contractivity

From the $\mathcal{L}$-optimal transportation construction in Theorem 5.5 one can [24] recover Perelman's monotonic quantities (involving both entropies and $\mathcal{L}$-length) which are central in his work on Ricci flow [18–20], with the case of Perelman's reduced volume appearing in [14]. We will highlight the case of Perelman's $\mathcal{W}$-entropy since it illustrates how entropies and $\mathcal{L}$-length interact and will solve the degeneration issue discussed in Section 5.3.2. But first we indicate how Theorem 5.5 generalises the main implication $(A) \implies (B_2)$ in Theorem 5.2.

As mentioned in (5.17), by inspection of the definition of $\hat{\mathcal{L}}$ in (5.16) we see that when τ_2 is only marginally larger than τ_1, the $\sqrt{\tau}|\gamma'(\tau)|^2$ part of the integrand dominates and

$$\lim_{\tau_2 \downarrow \tau_1} 2(\sqrt{\tau_2} - \sqrt{\tau_1})\hat{Q}(x, \tau_1; y, \tau_2) = d^2_{g(\tau_1)}(x, y).$$

This suggests that, for τ_2 only marginally larger than τ_1,

$$2(\sqrt{\tau_2} - \sqrt{\tau_1})\mathcal{D}(\nu_1, \tau_1; \nu_2, \tau_2) \simeq \left[W_2^{g(\tau_1)}(\nu_1, \nu_2)\right]^2, \qquad (5.25)$$

and since by Theorem 5.5 the function

$$s \mapsto 2(\sqrt{s\bar{\tau}_2} - \sqrt{s\bar{\tau}_1})\mathcal{D}(\nu_1(s\bar{\tau}_1), s\bar{\tau}_1; \nu_2(s\bar{\tau}_2), s\bar{\tau}_2) \qquad (5.26)$$

is decreasing, we are led to property (B_2):

Corollary 5.6 (McCann and Topping [15]). *Given two diffusions $\nu_1(\tau)$ and $\nu_2(\tau)$ (Definition 3) on a reverse Ricci flow $g(\tau)$, the function*

$$\tau \to W_2^{g(\tau)}(\nu_1(\tau), \nu_2(\tau))$$

is (weakly) decreasing in τ.

Unlike some of the heuristic arguments in these lectures, the one above is relatively easy to make rigorous [24].

5.5.5.1 Recovering Perelman's $\mathcal{W}$-entropy

We now want to take the limit $\bar{\tau}_2 \downarrow \bar{\tau}_1$ again in Theorem 5.5, but this time with $\nu_1(\tau) = \nu_2(\tau) =: \nu(\tau)$ the same diffusion. In this case, by (5.25), the monotonic quantity of (5.26) is zero to leading order in $\bar{\tau}_2 - \bar{\tau}_1$. The trick is to look at the next order term in the expansion, which will also be monotonic.

We will need to consider the infinitesimal version of the $\mathcal{L}$-Wasserstein distance $\mathcal{D}$ implied by the following lemma, which is a variation on a well-known infinitesimal description of the usual W_2 distance as an H^{-1} norm (see [25]). Given a smooth family of positive probability measures $\nu(\tau)$ on a closed manifold $\mathcal{M}$, for τ in some neighbourhood of τ_1, we call a vector field $X \in \Gamma(T\mathcal{M})$ an *advection* field for $\nu(\tau)$ at $\tau = \tau_1$ if there exists a smooth family of diffeomorphisms $\psi_\tau : \mathcal{M} \to \mathcal{M}$, for τ in a neighbourhood of τ_1, with ψ_{τ_1} the identity, and such that $(\psi_\tau)_{\#}\nu(\tau_1) = \nu(\tau)$ and $X = \frac{\partial \psi}{\partial \tau}|_{\tau=\tau_1}$.

Lemma 5.7 (Lemma 1.3 from [24]). *Suppose $g(\tau)$ is a (reverse) Ricci flow, and $\nu(\tau)$ is a smooth family of positive probability measures on a closed manifold $\mathcal{M}$, for τ in some neighbourhood of $\tau_1 \in \mathbb{R}$. Then as $\tau_2 \downarrow \tau_1$,*

$$\mathcal{D}(\nu(\tau_1), \tau_1; \nu(\tau_2), \tau_2)$$
$$= (\tau_2 - \tau_1)\left[\inf_X \sqrt{\tau_1} \int_{\mathcal{M}} \left(R(\cdot, \tau_1) + |X|^2_{g(\tau_1)} - \frac{n}{2\tau_1}\right) d\nu(\tau_1)\right] + o(\tau_2 - \tau_1),$$

$$(5.27)$$

where the infimum is taken over all advection fields X for $\nu(\tau)$ at $\tau = \tau_1$.

We call the coefficient of $(\tau_2 - \tau_1)$ in (5.27) (the part within square brackets) the infinitesimal $\mathcal{L}$-Wasserstein distance, or $\mathcal{L}$-Wasserstein *speed* of $\nu(\tau)$, with respect to $g(\tau)$, at $\tau = \tau_1$.

One can give an expression for X as the gradient of the solution to a certain elliptic equation (see [24]), but here it suffices to note that, by considering Hodge decompositions, we find that there can be only one advection field which is the gradient of a function, and that must be optimal. But if $v(\tau)$ is a diffusion, then Fourier's law tells us that $\nabla \ln u$ is an advection field, so it must be optimal.

We now combine these considerations with Theorem 5.5 specialised to the situation that $v_1(\tau) = v_2(\tau) =: v(\tau)$. Writing $\bar\tau_2 = (1 + \eta)\bar\tau_1$ (for small η) we can express the monotonic quantity from Theorem 5.5 as

$$s^{\frac{1}{2}}\mathcal{D}(v_1(s\bar\tau_1), s\bar\tau_1; v_2(s\bar\tau_2), s\bar\tau_2)$$

$$= \eta\left(\tau_1^2 \int_M \left(R + |\nabla \ln u|^2 - \frac{n}{2\tau_1}\right)dv(\tau_1)\right)\bar\tau_1^{-1/2} + o(\eta) \quad (5.28)$$

$$= \eta\left(\tau_1^2 \mathcal{F}(\tau_1) - \frac{n\tau_1}{2}\right)\bar\tau_1^{-1/2} + o(\eta),$$

where $\tau_1 := s\bar\tau_1$ and where, for each τ,

$$\mathcal{F} = \int_M (R + |\nabla \ln u|^2)u\,d\mu$$

is Perelman's $\mathcal{F}$-information functional [18] and [23, Section 6.2]. Theorem 5.5 then tells us that

$$\left(\tau^2\mathcal{F}(\tau) - \frac{n\tau}{2}\right) \text{ is weakly decreasing in } \tau. \quad (5.29)$$

In terms of the normalised Boltzmann entropy

$$E(\tau) := -\int_M u \ln u\,d\mu - \frac{n}{2}(1 + \ln(4\pi\tau))$$

(normalised so that the heat kernel in Euclidean space would have zero entropy for all time), the decreasing quantity in (5.29) can be written $\tau^2\frac{dE}{d\tau}$. Therefore,

$$0 \geq \frac{d}{d\tau}\left(\tau^2\frac{dE}{d\tau}\right) = \tau\frac{d^2}{d\tau^2}(\tau E)$$

and we see that (τE) is concave in τ and, in particular, the quantity

$$W(\tau) := \frac{d}{d\tau}(\tau E)$$

is weakly decreasing in τ. A short calculation gives the explicit formula

$$W(\tau) = \int_M \left[\tau(|\nabla \ln u|^2 + R) - \ln u\right]u\,d\mu - n - \frac{n}{2}\ln(4\pi\tau),$$

which is the entropy quantity discovered by Perelman [18].

5.5.5.2 No local collapsing

In this section we sketch how the monotonicity of the previous section rules out the local collapsing that was discussed in Section 5.3.2. Recall, we have a Ricci flow defined on some time interval $[0, T)$, and we have to show that there exists $\kappa > 0$ such that for any $\bar{t} \in [0, T)$ and $r \in (0, 1]$, if the sectional curvatures of $g(\bar{t})$ are bounded in norm by r^{-2} in some geodesic ball $B(x, r)$ then the volume of $B(x, r)$ is bounded below by κr^n.

The main step is to prove the following log-Sobolev inequality:

Proposition 5.8. *For all* $\bar{t} \in [0, T)$, $r \in (0, 1]$ *and* $\varphi \in C^{0,1}(\mathcal{M})$, *there holds on* $(\mathcal{M}, g(\bar{t}))$

$$\int_{\mathcal{M}} \varphi^2 \ln \varphi^2 - \left(\int_{\mathcal{M}} \varphi^2 \right) \ln \left(\int_{\mathcal{M}} \varphi^2 \right) \leq r^2 \int_{\mathcal{M}} \left(4|\nabla \varphi|^2 + R\varphi^2 \right)$$
$$+ (C - \ln(r^n)) \left(\int_{\mathcal{M}} \varphi^2 \right), \quad (5.30)$$

where C depends only on n, T and $(\mathcal{M}, g(0))$.

Proof. (Sketch. For more details, see [23, Chapter 8].) If we view Perelman's entropy from the previous section as a functional[5]

$$\mathcal{W}(g, u, \tau) := \int_{\mathcal{M}} \left[\tau(|\nabla \ln u|^2 + R) - \ln u \right] u \, d\mu - n - \frac{n}{2} \ln(4\pi\tau),$$

on probability densities u and numbers $\tau > 0$, for a given metric g, then as we shall see, an inequality

$$\mathcal{W}(g, \cdot, \tau) \geq -C$$

can be viewed as a log-Sobolev inequality on $(\mathcal{M}, g)$ at 'scale' τ.

All closed manifolds admit some log-Sobolev inequality, but if the manifold evolves, then that log-Sobolev inequality could degenerate; that is, the constants in the inequality could just get worse and worse as we approach a singular time, and the inequality would become unusable. The monotonicity from the previous section will rule out that possibility.

To analyse the flow at our time $\bar{t} \in [0, T)$, we define a reverse time parameter $\tau = \bar{t} + r^2 - t$ and look at the flow over the range $\tau \in [r^2, \bar{t} + r^2]$.

Let us accept the log-Sobolev inequality at $t = 0$

$$\mathcal{W}(g(\tau = \bar{t} + r^2), \cdot, \bar{t} + r^2) \geq -C_0$$

(noting that $\tau = \bar{t} + r^2$ corresponds to $t = 0$), where C_0 depends on T (i.e. an upper bound for $\bar{t}$) and $(\mathcal{M}, g(t = 0))$. In particular, if we let $u : \mathcal{M} \times [r^2, \bar{t} + r^2] \to [0, \infty)$ be the solution of the conjugate heat equation (5.8)

---

[5] Note the distinction between W and $\mathcal{W}$!

with $u(\cdot, r^2) = \left(\int \varphi^2\right)^{-1} \varphi^2$, then we have $W(\bar{t} + r^2) \geq -C_0$. Appealing to the monotonicity of W, we then also have $W(r^2) \geq -C_0$. Unravelling the definition of $W(r^2)$ using the fact that we know $u(\cdot, r^2)$ gives the proposition. $\square$

To rule out local collapsing using (5.30) is essentially a question of finding the right function φ to plug in to the inequality. Given a ball $B(x, r)$ at time $\bar{t}$, with volume V, satisfying the assumption that all sectional curvatures are less than r^{-2}, we should take φ to be the function identically equal to 1 on $B(x, r/2)$, equal to zero outside $B(x, r)$, and equal to $2 - 2d(\cdot, x)/r$ on the remaining 'annulus'.

If we check the order of magnitude of each term in (5.30) (we need the fact that $B(x, r/2)$ will also have volume of order V, which follows from a theorem of Bishop and Gromov from comparison geometry [8] that we do not discuss here) we find that

$$\int_{\mathcal{M}} \varphi^2 \sim V; \qquad -\int_{\mathcal{M}} \varphi^2 \ln \varphi^2 \lesssim V; \qquad r^2 \int_{\mathcal{M}} \left(4|\nabla\varphi|^2 + R\varphi^2\right) \lesssim V,$$

and thus (5.30) gives a positive lower bound for $\frac{V}{r^n}$ as desired.

More refined and complete arguments along the lines of this section can be found in [23] and [22].

5.6 Harnack inequalities

5.6.1 The canonical expanding soliton

Up to now, we have shown how a reasonable-sounding result combining optimal transportation and Ricci flow (Theorem 5.2) can be applied to the canonical solitons to give more exotic results about Ricci flow, such as Theorem 5.5. In fact, the theory developed a little differently: after our work [15] with McCann, we proved Theorem 5.5 as an extension, and then derived the canonical shrinking solitons with Cabezas-Rivas [4] as the flows to which we should apply Theorem 5.2 in order to get Theorem 5.5. In this way we can see the canonical solitons as an *application* of optimal transportation theory.

In this final section, we describe a slight variation on the canonical shrinking soliton construction: we introduce the canonical expanding solitons, and show how a quite different part of Ricci flow theory arises naturally from them. A more complete explanation of this side of the theory is to be found in [5].

Theorem 5.9 ([5]). *Suppose $g(t)$ is a Ricci flow, i.e. a solution of $\frac{\partial g}{\partial t} = -2\operatorname{Ric}(g(t))$, defined on a manifold $\mathcal{M}$ of dimension $n \in \mathbb{N}$, for t within a time interval $[0, T]$, with uniformly bounded curvature. Suppose $N > 0$, and define* *a metric $\check{g}_N$ (which we often write simply as $\check{g}$) on $\check{\mathcal{M}} := \mathcal{M} \times (0, T]$ by*

$$\check{g}_{ij} = \frac{g_{ij}}{t}; \qquad \check{g}_{00} = \frac{N}{2t^3} + \frac{R}{t} + \frac{n}{2t^2}; \qquad \check{g}_{0i} = 0,$$

where i, j are coordinate indices on the $\mathcal{M}$ factor, 0 represents the index of the time coordinate $t \in (0, T]$, and the scalar curvature of g is written as R.

Then, up to errors of order $\frac{1}{N}$, the metric $\check{g}$ is a gradient expanding Ricci soliton on the higher dimensional space $\check{\mathcal{M}}$:

$$E_N := \text{Ric}(\check{g}) + \text{Hess}_{\check{g}}\left(-\frac{N}{2t}\right) + \frac{1}{2}\check{g} \simeq 0, \tag{5.31}$$

by which we mean that for any $k \in \{0, 1, 2, \ldots\}$ the quantity

$$N\left[\check{\nabla}^k E_N\right] \tag{5.32}$$

is bounded uniformly locally on $\check{\mathcal{M}}$ (independently of N) where $\check{\nabla}$ is the Levi–Civita connection corresponding to $\check{g}$.

One point that should be clarified is why $\check{g}$ is positive definite, and in particular why $\check{g}_{00} \geq 0$. By refining the maximum principle argument that shows in Section 5.3.1 that weakly positive scalar curvature is preserved, one can show (see [23, Corollary 3.2.5] and [5]) that in fact $R + \frac{n}{2t} \geq 0$ for every Ricci flow starting at $t = 0$, which implies $\check{g}_{00} \geq 0$ as desired.

One sees that this result is close in form to Theorem 5.3. Various signs have changed, and every τ has been replaced with a t. We have also asserted that we have a soliton in a smoother sense in the limit $N \to \infty$, which makes the construction more applicable in rigorous proofs. Finally, we only consider Ricci flows starting at time $t = 0$, as is natural when deriving Harnack inequalities. In fact, it is possible to improve considerably the sense in which $\check{g}$ is a soliton near $t = 0$, but we do not pursue that here.

It will also be important to get a feel for the geometry of the canonical expanding solitons near $t = 0$. In this region the dominant term in the definition of $\check{g}_{00}$ is emphatically $\frac{N}{2t^3}$. If we neglect the other terms for the moment, and change variables from t to $r := t^{-1/2}$, then for large r (small t) $\check{g}$ can be written approximately as

$$\begin{aligned}\check{g}_N &\simeq \frac{g(t)}{t} + \frac{N}{2t^3}dt^2 \\ &= r^2 g(r^{-2}) + 2Ndr^2\end{aligned} \tag{5.33}$$

and we see that asymptotically the canonical soliton opens like the cone

$$\Sigma_N := (\mathcal{M} \times (0, \infty), \tilde{r}^2 g(0) + 2Nd\tilde{r}^2)$$

(using coordinates $(x, \tilde{r})$ on $\mathcal{M} \times (0, \infty)$) with shallow cone angle for large N.

A precise way of saying this, involving the notion of Cheeger–Gromov convergence [23], would be that for any $x \in \mathcal{M}$ and sequence $t_i \downarrow 0$ we have convergence of pointed rescalings of the canonical expanding soliton:

$$(\check{\mathcal{M}}, t_i \check{g}_N, (x, t_i)) \rightarrow (\Sigma_N, (x, 1)).$$

It is essentially equivalent to consider the geometry of the corresponding canonical expanding soliton flow

$$G(s) := s\psi_s^*(\check{g})$$

near $s = 0$, where the maps $\psi_s : \check{\mathcal{M}} \rightarrow \check{\mathcal{M}}$, defined by setting $\psi_1 : \check{\mathcal{M}} \rightarrow \check{\mathcal{M}}$ to be the identity and then (following the construction of Section 5.3.3) obtaining ψ_s for $s \in (0, 1]$ by integrating the vector field $X_s := \frac{1}{s}\check{\nabla}\left(-\frac{N}{2t}\right) \simeq \frac{t}{s}\frac{\partial}{\partial t}$ to give

$$\psi_s(x, t) \simeq (x, st)$$

(cf. (5.23) of Section 5.5.4, though note that s is now a *forwards* time parameter). We find that for small $s > 0$, and r of order 1,

$$G(s) \simeq r^2 g(0) + 2N dr^2.$$

Thus, $G(s)$ converges to the cone Σ_N as $s \downarrow 0$ even at the level of convergence of tensors (i.e. one does not need the diffeomorphisms of Cheeger–Gromov convergence), and for large N this cone looks like a cylinder $(\mathcal{M}, g(0)) \times \mathbb{R}$.

See [5] for a more detailed and accurate presentation of these issues.

5.6.2 Harnack estimates generated by the canonical expanding soliton

We saw in Section 5.3 that various positivity-of-curvature conditions are preserved under Ricci flow. For example, if $g(t)$ is a Ricci flow for which $g(0)$ has a (weakly) positive curvature operator, then $g(t)$ has a (weakly) positive curvature operator for all $t \geq 0$. It will help us later to rephrase this slightly, and consider the condition 'weakly positive curvature form', by which we mean the bilinear form in $Sym^2(\Lambda^2 T^* \mathcal{M})$ induced by the curvature operator and the metric is weakly positive definite.

In this section we show in a manner reminiscent of Section 5.5 that one can get much further by applying such a Ricci flow result not to a given Ricci flow, but to its canonical soliton.

To facilitate this, we note that for the Ricci flow above, $(\mathcal{M}, g(0)) \times \mathbb{R}$ will have (weakly) positive curvature form, and thus (for large N) this will almost be true for $G(s)$ in the limit $s \downarrow 0$ by the previous section.

If we make the leap of faith that (approximately) positive curvature form is preserved under the (approximate) Ricci flow $G(s)$, then we see that this condition then holds for $G(s)$ for all $s \in (0, 1]$, and in particular for $\check{g} = G(1)$.

One might also hope that, as we let $N \to \infty$, this statement becomes less and less approximate. In fact, this is exactly what happens:

Theorem 5.10. *Suppose that $g(t)$ is a Ricci flow on a manifold $\mathcal{M}$ for $t \in [0, T]$ such that $g(t)$ is complete for each $t \geq 0$ and has uniformly bounded curvature, and suppose that $(\mathcal{M}, g(0))$ has weakly positive curvature form. For each N, let $\check{g}_N$ be the canonical expanding soliton associated with $g(\cdot)$, and $\mathcal{R}_N$ the curvature form associated with $\check{g}_N$. Then there exists $\mathcal{R}_\infty$ such that*

$$\mathcal{R}_N \to \mathcal{R}_\infty$$

pointwise on $\check{\mathcal{M}}$ as $N \to \infty$. Moreoever, $\mathcal{R}_\infty$ is weakly positive definite.

In fact, the curvature *operators* of the metrics $\check{g}_N$ would also converge, but because the metrics $\check{g}_N$ are degenerating as $N \to \infty$, we get more information from considering curvature *forms*.

The final assertion of Theorem 5.10 turns out to be precisely equivalent to Hamilton's matrix Harnack estimates [10] and was thus known to hold rigorously long before most of the theory in these lectures was developed. It also arises as a corollary of the rigorous development of the ideas of this section, found in [5].

Our canonical expanding construction follows earlier geometric constructions [6, 7, 18] to which one can associate curvatures that are similar to Hamilton's Harnack expressions (typically modulo some missing terms or alternative signs). For the exact expression for the curvature forms $\mathcal{R}_N$ and the limit $\mathcal{R}_\infty$, see [5]. In these notes, it will suffice to detail the following special case:

Corollary 5.11. *Under the conditions of Theorem 5.10, let $\mathrm{Ric}_N \in \Gamma(Sym^2 T^* \check{\mathcal{M}})$ be the Ricci tensor associated to $\check{g}_N$. Then*

$$\mathrm{Ric}_N \to \mathrm{Ric}_\infty$$

pointwise as $N \to \infty$, where $\mathrm{Ric}_\infty \in \Gamma(Sym^2 T^* \check{\mathcal{M}})$ is determined by

$$\mathrm{Ric}_\infty \left(X + \frac{\partial}{\partial t}, X + \frac{\partial}{\partial t} \right) = \mathrm{Ric}^{g(t)}(X, X) + \langle X, \nabla R \rangle_{g(t)} + \frac{1}{2} \left(\frac{\partial R}{\partial t} + \frac{R}{t} \right),$$

for each $X \in T\mathcal{M}$. Moreoever, Ric_∞ is weakly positive definite.

This corollary follows immediately from the theorem (modulo our omission here of the exact expression for $\mathcal{R}_\infty$) because $\check{g}_N$ viewed as a metric on the cotangent bundle converges to a (degenerate) limit metric $\check{g}_\infty$ as $N \to \infty$, and so Ric_∞ is simply the appropriate trace of $\mathcal{R}_\infty$ with respect to $\check{g}_\infty$.

Once this intuition outlined above is in place, one can uncover alternative Harnack results not included in Hamilton's theory. At least at the heuristic level, these arguments suggest the general Harnack principle:

Suppose $g(t)$ is a Ricci flow on $\mathcal{M}^n$ for $t \in [0, T]$ and that the curvature tensor of $(\mathcal{M}, g(0)) \times \mathbb{R}$ lies in a cone C of curvature tensors which is invariant under Ricci flow in $(n + 1)$-dimensions. Then $\mathcal{R}_\infty$ lies in C.

To clarify, an invariant cone here is a cone C within the linear space of algebraic curvature tensors such that if the curvature tensor at each point of an initial metric on an $(n + 1)$-dimensional closed manifold lies in C, then we know that the curvature tensor at each point of a subsequent Ricci flow must lie in C. An example would be the cone of positive definite curvature forms.

In practice, we can turn this principle into a rigorous theorem for appropriate C, as outlined in [5]. One example which goes beyond Hamilton's Theorem 5.10 is when one takes C to be the cone of so-called WPIC2 curvature tensors (equivalently, weakly positive complex sectional curvature). A manifold $(\mathcal{N}, h)$ has WPIC2 curvature tensor if $(\mathcal{N}, h) \times \mathbb{R}^2$ has weakly positive isotropic curvature. In this way, one recovers the recent Harnack estimate of Brendle [2]. We refer to [5] for more details, and for other curvature cones C which provide completely new Harnack estimates. The analogue of this theory for mean curvature flow can be found in [12].

5.6.3 Applications to understanding blown-up Ricci flows

We want to conclude the lectures by applying the estimates from Section 5.6.2 in order to establish a relationship between the scalar curvature at different points of a Ricci flow which has arisen after blowing up a singularity of a three-dimensional Ricci flow as discussed in Section 5.3.2.

More precisely, we consider a complete bounded curvature Ricci flow $g(t)$ for $t \in (-\infty, 0)$ with weakly positive curvature operator, but non-flat (so $R > 0$) and wish to show that for $x_1, x_2 \in \mathcal{M}$ and times $t_1 < t_2 < 0$ we have

$$\frac{R(x_2, t_2)}{R(x_1, t_1)} \geq \exp\left(-\frac{d^2_{g(t_1)}(x_1, x_2)}{2(t_2 - t_1)}\right).$$

This estimate was first derived by Hamilton [11] following Li and Yau [13].

Proof. Because $\mathrm{Ric} \geq 0$, we can estimate $\mathrm{Ric}(X, X) \leq R|X|^2$, and so Corollary 5.11 applied to $g(t)$ for $t \in [\bar{t}, 0]$ for some $\bar{t} < 0$ gives

$$R|X|^2 + \langle X, \nabla R \rangle + \frac{1}{2}\left(\frac{\partial R}{\partial t} + \frac{R}{t - \bar{t}}\right) \geq 0.$$

By allowing $\bar{t}$ to tend to $-\infty$ while keeping t fixed, we then see that in fact

$$R|X|^2 + \langle X, \nabla R \rangle + \frac{1}{2}\frac{\partial R}{\partial t} \geq 0.$$

If we divide through by R and set $X = -\frac{1}{2}\nabla \ln R$ in order to minimise the left-hand side over X, we get the estimate

$$\frac{\partial}{\partial t} \ln R - \frac{1}{2}|\nabla \ln R|^2 \geq 0.$$

We now wish to integrate this in the spirit of Li and Yau [13]. If $\gamma : [t_1, t_2] \to \mathcal{M}$ satisfies $\gamma(t_1) = x_1$ and $\gamma(t_2) = x_2$, then we can compute

$$\frac{d}{dt} \ln R(\gamma(t), t) = \frac{\partial}{\partial t} \ln R + \langle \nabla \ln R, \dot{\gamma} \rangle$$

$$\geq \frac{1}{2}|\nabla \ln R|^2 - \frac{1}{2}(|\nabla \ln R|^2 + |\dot{\gamma}|^2) \qquad (5.34)$$

$$= -\frac{1}{2}|\dot{\gamma}|^2_{g(t)}.$$

By inspection of the Ricci flow equation (5.1) and the fact that Ric ≥ 0, we must have $g(t) \leq g(t_1)$, and so

$$\frac{d}{dt} \ln R(\gamma(t), t) \geq -\frac{1}{2}|\dot{\gamma}|^2_{g(t_1)}.$$

If we integrate this inequality, we find that

$$\ln R(x_2, t_2) - \ln R(x_1, t_1) \geq -\frac{1}{2}\int_{t_1}^{t_2}|\dot{\gamma}(t)|^2 dt,$$

and minimising over all valid paths γ (i.e. choosing γ to be a minimising geodesic with respect to the metric $g(t_1)$) we find that

$$\ln \frac{R(x_2, t_2)}{R(x_1, t_1)} \geq -\frac{d^2_{g(t_1)}(x_1, x_2)}{2(t_2 - t_1)},$$

which can be exponentiated to give the desired inequality. $\qquad\square$

Acknowledgements

These lectures were given at the summer school on 'Optimal Transportation: Theory and Applications' at Grenoble in June 2009, and I thank Yann Ollivier, Hervé Pajot and Cédric Villani for their invitation. I would also like to thank Esther Cabezas-Rivas and Dan Jane for many useful comments.

References

[1] M. Arnaudon, A. Coulibaly and A. Thalmaier, Horizontal diffusion in C^1 path space. *Séminaire de Probabilités XLIII*, pp. 73–94, Lecture Notes in Mathematics 2006, Springer (2010).

[2]   S. Brendle, A generalization of Hamilton's differential Harnack inequality for the Ricci flow. *J. Differential Geom.*, **82** (2009), 207–227.

[3]   S. Brendle and R. Schoen, Manifolds with 1/4-pinched curvature are space forms. *J. Am. Math. Soc.*, **22** (2009) 287–307.

[4]   E. Cabezas-Rivas and P.M. Topping, The canonical shrinking soliton associated to a Ricci flow. *Calc. Var. PDE*, **43** (2012) 173–184.

[5]   E. Cabezas-Rivas and P.M. Topping, The canonical expanding soliton and Harnack inequalities for Ricci flow. *Trans. Am. Math. Soc.*, **364** (2012) 3001–3021.

[6]   B. Chow and S.-C. Chu, A geometric interpretation of Hamilton's Harnack inequality for the Ricci flow. *Math. Res. Lett.*, **2** (1995) 701–718.

[7]   B. Chow and D. Knopf, New Li–Yau–Hamilton inequalities for the Ricci flow via the space-time approach. *J. Differential Geom.*, **60** (2002), 1–54.

[8]   S. Gallot, D. Hulin and J. Lafontaine, *Riemannian Geometry* (second edition), Springer-Verlag (1993).

[9]   R.S. Hamilton, Three-manifolds with positive Ricci curvature. *J. Differential Geom.*, **17** (1982) 255–306.

[10] R.S. Hamilton, The Harnack estimate for the Ricci flow. *J. Differential Geom.*, **37** (1993) 225–243.

[11] R.S. Hamilton, The formation of singularities in the Ricci flow. *Surveys in Differential Geometry*, Vol. II (Cambridge, MA, 1993), pp. 7–136, International Press, Cambridge, MA, 1995.

[12] S. Helmensdorfer and P.M. Topping, The geometry of differential Harnack estimates. *Act. Sémin. Théor. Spectr. Géom. [Grenoble 2011–2012]*, **30** (2013) 77–89.

[13] P. Li and S.-T. Yau, On the parabolic kernel of the Schrödinger operator. *Acta Math.*, **156** (1986) 153–201.

[14] J. Lott, Optimal transport and Perelman's reduced volume. *Calc. Var. Partial Dif. Equations*, **36** (2009) 49–84.

[15] R.J. McCann and P.M. Topping, Ricci flow, entropy and optimal transportation. *Am. J. Math.*, **132** (2010) 711–730.

[16] H. Nguyen, Invariant curvature cones and the Ricci flow. PhD thesis, Australian National University (2007).

[17] F. Otto and C. Villani, Generalization of an inequality by Talagrand and links with the logarithmic Sobolev inequality. *J. Funct. Anal.*, **173** (2000) 361–400.

[18] G. Perelman, The entropy formula for the Ricci flow and its geometric applications. http://arXiv.org/abs/math/0211159v1 (2002).

[19] G. Perelman, Ricci flow with surgery on three-manifolds. http://arxiv.org/abs/math/0303109v1 (2003).

[20] G. Perelman, Finite extinction time for the solutions to the Ricci flow on certain three-manifolds. http://arXiv.org/abs/math/0307245v1 (2003).

[21] K.-T. Sturm and M.-K. von Renesse, Transport inequalities, gradient estimates, entropy and Ricci curvature. *Commun. Pure Appl. Math.*, **58** (2005) 923–940.

[22] P.M. Topping, Diameter control under Ricci flow. *Commun. Anal. Geom.*, **13** (2005) 1039–1055.

[23] P.M. Topping, 'Lectures on the Ricci flow.' L.M.S. Lecture Note Series **325**, C.U.P. (2006) http://www.warwick.ac.uk/~maseq/RFnotes.html

[24] P.M. Topping, $\mathcal{L}$-optimal transportation for Ricci flow. *J. Reine Angew. Math.*, **636** (2009) 93–122.

[25] C. Villani, *Topics in Optimal Transportation*, Graduate Studies in Mathematics, Vol. 58 American Mathematical Society (2003).

[26] Z.-H. Zhang, On the completeness of gradient Ricci solitons. *Proc. Am. Math. Soc.*, **137** (2009) 2755–2759.

6

Lecture notes on gradient flows and optimal transport

SARA DANERI AND GIUSEPPE SAVARÉ

Abstract

We present a short overview on the strongest variational formulation for
gradient flows of geodesically λ-convex functionals in metric spaces, with
applications to diffusion equations in Wasserstein spaces of probability
measures. These notes are based on a series of lectures given by the
second author for the Summer School "Optimal Transportation: Theory
and Applications" in Grenoble during the week of June 22–26, 2009.

6.1 Introduction

These notes are based on a series of lectures given by the second author for
the Summer School "Optimal Transportation: Theory and Applications" in
Grenoble during the week of June 22–26, 2009.

We try to summarize some of the main results concerning gradient flows of
geodesically λ-convex functionals in metric spaces and applications to diffusion partial differential equations (PDEs) in the Wasserstein space of probability
measures. Due to obvious space constraints, the theory and the references presented here are largely incomplete and should be intended as an oversimplified
presentation of a quickly evolving subject. We refer to the books [3, 68] for a
detailed account of the large literature available on these topics.

In the Section 6.2 we collect some elementary and well-known results
concerning gradient flows of smooth convex functions in $\mathbb{R}^d$. We selected
just a few topics, which are well suited for a "metric" formulation and provide
a useful guide for the more abstract developments.

In the Section 6.3 we present the main (and strongest) notion of gradient flow in metric spaces characterized by the solution of a metric *evolution variational inequality*: the aim here is to show the consequence of this

definition, without any assumptions on the space and on the functional (except completeness and lower semicontinuity); we shall see that solutions to evolution variational inequalities enjoy nice stability, asymptotic, and regularization properties. We also investigate the relationships with two different approaches, *curves of maximal slope* and *minimizing movements*, and we discuss a first stability result with respect to perturbations of the generating functional with respect to Γ-convergence.

Section 6.4 is devoted to some fundamental generation results for gradient flows of geodesically λ-convex functionals; here we adopt the method of *minimizing movement* to construct suitable families of discrete approximating solutions and we show three basic convergence results.

Apart from Sections 6.3.6 (stability of gradient flows with respect to Γ-convergence of the functionals) and 6.4.1 (existence of curves of maximal slope), we made a substantial effort to avoid any compactness argument in the theory, which is mainly focused on purely metric arguments. So we will present a slightly relaxed version of the minimizing movement scheme, which is always solvable by invoking Ekeland's variational principle, and the main existence and generation results for λ-gradient flows rely on refined Cauchy estimates and crucial geometric assumptions on the distance of the metric space.

Section 6.5 is devoted to applications of the metric theory to evolution equations in the so called "Wasserstein spaces" $\mathscr{P}_2(X)$ of probability measures. We recall a few basic facts about such spaces, the characterization of geodesics and absolutely continuous curves, and some geometric properties of the Wasserstein distance. Three basic examples of (or, better, displacement-) λ-convex functionals in $\mathscr{P}_2(\mathbb{R}^d)$ are presented, together with the evolutionary PDEs they are associated with. A short account of possible extensions of the theory to measure-metric spaces concludes the notes.

6.2 Gradient flows for smooth λ-convex functions in the Euclidean space

In this section we recall some simple properties of the gradient flow of a C^2 function $\varphi : \mathbb{R}^d \to \mathbb{R}$ satisfying the global lower bound $D^2\varphi \geq \lambda I$ for some $\lambda \in \mathbb{R}$. We will focus on those aspects which rely just on the "metric" structure of $\mathbb{R}^d$ and therefore could make sense in more general metric spaces. We denote by $\mathsf{d}(u, v) = |u - v|$ the Euclidean distance on $\mathbb{R}^d$ induced by the scalar product $\langle \cdot, \cdot \rangle$.

Remark 1 (a few basic facts about λ-convex functions). We will extensively use the following well-known equivalent characterizations of a λ-convex function $\varphi : \mathbb{R}^d \to \mathbb{R}$ (here x, x_0, x_1 are arbitrary points in $\mathbb{R}^d$):

Hessian inequality

$$D^2\varphi(x) \geq \lambda I, \quad \text{i.e.} \quad \langle D^2\varphi(x)\xi, \xi \rangle \geq \lambda|\xi|^2 \quad \text{for every } \xi \in \mathbb{R}^d.$$
(6.1a)

λ-monotonicity of $\nabla\varphi$

$$\langle \nabla\varphi(x_0) - \nabla\varphi(x_1), x_0 - x_1 \rangle \geq \lambda|x_0 - x_1|^2.$$
(6.1b)

λ-convexity inequality

$$\varphi(x_\theta) \leq (1 - \theta)\varphi(x_0) + \theta\varphi(x_1) - \frac{\lambda}{2}\theta(1 - \theta)|x_0 - x_1|^2$$
$$x_\theta := (1 - \theta)x_0 + \theta x_1, \theta \in [0, 1].$$
(6.1c)

Sub-gradient inequality

$$\langle \nabla\varphi(x_1), x_1 - x_0 \rangle - \frac{\lambda}{2}|x_1 - x_0|^2$$
$$\geq \varphi(x_1) - \varphi(x_0)$$
$$\geq \langle \nabla\varphi(x_0), x_1 - x_0 \rangle + \frac{\lambda}{2}|x_1 - x_0|^2.$$
(6.1d)

Notice that

$$\varphi \text{ is } \lambda\text{-convex if and only if } \tilde{\varphi}(x) := \varphi(x) - \frac{\lambda}{2}|x|^2 \text{ is convex.}$$
(6.1e)

In particular, there exist constants $a \in \mathbb{R}$, $\boldsymbol{b} \in \mathbb{R}^d$ such that

$$\varphi(x) \geq a + \langle \boldsymbol{b}, x \rangle + \frac{\lambda}{2}|x|^2.$$
(6.1f)

Definition 1 (gradient flow). The gradient flow of φ is the family of maps

$$S_t : \mathbb{R}^d \to \mathbb{R}^d, \quad t \in [0, +\infty),$$

characterized by the following property: for every $u_0 \in \mathbb{R}^d$, $S_0(u_0) := u_0$ and the curve $u_t := S_t(u_0)$, $t \in (0, +\infty)$, is the unique C^1 solution of the Cauchy problem

$$\frac{d}{dt}u_t = -\nabla\varphi(u_t) \quad \text{in } (0, +\infty), \quad \lim_{t\downarrow 0} u_t = u_0.$$
(6.2)

By the standard Cauchy–Lipschitz theory and the *a priori* estimates we will show in the next theorem, for every initial datum $u_0 \in \mathbb{R}^d$, Equation (6.2) admits a unique global solution so that the family S_t, $t \in [0, +\infty)$, is a continuous semigroup of Lipschitz maps, thus satisfying

$$S_{t+h}(u_0) = S_t(S_h(u_0)), \quad \lim_{t \downarrow 0} S_t(u_0) = S_0(u_0) = u_0 \quad \text{for every } u_0 \in \mathbb{R}^d.$$
$$(6.3)$$

6.2.1 Basic estimates

Theorem 6.1 (basic differential estimates). *Let us assume that* $\varphi \in C^2(\mathbb{R}^d)$ *is λ-convex; if* $u : [0, +\infty) \to \mathbb{R}^d$ *is a solution of (6.2) then*

$$\frac{\mathrm{d}}{\mathrm{d}t} \frac{1}{2} |u_t - v|^2 + \frac{\lambda}{2} |u_t - v|^2$$

$$= e^{-\lambda t} \frac{\mathrm{d}}{\mathrm{d}t} \left(e^{\lambda t} \frac{1}{2} |u_t - v|^2 \right) \le \varphi(v) - \varphi(u_t) \quad \text{for every } v \in \mathbb{R}^d, \quad (\text{EVI}_\lambda)$$

$$\frac{\mathrm{d}}{\mathrm{d}t} \varphi(u_t) = -|u_t'|^2 = -|\nabla\varphi(u_t)|^2 \le 0, \quad (\text{EI})$$

$$\frac{\mathrm{d}}{\mathrm{d}t} \left(e^{2\lambda t} |\nabla\varphi(u_t)|^2 \right) = \frac{\mathrm{d}}{\mathrm{d}t} \left(e^{2\lambda t} |u_t'|^2 \right) \le 0; \quad (\text{SI}_\lambda)$$

moreover, if v is another solution to (6.2) then

$$\frac{\mathrm{d}}{\mathrm{d}t} \left(e^{\lambda t} |u_t - v_t| \right) \le 0. \quad (\text{Cont}_\lambda)$$

Proof. We sketch here the easy calculations.
For the *evolution variational inequality* (EVI_λ):

$$\frac{\mathrm{d}}{\mathrm{d}t} \frac{1}{2} |u_t - v|^2$$

$$= \langle u_t', u_t - v \rangle \overset{(6.2)}{=} \langle \nabla\varphi(u_t), v - u_t \rangle \overset{(6.1d)}{\le} \varphi(v) - \varphi(u_t) - \frac{\lambda}{2} |u_t - v|^2.$$

The *energy identity* (EI):

$$\frac{\mathrm{d}}{\mathrm{d}t} \varphi(u_t) = \langle \nabla\varphi(u_t), u_t' \rangle \overset{(6.2)}{=} -|\nabla\varphi(u_t)|^2 \overset{(6.2)}{=} -|u_t'|^2.$$

The *slope inequality* (SI_λ):

$$\frac{\mathrm{d}}{\mathrm{d}t} |\nabla\varphi(u_t)|^2 = 2\langle \mathrm{D}^2\varphi(u_t)\nabla\varphi(u_t), u_t' \rangle$$

$$\overset{(6.2)}{=} -2\langle \mathrm{D}^2\varphi(u_t)\nabla\varphi(u_t), \nabla\varphi(u_t) \rangle \overset{(6.1a)}{\le} -2\lambda |\nabla\varphi(u_t)|^2.$$

The λ-*contraction* property (Cont$_\lambda$):

$$\frac{\mathrm{d}}{\mathrm{d}t}|u_t - v_t|^2 = 2\langle u'_t - v'_t, u_t - v_t\rangle$$

$$\stackrel{(6.2)}{=} -2\langle\nabla\varphi(u_t) - \nabla\varphi(v_t), u_t - v_t\rangle \stackrel{(6.1b)}{\leq} -2\lambda|u_t - v_t|^2.$$

$\square$

In order to write in a simple way suitable integrated versions of the previous inequalities, we set

$$\mathsf{E}_\lambda(t) := \int_0^t \mathrm{e}^{\lambda r}\,\mathrm{d}r = \begin{cases} \frac{\mathrm{e}^{\lambda t}-1}{\lambda} & \text{if } \lambda \neq 0, \\ t & \text{if } \lambda = 0. \end{cases} \tag{6.4}$$

Corollary 6.2 (pointwise and integral inequalities). *If $u : [0, +\infty) \to \mathbb{R}^d$ is a solution to (6.2), then*

$$\frac{\mathrm{e}^{\lambda t}}{2}|u_t - v|^2 + \mathsf{E}_\lambda(t)\big(\varphi(u_t) - \varphi(v)\big) + \frac{\big(\mathsf{E}_\lambda(t)\big)^2}{2}|\nabla\varphi(u_t)|^2 \tag{6.5}$$

$$\leq \frac{1}{2}|u_0 - v|^2 \quad \text{for every } v \in \mathbb{R}^d,$$

$$\varphi(u_t) + \frac{1}{2}\int_0^t \Big(|u'_r|^2 + |\nabla\varphi(u_r)|^2\Big)\,\mathrm{d}r = \varphi(u_0), \tag{6.6}$$

$$|\nabla\varphi(u_t)| \leq \mathrm{e}^{-\lambda t}|\nabla\varphi(u_0)|; \tag{6.7}$$

moreover, if v is another solution to (6.2), then

$$|u_t - v_t| \leq \mathrm{e}^{-\lambda t}|u_0 - v_0|. \tag{6.8}$$

In particular, when $\lambda > 0$, φ admits a unique minimum point $\bar{u}$ and

$$\frac{\lambda}{2}|u_t - \bar{u}|^2 \leq \varphi(u_t) - \varphi(\bar{u}) \leq \frac{1}{2\lambda}|\nabla\varphi(u_t)|^2 \tag{6.9}$$

$$|u_t - \bar{u}| \leq \mathrm{e}^{-\lambda t}|u_0 - \bar{u}|, \quad \varphi(u_t) - \varphi(\bar{u}) \leq \mathrm{e}^{-2\lambda t}\big(\varphi(u_0) - \varphi(\bar{u})\big). \tag{6.10}$$

Proof. We have just to check (6.5): if A_t denotes the quantity in the left-hand side, we show that A_t is nonincreasing. A differentiation in time yields

$$\frac{d}{dt} A_t = e^{\lambda t} \left(\frac{\lambda}{2} |u_t - v|^2 + \frac{d}{dt} \frac{1}{2} |u_t - v|^2 + \varphi(u_t) - \varphi(v) + \mathsf{E}_\lambda(t) |\nabla\varphi(u_t)|^2 \right)$$

$$+ \mathsf{E}_\lambda(t) \frac{d}{dt} \varphi(u_t) + \frac{(\mathsf{E}_\lambda(t))^2}{2} \frac{d}{dt} |\nabla\varphi(u_t)|^2$$

$$\overset{\text{(EVI}_\lambda)}{\leq} \mathsf{E}_\lambda(t) \left(e^{\lambda t} |\nabla\varphi(u_t)|^2 + \frac{d}{dt} \varphi(u_t) + \frac{\mathsf{E}_\lambda(t)}{2} \frac{d}{dt} |\nabla\varphi(u_t)|^2 \right)$$

$$\overset{\text{(EI)}}{=} \mathsf{E}_\lambda(t) \left((e^{\lambda t} - 1) |\nabla\varphi(u_t)|^2 + \frac{\mathsf{E}_\lambda(t)}{2} \frac{d}{dt} |\nabla\varphi(u_t)|^2 \right)$$

$$\overset{\text{(SI}_\lambda)}{\leq} \mathsf{E}_\lambda(t) \left((e^{\lambda t} - 1 - \lambda\mathsf{E}_\lambda(t)) |\nabla\varphi(u_t)|^2 \right) \overset{(6.4)}{=} 0.$$

$\square$

In terms of the maps S_t, (6.7) yields the λ-contraction estimate

$$\mathsf{d}(\mathsf{S}_t(u_0), \mathsf{S}_t(v_0)) \leq e^{-\lambda t} \mathsf{d}(u_0, v_0) \quad \text{for every } u_0, v_0 \in \mathbb{R}^d, \ t \geq 0, \qquad (6.11)$$

thus showing the Lipschitz property of S_t and the uniqueness and continuous dependence w.r.t. the initial data of the solutions of (6.2).

6.2.2 Approximation by the implicit Euler scheme

One of the simplest but very useful ways to construct discrete approximations of the solution to (6.2) (and to show its existence by a limiting process) is given by the *implicit Euler scheme*.

For a given time step $\tau > 0$ we consider the associated uniform partition of $[0, +\infty)$

$$\mathcal{P}_\tau := \{0 = t_\tau^0 < t_\tau^1 < \cdots < t_\tau^n < \cdots\}, \quad t_\tau^n := n\tau, \qquad (6.12)$$

and we look for a discrete sequence $(U_\tau^n)_{n \in \mathbb{N}}$ whose value U_τ^n should provide an effective approximation of $u(t_\tau^n)$. U_τ^n are defined recursively, starting from a suitable choice of $U_\tau^0 \approx u_0$, by solving at each step the equation of the Euler scheme

$$\frac{U_\tau^n - U_\tau^{n-1}}{\tau} = -\nabla\varphi(U_\tau^n) \quad n = 1, 2, \ldots, \qquad (6.13)$$

or, equivalently,

$$U_\tau^n = J_\tau(U_\tau^{n-1}), \quad J_\tau := (I + \tau\nabla\varphi)^{-1}. \qquad (6.14)$$

Existence of a discrete approximating solution can be easily obtained by looking for the minimizers of the function

$$U \mapsto \Phi(\tau, U_\tau^{n-1}; U) := \frac{1}{2\tau}|U - U_\tau^{n-1}|^2 + \varphi(U). \qquad (6.15)$$

In fact, it is immediate to check that any minimizer U_τ^n of (6.15) solves (6.13); moreover, the function defined by (6.15) is $(\tau^{-1} + \lambda)$-convex and therefore it admits a unique minimizer whenever $\tau^{-1} > -\lambda$.

Denoting by $U_\tau : [0, +\infty) \to \mathbb{R}^d$ the piecewise linear interpolant of the discrete values $(U_\tau^n)_{n \in \mathbb{N}}$ on the grid $\mathcal{P}_\tau$, defined by

$$U_\tau(t) := \frac{t - t_\tau^{n-1}}{\tau} U_\tau^{n-1} + \frac{t_\tau^n - t}{\tau} U_\tau^n \quad \text{if } t \in [t_\tau^{n-1}, t_\tau^n], \qquad (6.16)$$

one expects that $U_\tau(t)$ converges to the solution u_t to (6.2) as $\tau \downarrow 0$.

Theorem 6.3. *If $\lim_{\tau \downarrow 0} U_\tau^0 = u_0$ then the family of piecewise linear interpolants $(U_\tau)_{\tau > 0}$ satisfies the Cauchy condition as $\tau \downarrow 0$ with respect to the uniform convergence on each compact interval $[0, T]$, $T > 0$; its unique limit is the solution u_t of (6.2). Moreover, for every $T > 0$ there exists a universal constant $C(\lambda, T)$ such that*

$$\sup_{t \in [0,T]} |u_t - U_\tau(t)| \le |u_0 - U_\tau^0| + C(\lambda, T)|\nabla\varphi(u_0)|\,\tau. \qquad (6.17)$$

In particular, when $\lambda = 0$ we can choose $C = \frac{1}{\sqrt{2}}$, independent of T.

Remarks about the proof. In the present finite-dimensional smooth setting, the proof of the convergence of U_τ is not difficult: e.g. considering the case $\lambda = 0$, we can apply the contraction property of the map J_τ defined by (6.14)

$$|J_\tau(x) - J_\tau(y)| \le |x - y| \quad \text{for every } x, y \in \mathbb{R}^d \qquad (6.18)$$

to obtain the uniform bound

$$\tau^{-1}|U_\tau^n - U_\tau^{n-1}| = |\nabla\varphi(U_\tau^n)| \le |\nabla\varphi(U_\tau^{n-1})| \quad \text{for every } n \ge 1, \qquad (6.19)$$

so that

$$|U_\tau'(t)| \le \sup_{n \in \mathbb{N}} \tau^{-1}|U_\tau^n - U_\tau^{n-1}| = \tau^{-1}|U_\tau^1 - U_\tau^0| \le |\nabla\varphi(U_\tau^0)|$$
$$\text{for every } t \in [0, +\infty) \setminus \mathcal{P}_\tau. \qquad (6.20)$$

Since $\lim_{\tau \downarrow 0} |\nabla\varphi(U_\tau^0)| = |\nabla\varphi(u_0)|$ it follows that $(U_\tau)_{\tau > 0}$ satisfies a uniform Lipschitz condition and therefore it admits a suitable subsequence uniformly converging to a Lipschitz curve u in each compact interval $[0, T]$. Denoting by

$\bar{U}_\tau(t)$ the piecewise constant interpolant

$$\bar{U}_\tau(t) := U_\tau^n \quad \text{if } t \in (t_\tau^{n-1}, t^n], \tag{6.21}$$

the same estimate (6.20) shows that

$$\sup_{t \in (0, +\infty)} |U_\tau(t) - \bar{U}_\tau(t)| \leq \tau |\nabla \varphi(U_\tau^0)|, \tag{6.22}$$

so that $\bar{U}_\tau$ has the same limit points as U_τ. On the other hand, (6.13) yields

$$U_\tau'(t) = -\nabla \varphi(\bar{U}_\tau(t)) \quad \text{in } [0, +\infty) \setminus \mathcal{P}_\tau, \tag{6.23}$$

and we can pass to the limit in an integrated form of (6.23) thus showing that u solves (6.2).

The uniform error estimate (6.17) is subtler: a simple derivation in the case $\lambda = 0$ can be found in [51]; see also [59,61]. Its main functional interest relies on the fact that it involves just the lower bound on the Hessian of φ but not its upper bound (and, therefore, it does not require a uniform Lipschitz condition on $\nabla \varphi$). $\qquad\square$

6.2.3 Metric characterization of gradient flows in $\mathbb{R}^d$

The energy identity (EI) (with its integrated version (6.6)) and the evolution variational inequality (EVI_λ) not only provide important estimates on the solution to (6.2) but can also be used to characterize it.

Concerning (EI), we can even relax the identity, as the following proposition shows.

Proposition 6.4 (curves of maximal slope). *A C^1 curve $u : [0, +\infty) \to \mathbb{R}^d$ is a solution to (6.2) if and only if it satisfies the* energy dissipation inequality

$$\frac{d}{dt} \varphi(u_t) \leq -\frac{1}{2} |u_t'|^2 - \frac{1}{2} |\nabla \varphi(u_t)|^2 \quad \text{in } (0, +\infty) \tag{EDI}$$

or its weaker integrated form

$$\varphi(u_t) + \frac{1}{2} \int_0^t \left(|u_r'|^2 + |\nabla \varphi(u_r)|^2 \right) dr \leq \varphi(u_0) \quad \text{for every } t \in (0, +\infty). \tag{EDI'}$$

Proof. If u is a C^1 curve the chain rule yields

$$\varphi(u_t) = \varphi(u_0) + \int_0^t \langle \nabla \varphi(u_r), u_r' \rangle \, dr, \tag{6.24}$$

so that (EDI$'$) yields

$$\frac{1}{2}\int_0^t |u_r' + \nabla\varphi(u_r)|^2 \, dr$$
$$= \frac{1}{2}\int_0^t \left(|u_r'|^2 + |\nabla\varphi(u_r)|^2\right) dr + \int_0^t \langle\nabla\varphi(u_r), u_r'\rangle \, dr \le 0,$$

and therefore $u_r' = -\nabla\varphi(u_r)$ for $\mathcal{L}^1$-a.e. $r \in (0,t)$. Since t is arbitrary and $u \in C^1$, u solves (6.2). □

Notice that in the previous formulation we did not use the λ-convexity of φ: the argument only relies on the chain rule.

In the following proposition we show that also the evolution variational inequality (EVI$_\lambda$) characterizes a solution of (6.2). In fact, if (EVI$_\lambda$) admits a solution *for every initial datum* u_0, then φ is λ-convex.

Proposition 6.5 (characterization of gradient flows through the EVI). *If $u : [0, +\infty) \to \mathbb{R}^d$ is a C^1 curve solving (EVI$_\lambda$) then u is a solution to (6.2).*

Proof. Applying the chain rule for the squared distance function $\frac{1}{2}|\cdot - v|^2$ we easily have

$$\langle u_t', u_t - v\rangle \le \varphi(v) - \varphi(u_t) - \frac{\lambda}{2}|u_t - v|^2 \quad \text{for every } v \in \mathbb{R}^d, \ t > 0.$$
$$(6.25)$$

Choosing $v := u_t + \varepsilon\xi$, for $\varepsilon > 0$ and $\xi \in \mathbb{R}^d$ and dividing by ε we obtain

$$-\langle u_t', \xi\rangle \le \varepsilon^{-1}\left(\varphi(u_t + \varepsilon\xi) - \varphi(u_t)\right) - \frac{\lambda\varepsilon}{2}|\xi|^2 \quad \text{for every } \xi \in \mathbb{R}^d.$$

Passing to the limit as $\varepsilon \downarrow 0$ we eventually get

$$-\langle u_t', \xi\rangle \le \langle\nabla\varphi(u_t), \xi\rangle \quad \text{for every } \xi \in \mathbb{R}^d,$$

so that $-u_t' = \nabla\varphi(u_t)$. □

Proposition 6.6. *Let us suppose that there exists a C^1 semigroup $\tilde{S}_t : \mathbb{R}^d \to \mathbb{R}^d$, $t \ge 0$, of smooth maps such that for every $u_0 \in \mathbb{R}^d$ the curve $u_t := \tilde{S}_t(u_0)$ satisfies (EVI$_\lambda$). Then φ is λ-convex.*

Proof. We consider for simplicity the case $\lambda = 0$; for arbitrary $u^0, u^1 \in \mathbb{R}^d$ we set

$$u^s := (1-s)u^0 + su^1, \quad u_t^s := \tilde{S}_t(u^s)$$

and we want to show that

$$\frac{d}{ds}\varphi(u^s)|_{s=0} \leq \frac{d}{ds}\varphi(u^s)|_{s=1}.$$

We get

$$\frac{d}{ds}\varphi(u^s)|_{s=0} = \langle \nabla\varphi(u^0), u^1 - u^0 \rangle \overset{(6.2)}{=} -\left\langle \frac{d}{dt}u_t^0|_{t=0}, u^1 - u^0 \right\rangle$$

$$= \frac{d}{dt}\left(\frac{1}{2}|u_t^0 - u^1|^2\right)|_{t=0} \overset{(EVI_\lambda)}{\leq} \varphi(u^1) - \varphi(u^0)$$

$$\overset{(EVI_\lambda)}{\leq} -\frac{d}{dt}\left(\frac{1}{2}|u^0 - u_t^1|^2\right)|_{t=0}$$

$$\overset{(6.2)}{\leq} \langle \nabla\varphi(u^1), u^1 - u^0 \rangle = \frac{d}{ds}\varphi(u^s)|_{s=1}$$

$\square$

6.2.4 Extensions to more general functional settings

The simple finite-dimensional theory for smooth functionals has been extended in various directions; without claiming any completeness, we quote here four different points of view.

1. *The theory of differential inclusions and maximal monotone operators in Hilbert spaces*, developed in the 1970s by Kōmura [39], Crandall and Pazy [25], Crandall and Liggett [24], Brézis [13], Bénilan [10], Lions [40]; we refer to the monographs [8, 14, 40]. In this framework one considers the gradient flow generated by a proper lower semicontinuous λ-convex functional ϕ : $H \to (-\infty, +\infty]$, where H is a separable Hilbert space. By using tools of convex analysis, clever regularization techniques, and replacing $\nabla\varphi$ with the multivalued subdifferential operator $\partial\phi$, one can basically reproduce all the estimates and results we briefly discussed in the finite-dimensional setting which just depend on the lower bound of the Hessian of φ, avoiding any strong compactness assumptions.

In this framework, the resolvent operator $J_\tau := (I + \tau\partial\phi)^{-1}$ is single valued and non-expansive, i.e.

$$d(J_\tau[u], J_\tau[v]) \leq d(u, v) \quad \text{for every } u, v \in H, \quad \tau > 0. \tag{6.26}$$

This property is the key ingredient to prove, as in the Crandall and Liggett generation theorem [24], uniform convergence of the exponential formula

$$u_t = \lim_{n\to+\infty} (J_{t/n})^n[u_0], \quad d(u_t, (J_{t/n})^n[u_0]) \leq \frac{2|\partial\phi|(u_0)t}{\sqrt{n}}, \tag{6.27}$$

and therefore to define a contraction semigroup on $\overline{D(\phi)}$.

Being generated by a convex functional, this semigroup exhibits a nice regularization effect, since $u_t \in D(\partial\phi)$ even if $u_0 \in \overline{D(\phi)}$. Moreover, the curve u_t can be characterized as the unique solution of the evolution variational inequality (EVI$_\lambda$), whose formulation goes back to [41]. Optimal error estimates for the implicit Euler discretization in the spirit of (6.17) have been obtained by [7,51,59,61].

2. *The theory of the curves of maximal slope in metric spaces*, developed in the 1980s by De Giorgi, Degiovanni, Marino, and Tosques in a series of papers originating from [29, 30], and culminating in [31, 44] (but see also the more recent [16] and the presentation of [2, 3]). Here, $\phi : X \to (-\infty, +\infty]$ is a proper and lower semicontinuous functional defined in the complete metric space X and one looks for absolutely continuous curves satisfying a suitable form of the energy dissipation inequality (EDI), where $|u'|$ should be interpreted as the *metric velocity of the curve u* and $|\nabla\varphi(u)|$ should be replaced by the *metric slope of ϕ*. The theory is usually based on local compactness of the sublevels of ϕ and various kinds of assumptions on its slope, yielding in particular its lower semicontinuity and the possibility to write a weak form of the chain rule. The advantage of this approach relies on its flexibility, but in general metric spaces uniqueness and stability properties of curves of maximal slope are not known.

3. *Limits of discrete solutions, generalized minimizing movements*: this is the weakest approach, which has been clarified in [28] and independently applied to different kinds of problems (e.g. see [35], [36], [43], [48]). It just provides a general approximating scheme which is quite useful to construct some limit curves by compactness arguments, but one can hardly deduce refined properties of these curves from general metric results, and each example deserves a careful ad hoc investigation.

4. *Evolution variational inequalities in metric spaces*: this is the strongest point of view, which is related to the metric evolution variational inequality (EVI$_\lambda$) and goes back to Bénilan's [10] notion of integral solutions to evolution equations in Banach spaces. Its application to gradient flows in metric spaces has been developed in [3] and it will be adopted in these notes.

6.3 Gradient flows and evolution variational inequalities in metric spaces

The aim of this section is to study the metric notion of gradient flows associated with the (metric formulation of the) evolution variational inequality (EVI$_\lambda$).

Throughout the rest of these notes, (X, d) will be a *complete and separable metric space* and $\phi : X \to (-\infty, +\infty]$ a proper and l.s.c. functional on X with non-empty domain $D(\phi) = \{v \in X : \phi(v) < +\infty\}$. We will look for curves $u : [0, +\infty) \to X$ which satisfy properties that depend only on the metric structure of X and that in the case of a smooth function $\phi = \varphi$ on $X = \mathbb{R}^d$ satisfy the ordinary differential equation (6.2).

6.3.1 A few metric concepts

Let us first recall the notion of *metric velocity* and *metric slope* (e.g. see [3]).

Definition 2 (absolutely continuous curves). We say that a curve $v : (a, b) \subset \mathbb{R} \to X$ belongs to $AC^p_{(\text{loc})}(a, b; X)$ for some $p \in [1, +\infty]$ if there exists $m \in L^p_{(\text{loc})}(a, b)$ such that

$$d(v_s, v_t) \leq \int_s^t m(r)\, dr \qquad \text{for every } a < s \leq t < b. \tag{6.28}$$

If $p = 1$ we say that v is a *(locally) absolutely continuous curve*.

Theorem 6.7 (metric derivative). If $v : (a, b) \to X$ is an absolutely continuous curve then the limit

$$|v'|(t) = \lim_{s \to t} \frac{d(v_s, v_t)}{|t - s|} \tag{6.29}$$

exists for $\mathcal{L}^1$-a.e. $t \in (a, b)$ and it is called the *metric derivative* of v at the point t. Moreover, the function $t \mapsto |v'|(t)$ belongs to $L^1(a, b)$, it is an admissible integrand for the right-hand side of (6.28), and it is minimal in the following sense:

$$|v'|(t) \leq m(t) \quad \text{for } \mathcal{L}^1\text{-a.e. } t \in (a, b), \text{ for each function } m \text{ satisfying (6.28).}$$

Definition 3 (metric slope). The metric slope of ϕ at a point $v \in X$ is given by

$$|\partial\phi|(v) = \begin{cases} +\infty & \text{if } v \notin D(\phi), \\ 0 & \text{if } v \in D(\phi) \text{ is isolated,} \\ \limsup\limits_{w \to v} \dfrac{(\phi(v) - \phi(w))^+}{d(v, w)} & \text{otherwise.} \end{cases} \tag{6.30}$$

6.3.2 Structural properties of solutions to evolution variational inequalities

The next (quite restrictive) definition is modeled on the case of λ-convex functionals in Euclidean-like spaces and has been introduced and discussed in [3, Chapter 4].

Definition 4 (EVI and gradient flow). A solution of the evolution variational inequality $\mathrm{EVI}_\lambda(X, \mathsf{d}, \phi)$, $\lambda \in \mathbb{R}$, is a locally absolutely continuous curve $u : t \in (0, +\infty) \mapsto u_t \in D(\phi)$ such that

$$\frac{1}{2}\frac{d}{dt}\mathsf{d}^2(u_t, v) + \frac{\lambda}{2}\mathsf{d}^2(u_t, v)$$

$$\leq \phi(v) - \phi(u_t) \quad \mathscr{L}^1\text{-a.e. in } (0, +\infty), \quad \text{for every } v \in D(\phi). \quad (\mathrm{EVI}_\lambda)$$

A λ-gradient flow of ϕ is a family of continuous maps $\mathsf{S}_t : \overline{D(\phi)} \to D(\phi)$, $t > 0$, such that for every $u \in \overline{D(\phi)}$

$$\lim_{t\downarrow 0} \mathsf{S}_t(u) = u =: \mathsf{S}_0(u), \qquad \mathsf{S}_{t+h}(u) = \mathsf{S}_h(\mathsf{S}_t(u)) \quad \text{for every } t, h \geq 0, \tag{6.31a}$$

$$\text{the curve } t \mapsto \mathsf{S}_t(u) \quad \text{is a solution of } \mathrm{EVI}_\lambda(X, \mathsf{d}, \phi). \tag{6.31b}$$

The next result shows that (EVI_λ) can be formulated avoiding differentiation and without assuming the absolute continuity of u (see [60] for the proof).

Theorem 6.8 (derivative free characterization of solutions to (EVI_λ)). *A curve $u : (0, +\infty) \to \overline{D(\phi)}$ is a solution of $\mathrm{EVI}_\lambda(X, \mathsf{d}, \phi)$ according to Definition 4 if and only if for every $s, t \in (0, +\infty)$ with $s < t$ and $v \in D(\phi)$*

$$\frac{e^{\lambda(t-s)}}{2}\mathsf{d}^2(u_t, v) - \frac{1}{2}\mathsf{d}^2(u_s, v) \leq \mathsf{E}_\lambda(t - s)\Big(\phi(v) - \phi(u_t)\Big). \quad (\mathrm{EVI}'_\lambda)$$

Notice that (EVI'_λ) yields the pointwise right-upper differential inequality

$$\frac{1}{2}\frac{d^+}{dt}\mathsf{d}^2(u_t, v) + \frac{\lambda}{2}\mathsf{d}^2(u_t, v) \leq \phi(v) - \phi(u_t) \quad \text{for every } v \in D(\phi), \tag{6.32}$$

at *every* time $t > 0$; here, $\frac{d^+}{dt}\zeta$ denotes the right-upper Dini derivative $\limsup_{h\downarrow 0} h^{-1}(\zeta(t + h) - \zeta(t))$.

The next result collects many useful properties of solutions to $\mathrm{EVI}_\lambda(X, \mathsf{d}, \phi)$ (see [60] and an analogous result of [4] in the Wasserstein framework): they reproduce in the metric framework the estimates of the previous section and show that (EVI_λ) contains all the information concerning the gradient flow of ϕ.

Theorem 6.9 (properties of solutions to (EVI$_\lambda$)). *Let $u, u^1, u^2 : [0, +\infty) \to X$ be solutions of* $\mathrm{EVI}_\lambda(X, \mathsf{d}, \phi)$.

  λ-contraction and uniqueness:

$$\mathsf{d}(u^1_t, u^2_t) \leq e^{-\lambda(t-s)}\mathsf{d}(u^1_s, u^2_s) \quad \text{for every } 0 \leq s < t < +\infty. \quad (6.33)$$

  In particular, for every $u_0 \in \overline{D(\phi)}$ there is at most one solution u of $\mathrm{EVI}_\lambda(X, \mathsf{d}, \phi)$ satisfying the initial condition $\lim_{t\downarrow 0} u_t = u_0$.

Regularizing effects: *u is locally Lipschitz continuous in $(0, +\infty)$ and $u_t \in D(|\partial\phi|) \subset D(\phi)$ for every $t > 0$. Moreover, in the time interval $[0, +\infty)$,*

$$\text{the map } t \mapsto \phi(u_t) \text{ is non-increasing and (locally semi-, if } \lambda < 0) \text{ convex,} \tag{6.34}$$

$$\text{the map } t \mapsto e^{\lambda t}|\partial\phi|(u_t) \text{ is non-increasing and right continuous,} \tag{6.35}$$

  the following regularization/a priori estimate holds

$$\frac{e^{\lambda t}}{2}\mathsf{d}^2(u_t, v) + \mathsf{E}_\lambda(t)\Big(\phi(u_t) - \phi(v)\Big) + \frac{\big(\mathsf{E}_\lambda(t)\big)^2}{2}|\partial\phi|^2(u_t) \leq \frac{1}{2}\mathsf{d}^2(u_0, v) \tag{6.36}$$

  for every $v \in D(\phi)$; in particular

$$\phi(u_t) \leq \phi(v) + \frac{1}{2\mathsf{E}_\lambda(t)}\mathsf{d}^2(u_0, v), \tag{6.37}$$

$$|\partial\phi|^2(u_t) \leq \frac{1}{2e^{\lambda t} - 1}|\partial\phi|^2(v) + \frac{1}{(\mathsf{E}_\lambda(t))^2}\mathsf{d}^2(u_0, v) \quad \text{if } -\lambda t < \log 2. \tag{6.38}$$

Asymptotic expansion for $t \downarrow 0$: *If $u_0 \in D(|\partial\phi|)$ and $\lambda \leq 0$ then for every $v \in D(\phi)$ and $t \geq 0$*

$$\frac{e^{2\lambda t}}{2}\mathsf{d}^2(u_t, v) - \frac{1}{2}\mathsf{d}^2(u_0, v) \leq \mathsf{E}_{2\lambda}(t)\big(\phi(v) - \phi(u_0)\big) + \frac{t^2}{2}|\partial\phi|^2(u_0). \tag{6.39}$$

Right and left limits, energy identity: *For every $t > 0$ the right limits*

$$|\dot{u}_{t+}| := \lim_{h\downarrow 0}\frac{\mathsf{d}(u_t, u_{t+h})}{h}, \quad \frac{\mathrm{d}}{\mathrm{d}t}\phi(u_{t+}) := \lim_{h\downarrow 0}\frac{\phi(u_{t+h}) - \phi(u_t)}{h} \tag{6.40}$$

  exist, they satisfy

$$\frac{\mathrm{d}}{\mathrm{d}t}\phi(u_{t+}) = -|\dot{u}_{t+}|^2 = -|\partial\phi|^2(u_t), \tag{6.41}$$

and they define a right continuous map. (6.40) and (6.41) hold at $t = 0$ iff $u_0 \in D(|\partial\phi|)$. Moreover, there exists an at most countable set $\mathcal{C} \subset (0, +\infty)$ such that the analogous identities for the left limits *hold for every $t \in (0, +\infty) \setminus \mathcal{C}$.*

Asymptotic behavior: *If $\lambda > 0$, then ϕ admits a unique minimum point $\bar{u}$ and for every $t \geq t_0 \geq 0$ we have*

$$\frac{\lambda}{2}d^2(u_t, \bar{u}) \leq \phi(u_t) - \phi(\bar{u}) \leq \frac{1}{2\lambda}|\partial\phi|^2(u_t), \tag{6.42a}$$

$$d^2(u_t, \bar{u}) \leq d^2(u_{t_0}, \bar{u})e^{-\lambda(t-t_0)}, \tag{6.42b}$$

$$\phi(u_t) - \phi(\bar{u}) \leq \left(\phi(u_{t_0}) - \phi(\bar{u})\right)e^{-2\lambda(t-t_0)},$$

$$\phi(u_t) - \phi(\bar{u}) \leq \frac{1}{2}e^{\lambda(t-t_0)}d^2(u_{t_0}, \bar{u}), \tag{6.42c}$$

$$|\partial\phi|(u_t) \leq |\partial\phi|(u_{t_0})e^{-\lambda(t-t_0)}, \quad |\partial\phi|(u_t) \leq \frac{1}{e^{\lambda(t-t_0)}d(u_{t_0}, \bar{u})}. \tag{6.42d}$$

If $\lambda = 0$ and $\bar{u}$ is any minimum point of ϕ then we have

$$|\partial\phi|(u_t) \leq \frac{d^2(u_0, \bar{u})}{t}, \quad \phi(u_t) - \phi(\bar{u}) \leq \frac{d^2(u_0, \bar{u})}{2t}, \tag{6.43}$$

the map $\quad t \mapsto d^2(u_t, \bar{u}) \quad$ is not increasing.

Continuity of the energy and the slope: *If $u^n \in C^0([0, +\infty); X)$ are solutions of $\mathrm{EVI}_\lambda(X, d, \phi)$ such that $\lim_{n\uparrow+\infty} u_0^n = u_0$, then*

$$\lim_{n\uparrow\infty} \phi(u_t^n) = \phi(u_t) \qquad \text{for every } t > 0, \tag{6.44}$$

$$\lim_{n\uparrow\infty} |\partial\phi|(u_t^n) = |\partial\phi|(u_t) \qquad \text{for every } t \in (0, +\infty) \setminus \mathcal{C}. \tag{6.45}$$

We just sketch the proof of the contraction property (6.33). For a fixed $s \in (0, +\infty)$ we have that

$$\frac{\partial}{\partial t}\frac{1}{2}d^2(u_t^1, u_s^2) + \frac{\lambda}{2}d^2(u_t^1, u_s^2) \leq \phi(u_s^2) - \phi(u_t^1) \quad \text{for every } t \in (0, +\infty), \tag{6.46}$$

while for a fixed $t \in (0, +\infty)$

$$\frac{\partial}{\partial s}\frac{1}{2}d^2(u_t^1, u_s^2) + \frac{\lambda}{2}d^2(u_t^1, u_s^2) \leq \phi(u_t^1) - \phi(u_s^2) \quad \text{for every } s \in (0, +\infty). \tag{6.47}$$

Adding (6.46) and (6.47) we get

$$\frac{\partial}{\partial t}\frac{1}{2}d^2(u_t^1, u_s^2) + \frac{\partial}{\partial s}\frac{1}{2}d^2(u_t^1, u_s^2) + \lambda d^2(u_t^1, u_s^2) \leq 0.$$

Applying [3, Lemma 4.3.4] we obtain

$$\frac{d}{dt}d^2(u_t^1, u_t^2) \le -2\lambda d^2(u_t^1, u_t^2) \quad \mathscr{L}^1\text{-a.e. in } (0, +\infty)$$

and therefore we obtain (6.33). □

Theorem 6.9 concerns each single solution to (EVI$_\lambda$); when the λ-gradient flow S_t of ϕ exists we have further interesting properties, showing that the formulation by (EVI$_\lambda$) is really stronger than all the other metric approaches.

6.3.3 λ-gradient flows and λ-convexity along geodesics

Let us first recall the notion of (minimal, constant speed) geodesics in a metric space X and the related convexity.

Definition 5 (constant speed geodesics). A curve $\gamma : [0, 1] \to X$ is a *constant-speed geodesic* (or simply *geodesic*) if

$$d(\gamma_s, \gamma_t) = |t - s| \, d(\gamma_0, \gamma_1) \quad \text{for every } 0 \le s \le t \le 1. \tag{6.48}$$

A set $D \subset X$ is geodesically convex if every couple of points $x_0, x_1 \in D$ can be connected by a geodesic γ contained in D.

Definition 6 (λ-convexity along curves and geodesically λ-convex functionals). We say that $\phi : X \to (-\infty, +\infty]$ is λ-*convex* along a curve $\gamma : [0, 1] \to X$ if

$$\phi(\gamma_s) \le (1 - s)\phi(\gamma_0) + s\phi(\gamma_1) - \frac{\lambda}{2}s(1 - s)d^2(\gamma_0, \gamma_1) \quad \text{for every } s \in [0, 1]. \tag{6.49}$$

We say that ϕ is *geodesically λ-convex* if every couple of points $x_0, x_1 \in D(\phi)$ can be connected by a geodesic γ along which ϕ is λ-convex. If ϕ is geodesically convex and it is λ-convex along *every* geodesic connecting $x_0, x_1 \in D(\phi)$ in $\overline{D(\phi)}$ then we say that ϕ is *strongly* geodesically λ-convex.

Theorem 6.10 ([27]). *If the λ-gradient flow S_t of ϕ exists, then ϕ is λ-convex along any geodesic in $\overline{D(\phi)}$. In particular, if $\overline{D(\phi)}$ is geodesically convex, then ϕ is strongly geodesically λ-convex.*

Proof. Let $\gamma : s \in [0, 1] \mapsto \gamma^s \in \overline{D(\phi)}$ be a geodesic with $\gamma^0, \gamma^1 \in D(\phi)$ and let us set $\gamma_t^s := S_t(\gamma^s)$.

Applying (EVI$'_\lambda$) we have for every $s \in [0, 1]$ and $t > 0$

$$\frac{1}{2}e^{\lambda t}d^2(\gamma^s_t, \gamma^0) - \frac{1}{2}d^2(\gamma^s, \gamma^0) \le E_\lambda(t)\big(\phi(\gamma^0) - \phi(\gamma^s_t)\big), \qquad (6.50)$$

$$\frac{1}{2}e^{\lambda t}d^2(\gamma^s_t, \gamma^1) - \frac{1}{2}d^2(\gamma^s, \gamma^1) \le E_\lambda(t)\big(\phi(\gamma^1) - \phi(\gamma^s_t)\big). \qquad (6.51)$$

Multiplying (6.50) by $(1 - s)$ and (6.51) by s and adding the two inequalities we get

$$\frac{e^{\lambda t}}{2}\big((1 - s)d^2(\gamma^s_t, \gamma^0) + sd^2(\gamma^s_t, \gamma^1)\big) - \frac{1}{2}\big((1 - s)d^2(\gamma^s, \gamma^0) + sd^2(\gamma^s, \gamma^1)\big)$$
$$\le E_\lambda(t)\big((1 - s)\phi(\gamma^0) + s\phi(\gamma^1) - \phi(\gamma^s_t)\big). \qquad (6.52)$$

We now observe that the elementary inequality

$$(1 - s)a^2 + sb^2 \ge s(1 - s)(a + b)^2 \quad \text{for every } a, b \in \mathbb{R}, \quad s \in [0, 1], \qquad (6.53)$$

and the triangular inequality yield

$$(1 - s)d^2(\gamma^s_t, \gamma^0) + sd^2(\gamma^s_t, \gamma^1) \overset{(6.53)}{\ge} s(1 - s)\big(d(\gamma^s_t, \gamma^0) + d(\gamma^s_t, \gamma^1)\big)^2$$
$$\ge s(1 - s)d^2(\gamma^0, \gamma^1). \qquad (6.54)$$

On the other hand, since γ is a geodesic we have

$$(1 - s)d^2(\gamma^s, \gamma^0) + sd^2(\gamma^s, \gamma^1) = s(1 - s)d^2(\gamma^0, \gamma^1). \qquad (6.55)$$

Inserting (6.54) and (6.55) in (6.52) we get

$$\frac{e^{\lambda t} - 1}{2}s(1 - s)d^2(\gamma^0, \gamma^1) \le E_\lambda(t)\big((1 - s)\phi(\gamma^0) + s\phi(\gamma^1) - \phi(\gamma^s_t)\big). \quad (6.56)$$

Dividing then both sides of (6.56) by $E_\lambda(t)$ and passing to the limit as $t \downarrow 0$ we obtain

$$\phi(\gamma^s) \le (1 - s)\phi(\gamma^0) + s\phi(\gamma^1) - \frac{\lambda}{2}s(1 - s)d^2(\gamma^0, \gamma^1) \quad \text{for every } s \in [0, 1].$$

$$\square$$

6.3.4 λ-gradient flows and curves of maximal slope

Definition 7 (curves of maximal slope). We say that a curve $u \in AC^2_{\text{loc}}(0, +\infty; X)$ is a curve of maximal slope for the functional ϕ if the energy dissipation inequality

$$\frac{1}{2}\int_s^t |u'|^2(r)\,dr + \frac{1}{2}\int_s^t |\partial\phi|^2(u_r)\,dr \leq \phi(u_s) - \phi(u_t) \tag{6.57}$$

holds for all $0 < s \leq t < +\infty$.

The notion of curve of maximal slope was first introduced (in a slightly different form) by De Giorgi and provides a weak notion of gradient flow for nonsmooth functionals, also nonconvex. If ϕ admits a λ-gradient flow according to Definition 4, then these two definitions coincide.

Theorem 6.11. *Let us assume that the λ-gradient flow S of ϕ exists and let $u \in AC_{\mathrm{loc}}^2(0, +\infty; X)$ be satisfying (6.57) with $\lim_{t\downarrow 0} u_t = u_0 \in D(\phi)$. Then $u_t = S_t(u_0)$ for every $t \geq 0$ and (6.57) is in fact an identity for every $0 \leq s < t < +\infty$.*

6.3.5 λ-gradient flows and the minimizing movements variational scheme

A general variational method to approximate gradient flows (and often to prove their existence) is provided by the so-called *minimizing movements* variational scheme. In its original formulation (e.g. see [28]), the method consists in finding a discrete approximation $\overline{U}_\tau$ of the continuous gradient flow u by solving a recursive variational scheme, which is the natural generalization of (6.15) to a metric-space setting. If $\tau > 0$ denotes the step size of the uniform partition $\mathcal{P}_\tau$ (6.12), starting from a suitable approximation U_τ^0 of u_0 one looks at each step $((n-1)\tau, n\tau]$ for the minimizers of the functional

$$U \mapsto \Phi(\tau, U_\tau^{n-1}; U) := \frac{1}{2\tau}d^2(U, U_\tau^{n-1}) + \phi(U). \tag{6.58}$$

$\overline{U}_\tau$ thus takes a value $U_\tau^n \in \operatorname{argmin} \Phi(\tau, U^{n-1}; \cdot)$ on each interval $((n-1)\tau, n\tau]$.

Definition 8 (the minimizing movement variational scheme). Let us consider a time step $\tau > 0$ and a discrete initial datum $U_\tau^0 \in D(\phi)$. A τ-discrete minimizing movement starting from U_τ^0 is any sequence $(U_\tau^n)_{n\in\mathbb{N}}$ in $D(\phi)$ which satisfies

$$\Phi(\tau, U_\tau^{n-1}; U_\tau^n) \leq \Phi(\tau, U_\tau^{n-1}; V) \quad \text{for every } V \in X, \, n \in \mathbb{N}. \tag{6.59}$$

A discrete solution $\overline{U}_\tau$ is any piecewise constant interpolant of a τ-discrete minimizing movement on the grid $\mathcal{P}_\tau$ defined by

$$\overline{U}_\tau(0) = U_\tau^0, \quad \overline{U}_\tau(t) \equiv U_\tau^n \quad \text{if } t \in (t_\tau^{n-1}, t_\tau^n], n \geq 1. \tag{6.60}$$

The existence of a minimizing sequence $\{U_\tau^n\}_{n \in \mathbb{N}}$ is usually obtained by invoking the direct method of the calculus of variations, thus requiring that the functional (6.58) has compact sublevels with respect to some Hausdorff topology σ on X (e.g. see the setting of [3, Section 2.1]). In the next section we will discuss another possibility, still considered in [3], when the functional (6.58) satisfies a strong convexity assumption.

In a general setting it is also possible to avoid these restrictions by applying Ekeland's variational principle to the functional (6.58); this approach only requires the completeness of the metric space.

Definition 9 (a relaxed minimizing movement variational scheme). Let us consider a time step $\tau > 0$, a relaxation parameter $\eta \geq 0$, and a discrete initial datum $U_{\tau,\eta}^0 \in D(\phi)$. A (τ, η)-discrete minimizing movement starting from $U_{\tau,\eta}^0$ is any sequence $(U_{\tau,\eta}^n)_{n \in \mathbb{N}}$ in $D(\phi)$ which satisfies

$$\Phi(\tau, U_{\tau,\eta}^{n-1}; U_{\tau,\eta}^n) \leq \Phi(\tau, U_{\tau,\eta}^{n-1}; V) + \frac{\eta}{2} \, \mathsf{d}(U_{\tau,\eta}^n, U_{\tau,\eta}^{n-1}) \, \mathsf{d}(V, U_{\tau,\eta}^n)$$

$$\text{for every } V \in D(\phi), \quad (6.61a)$$

and the further condition

$$\Phi(\tau, U_{\tau,\eta}^{n-1}; U_{\tau,\eta}^n) = \frac{1}{2\tau} \mathsf{d}^2(U_{\tau,\eta}^n, U_{\tau,\eta}^{n-1}) + \phi(U_{\tau,\eta}^n) \leq \phi(U_{\tau,\eta}^{n-1}), \quad (6.61b)$$

for every $n \in \mathbb{N}$. A (τ, η)-discrete solution $\overline{U}_{\tau,\eta}$ is any piecewise constant interpolant of a (τ, η)-discrete minimizing movement on the grid $\mathcal{P}_\tau$, as in (6.60).

Notice that when $\eta = 0$ a solution to (6.61a) is a minimizer of (6.58) (and in particular satisfies (6.61b)), so that the usual discrete solutions arising from the minimizing movement scheme are included in this more general relaxed framework. The next result [60], which follows directly from Ekeland's variational principle, shows that the previous scheme is always solvable when $\eta > 0$.

Theorem 6.12. *Let us assume that X is complete and ϕ is quadratically bounded from below, i.e. for some $\kappa_o, \phi_o \in \mathbb{R}, o \in X$*

$$\phi(x) + \frac{\kappa_o}{2} \mathsf{d}^2(x, o) \geq \phi_o \quad \text{for every } x \in X. \quad (6.62)$$

Then for every $\eta > 0$, $\tau > 0$ with $\tau^{-1} > -\kappa_0$, and $U_{\tau,\eta}^0 \in D(\phi)$, the relaxed minimizing movement scheme admits at least a (τ, η)-discrete solution $(U_{\tau,\eta}^n)_{n \in \mathbb{N}}$.

Since under the general assumptions of Theorem 6.12 the relaxed minimizing movement scheme admits a (τ, η)-discrete solution $\overline{U}_{\tau,\eta}$ for fixed $\eta > 0$ and arbitrarily small step size τ, it is natural to ask what its limit is as $\tau \downarrow 0$. A first result in this direction is provided by the next theorem, which shows that the minimizing movement scheme is *consistent* with Definition 4 of λ-gradient flow. Notice that in Theorem 6.13 we will assume *a priori* that the λ-gradient flow of ϕ exists to get the convergence of $\overline{U}_{\tau,\eta}$; in Section 6.4 we will discuss how to remove this strong assumption. Still, it is sometimes useful to know that any λ-gradient flow, no matter how it has been constructed, admits a uniformly converging discrete approximation, which exhibits nice variational properties.

Theorem 6.13. *Let us assume that there exists the λ-gradient flow S_t of ϕ according to Definition 4 and that $\overline{D(\phi)}$ is geodesically convex. Let $\tau > 0, \eta \geq 0$ satisfy $\eta - \lambda < \frac{1}{2\tau}$, and let the sequence $(U^n_{\tau,\eta})_{n\in\mathbb{N}} \subset D(\phi)$ be a (τ, η)-discrete minimizing movement with $U^0_{\tau,\eta} \in D(|\partial\phi|)$. Setting $\alpha = \alpha_{\tau,\eta} := \frac{1}{2\tau}\log(1 + 2(\lambda - \eta)\tau)$ we have the* a priori *error estimate*

$$\mathsf{d}(\mathsf{S}_t(u_0), \overline{U}_{\tau,\eta}(t)) \leq \mathsf{d}(u_0, U^0_{\tau,\eta}) + e^{-\alpha T}\sqrt{T\tau}|\partial\phi|(U^0_{\tau,\eta}) \quad \textit{for every } t \in [0, T].$$
$$(6.63)$$

In particular, if for some $\eta \geq 0$ and every $\tau \in (0, \tau_0)$ $\overline{U}_{\tau,\eta}$ is a family of (τ, η)-discrete solutions with $U^0_{\tau,\eta} = u_0 \in D(|\partial\phi|)$, then $\lim_{\tau\downarrow 0}\overline{U}_{\tau,\eta}(t) = \mathsf{S}_t(u_0)$ uniformly on every compact interval.

Let us remark that η has been kept fixed in the previous convergence result, so that the coefficients $\alpha_{\tau,\eta} = \frac{1}{2\tau}\log(1 + 2(\lambda - \eta)\tau)$ in the estimate (6.63) are uniformly bounded from below as $\tau \downarrow 0$.

6.3.6 Stability of λ-gradient flows under Γ-convergence

We conclude this section by showing a simple stability property of gradient flows with respect to perturbations of the generating functional ϕ. Here we consider a coercive family of Γ-converging functionals $\phi^h : X \to (-\infty, +\infty]$, $h \in \bar{\mathbb{N}} = \mathbb{N} \cup \{+\infty\}$, which are quadratically bounded from below, uniformly w.r.t. h: for some $o \in X$, $\phi_o, \kappa_o \in \mathbb{R}$ they satisfy

$$\phi^h(x) + \frac{\kappa_o}{2}\mathsf{d}^2(x, o) \geq \phi_o \quad \text{for every } x \in X, h \in \mathbb{N}. \tag{6.64}$$

In Definition 10 we jointly recall the (sequential) notions of Γ-convergence and of coercivity [26, Definition 1.12]. For notational convenience, we will identify monotone subsequences $(h_n)_{n\in\mathbb{N}}$ with their unbounded image $H = \{h_n : n \in \mathbb{N}\} \subset \mathbb{N}$; expressions like $\lim_{h\in H}$, $\liminf_{h\in H}$ have an obvious meaning as limits for $h \uparrow +\infty, h \in H$.

Definition 10 (sequential $\Gamma(X, \mathsf{d})$-convergence of coercive functionals). We say that $(\phi^h)_{h\in\mathbb{N}}$ is a coercive family of functionals $\Gamma(X, \mathsf{d})$-sequentially converging to a proper functional $\phi^\infty : X \to (-\infty, +\infty]$ if the following two conditions are satisfied:

(1) For every infinite subset $H \subset \mathbb{N}$ and every bounded sequence $(x^h)_{h\in H}$ with $\sup_{h\in H} \phi^h(x^h) < +\infty$, there exists an infinite subsequence $H' \subset H$ such that $\lim_{h\in H'} x^h = x^\infty \in D(\phi^\infty)$ and

$$\liminf_{h\in H'} \phi^h(x^h) \geq \phi^\infty(x^\infty). \qquad (6.65)$$

(2) For every $\bar{x}^\infty \in D(\phi^\infty)$ there exists a sequence $(\bar{x}^h)_{h\in\mathbb{N}}$ such that

$$\lim_{h\uparrow+\infty} \mathsf{d}(\bar{x}^h, \bar{x}^\infty) = 0, \qquad \lim_{h\uparrow+\infty} \phi^h(\bar{x}^h) = \phi^\infty(\bar{x}^\infty). \qquad (6.66)$$

It is possible to prove that a coercive family of λ-convex functionals $(\phi^h)_{h\in\mathbb{N}}$ $\Gamma(X, \mathsf{d})$-converging to ϕ^∞ always satisfies the uniform lower bound (6.64).

Let us now state our first convergence result [60].

Theorem 6.14. *Let $(\phi^h)_{h\in\mathbb{N}}$ be a coercive family of functionals $\Gamma(X, \mathsf{d})$-converging to ϕ^∞ and let us assume that the λ-gradient flows S^h exist for every $h \in \mathbb{N}$. Then the functional ϕ^∞ admits a λ-gradient flow S^∞ and for every sequence $u_0^h \in \overline{D(\phi^h)}$ converging to $u_0^\infty \in \overline{D(\phi^\infty)}$ we have*

$$\lim_{h\uparrow+\infty} \mathsf{S}_t^h(u_0^h) = \mathsf{S}_t^\infty(u_0^\infty), \qquad \lim_{h\uparrow+\infty} \phi^h(\mathsf{S}_t^h(u_0^h)) = \phi^\infty(\mathsf{S}_t^\infty(u_0^\infty))$$
$$\text{for every } t > 0, \qquad\qquad\qquad\qquad\qquad\qquad (6.67)$$

locally uniformly on $(0, +\infty)$.

Proof. Here we consider the simpler case when (6.64) holds for $\kappa_o = 0$; it is not restrictive to assume $\lambda \leq 0$ and $\phi_o \geq 0$.

Step 1: uniform bounds. We set $u_t^h := \mathsf{S}_t^h(u_0^h)$ and we fix a compact time interval $[0, T]$, $T > 0$, a point $o^\infty \in D(\phi^\infty)$ and a corresponding sequence o^h as in (6.66). (EVI_λ') yields

$$\mathsf{d}^2(u_t^h, o^h) \leq \left(\mathsf{d}^2(u_0^h, o^h) + 2\mathsf{E}_\lambda(t)\,\phi^h(o^h)\right)e^{-\lambda t}; \qquad (6.68)$$

and therefore there exists a constant $C_1(T)$ independent of h such that

$$\mathsf{d}(u_t^h, o^h) \leq C_1(T) \quad \text{for every } t \in [0, T], \ h \in \mathbb{N}. \qquad (6.69)$$

The regularizing estimate (6.36) yields

$$\frac{e^{\lambda t}}{2}\mathsf{d}^2(u_t^h, o^h) + \mathsf{E}_\lambda(t)\phi^h(u_t^h) + \frac{\big(\mathsf{E}_\lambda(t)\big)^2}{2}|\partial\phi^h|^2(u_t^h)$$
$$\leq \frac{1}{2}\mathsf{d}^2(u_0^h, o^h) + \mathsf{E}_\lambda(t)\phi^h(o^h) \leq C_2(T) \qquad (6.70)$$

if $t \in (0, T]$, for a suitable constant $C_2(T)$ independent of h.

In particular, for every $0 < S < T$ there exists a constant $C(S, T)$ such that

$$\phi^h(u_t^h) \leq C(S, T), \quad |\partial\phi^h|(u_t^h) = |\dot{u}_{t+}^h| \leq C(S, T) \quad \text{for every } t \in [S, T].$$
$$(6.71)$$

Step 2: compactness. By the estimates of the previous point, the sequence $(u^h)_{h\in\mathbb{N}}$ is uniformly Lipschitz in each bounded interval $[S, T]$ of $(0, +\infty)$ and for every fixed t $\{u_t^h\}_{h\in\mathbb{N}}$ satisfies the assumptions of Definition 10, so that $(u_t^h)_{h\in\mathbb{N}}$ is relatively compact in X. Applying the Ascoli–Arzelà theorem we can find a subsequence $H = (h_n)_{n\in\mathbb{N}}$ such that u^{h_n} converge locally uniformly in time to a locally Lipschitz curve u^∞ in $(0, +\infty)$.

Step 3: characterization of the limit. Let us now fix an arbitrary point $v^\infty \in D(\phi^\infty)$ and a corresponding approximating sequence $v^h \in D(\phi^h)$ as in (6.66). By (EVI_λ') of Theorem 6.8 we know that

$$\frac{e^{\lambda(t-s)}}{2}\mathsf{d}^2(u_t^h, v^h) - \frac{1}{2}\mathsf{d}^2(u_s^h, v^h) \leq \mathsf{E}_\lambda(t-s)\Big(\phi^h(v^h) - \phi^h(u_t^h)\Big). \qquad (6.72)$$

We then pass to the limit in (6.72) as $h \uparrow +\infty, h \in H$, using the facts that u_t^h converges pointwise to u_t in X and applying (6.65) for u_t^h and (6.66) for v^h; we obtain

$$\frac{e^{\lambda(t-s)}}{2}\mathsf{d}^2(u_t^\infty, v^\infty) - \frac{1}{2}\mathsf{d}^2(u_s^\infty, v^\infty) \leq \mathsf{E}_\lambda(t-s)\Big(\phi(v^\infty) - \phi(u_t^\infty)\Big) \qquad (6.73)$$

for every $v^\infty \in D(\phi^\infty), 0 \leq s < t$. A further application of Theorem 6.8 shows that u^∞ solves $\text{EVI}_\lambda(X, \mathsf{d}, \phi^\infty)$.

In order to check that $\lim_{t\downarrow 0} u_t^\infty = u_0^\infty$ we use (6.73) at $s = 0$ and the lower semicontinuity of ϕ^∞, which yields

$$\limsup_{t\downarrow 0} \mathsf{d}^2(u_t^\infty, v^\infty) \leq \mathsf{d}^2(u_0^\infty, v^\infty) \quad \text{for every } v^\infty \in D(\phi^\infty); \qquad (6.74)$$

since $u_0^\infty \in \overline{D(\phi^\infty)}$ we conclude that $\lim_{t\downarrow 0} \mathsf{d}(u_t^\infty, u_0^\infty) = 0$.

Since the limit is the unique solution of $\text{EVI}_\lambda(X, \mathsf{d}, \phi^\infty)$ starting from u_0^∞, we conclude that the whole sequence u^h converges to u^∞.

Step 4: convergence of energy. We argue as in the proof of (6.44) and (6.45): for a fixed $t > 0$ and applying (6.66) to u_t^∞ we find a sequence $(\bar{u}_t^h)_{h\in\mathbb{N}}$ converging to u_t^∞ with $\lim_{h\uparrow+\infty} \phi^h(\bar{u}_t^h) = \phi^\infty(u_t^\infty)$. By estimate (6.70), the slope $|\partial\phi^h|(u_t^h)$ is uniformly bounded by a constant M_t so that

$$\phi^h(\bar{u}_t^h) \geq \phi^h(u_t^h) - M_t \mathsf{d}(\bar{u}_t^h, u_t^h) - \frac{\lambda}{2}\mathsf{d}^2(\bar{u}_t^h, u_t^h).$$

Passing to the limit as $h \uparrow +\infty$ we get $\limsup_{h\uparrow+\infty} \phi^h(u_t^h) \leq \phi^\infty(u_t^\infty)$, which combined with (6.65) yields the second identity of (6.67). ☐

6.4 Convergence of the minimizing movement method and generation results

We have seen in Theorem 6.13 a first convergence theorem for the relaxed minimizing movement method: it basically says that *if ϕ admits a λ-gradient flow* according to Definition 4 then any family of discrete solutions converges to the unique continuous solution of (EVI_λ) as the time step converges to 0.

In this section we revert to this point of view and we try to prove *the existence of the λ-gradient flow* when ϕ is geodesically λ-convex by studying the convergence of the (relaxed) minimizing movement method.

In the following we present three different results in this direction:

(1) A simpler convergence result when *the sublevels of ϕ are locally compact*: in this case we avoid any geometric restriction on the distance d of X and we do not need any Cauchy estimate. On the other hand, the (not necessarily unique) limit points of the discrete solutions are just curves of maximal slope, according to Definition 7: in general it is not possible to prove that they solve (EVI_λ).

(2) A first-generation result for λ-gradient flows, by assuming that the minimizing movement generating functional $\Phi(\tau, U; V)$ defined by (6.58) satisfies a suitable convexity property (which results from the combination of the convexity of d^2 and of ϕ).

(3) A second-generation result when $\mathsf{d}^2(\cdot, v)$ is semiconcave along geodesics and the metric space satisfies a *local angle condition* between a triple of geodesics emanating from the same point.

Differently from the first approach, the last ones provide explicit Cauchy estimates ensuring the convergence of the method and do not require any local compactness of the sublevels of ϕ.

6.4.1 Convergence of the variational scheme in the locally compact case

Let us first consider the case when ϕ is geodesically λ-convex and its sublevels are locally compact, i.e. $\exists\, o \in X$ s.t.

$$\left\{ x \in X : \phi(x) \le R \text{ and } d(x, o) \le R \right\} \quad \text{are compact in } X \text{ for every } R > 0.$$
$$(6.75)$$

Combining [3, Proposition 2.2.3, Corollary 2.4.11] we get:

Theorem 6.15 (limits of discrete minimizing movements are curves of maximal slope). *If ϕ is geodesically λ-convex and satisfies (6.75) then for every $\tau > 0$ satisfying $\tau^{-1} > -\lambda$ and $U_\tau^0 \in D(\phi)$ the minimizing movement variational scheme admits at least one solution $(U_\tau^n)_{n \in \mathbb{N}}$. If moreover*

$$\lim_{\tau \downarrow 0} U_\tau^0 = u_0, \quad \lim_{\tau \downarrow 0} \phi(U_\tau^0) = \phi(u_0), \qquad (6.76)$$

and $\overline{U}_\tau$ is a family of discrete solutions, any infinitesimal sequence of time steps $\tau_n \downarrow 0$ admits a convergent subsequence (still denoted by τ_n) and a limit curve $u \in AC_{loc}^2([0, +\infty); X)$ such that

$$\lim_{n \uparrow +\infty} \overline{U}_{\tau_n}(t) = u_t, \quad \lim_{n \uparrow +\infty} \phi(\overline{U}_{\tau_n}(t)) = \phi(u_t) \quad \text{for every } t \ge 0 \qquad (6.77)$$

uniformly in each compact interval $[0, T]$. u is a curve of maximal slope (see Definition 7), satisfying the energy identity

$$\frac{1}{2} \int_s^t |u'|^2(r)\, dr + \frac{1}{2} \int_s^t |\partial\phi|^2(u_r)\, dr$$
$$= \phi(u_s) - \phi(u_t) \quad \text{for all } 0 < s \le t < +\infty. \qquad (6.78)$$

Corollary 6.16 (existence of curves of maximal slope). *Under the same assumptions of the previous theorem, for every $u_0 \in D(\phi)$ there exists a curve of maximal slope $u \in AC_{loc}^2([0, +\infty); X)$ starting from u_0 and satisfying (6.78).*

6.4.2 Generation of λ-gradient flows by strong convexity of Φ

In the case when X is a Hilbert space and ϕ is an l.s.c. λ-convex functional, it is well known that the minimizing movement variational scheme admits a unique solution and the corresponding discrete solution U_τ converges to the solution of (EVI_λ). Applying this approximation scheme, it is then possible to show the existence of the λ-gradient flow of ϕ according to Definition 4.

Similar results for minimizing movements of convex functionals in Banach spaces do not always hold; indeed, the characterization of gradient flows through the EVI depends not only on the convexity of ϕ but also on structural properties of the distance d.

One fundamental property is the 1-convexity of the function $v \mapsto \frac{1}{2}\mathsf{d}(v, w)^2$, i.e.

$$\mathsf{d}^2(v_s, w) \le (1 - s)\mathsf{d}^2(v_0, w) + s\mathsf{d}^2(v_1, w) - s(1 - s)\mathsf{d}^2(v_0, v_1)$$

for every v_0, v_1, for every $[0, 1] \ni s \mapsto v_s$ geodesic between v_0 and v_1,

(6.79)

which in Banach spaces is equivalent to the fact that d is induced by a scalar product.

Equation (6.79) is satisfied by the geodesic distance on Riemannian manifolds of non-positive sectional curvature and characterizes the Aleksandrov *non-positively curved* (NPC) length spaces; e.g. see [15, 37].

Actually, using (6.79) and adapting a Crandall–Liggett argument, Mayer [46] was able to prove (6.26) and then (6.27) also for geodesically convex functionals on NPC spaces.

A crucial consequence of (6.79) and the λ-convexity of ϕ is that the generating functional $\Phi(\tau, V; U)$ of the minimizing movement scheme (6.58)

$$\Phi(\tau, V; U) := \frac{1}{2\tau}\mathsf{d}^2(U, V) + \phi(U) \quad \tau > 0, \ U, V \in X, \quad (6.80)$$

satisfies the $\tau^{-1} + \lambda$-convexity condition along geodesics, i.e.

the map $U \mapsto \Phi(\tau, V; U)$ is geodesically $(\tau^{-1} + \lambda)$-convex for every $V \in X$.

(6.81)

One of the main contributions of [3, Chapter 4] is to show that (6.81) can be relaxed, by assuming the $(\tau^{-1} + \lambda)$-convexity of $\Phi(\tau, V; \cdot)$ along more general families of curves in X connecting two arbitrary points in $D(\phi)$.

This improvement has been essential to apply the generation result in Wasserstein spaces, which do not satisfy (6.79) except for the one-dimensional case.

Theorem 6.17 (convergence of the minimizing movement scheme and generation result [3]). *Let us assume that the functional Φ defined in (6.80) satisfies the following property: for every $V, U_0, U_1 \in D(\phi)$ there exists a curve $\gamma_s : [0, 1] \to X$ with $\gamma_0 = U_0$ and $\gamma_1 = U_1$, such that*

$$U \mapsto \Phi(\tau, V; U) \text{ is } \left(\frac{1}{\tau} + \lambda\right)\text{-convex on } \gamma \text{ for each } 0 < \tau < \frac{1}{\lambda^-}, \quad (6.82)$$

i.e.

$$\Phi(\tau,V;\gamma_s) \le (1-s)\Phi(\tau,V;U_0) + s\,\Phi(\tau,V;U_1) - \frac{1+\lambda\tau}{2\tau}s(1-s)\mathrm{d}^2(U_0,U_1).$$
(6.83)

(1) *For every* $U_\tau^0 = u_0 \in \overline{D(\phi)}$ *and* $\tau > 0$ *with* $1 + \tau\lambda > 0$ *the minimizing movement method in Definition 8 admits a unique solution* $(U_\tau^n)_{n\in\mathbb{N}} \subset D(\phi)$.

(2) *The corresponding discrete solutions* $\bar{U}_\tau$ *converge to* u *as* $\tau \downarrow 0$ *uniformly on compact intervals.*

(3) *The limit* u *is the unique solution of* (EVI$_\lambda$). *In particular,* ϕ *admits a* λ-*gradient flow according to Definition 4, thus satisfying all the properties stated in Theorem 6.9.*

(4) *There exist universal constants* $C_{\lambda,T}$ *such that if* $u_0 \in D(|\partial\phi|)$ *the optimal error estimate holds:*

$$\mathrm{d}(u(t),\bar{U}_\tau(t)) \le C_{\lambda,T}|\partial\phi|(u_0)\,\tau \quad \text{for every } t \in [0,T].$$
(6.84)

The main arguments of the proof of Theorem 6.17 in a simplified setting can be found in [62]. Sub-optimal convergence estimates, inspired by the Crandall–Ligget approach, have also been obtained in a different way in [5] and in [23].

6.4.3 Generation results for geodesically convex functionals in spaces with a semiconcave squared distance

In this subsection we consider a geodesically λ-convex functional in a complete metric space (X, d) whose squared distance satisfies a *semi-concavity* condition.

Definition 11 (semi-concavity of the squared distance function). We say that $D(\phi) \subset X$ is a K-SC (semi-concave) space if for every geodesic $[0, 1] \ni s \mapsto v_s \in D(\phi)$ and for every $w \in D(\phi)$ we have

$$\mathrm{d}^2(v_s, w) \ge (1-s)\mathrm{d}^2(v_0, w) + s\,\mathrm{d}^2(v_1, w) - Ks(1-s)\mathrm{d}^2(v_0, v_1)$$

$$\text{for every } s \in [0, 1]. \quad (6.85)$$

6.4.3.1 Examples of K-SC spaces

- *PC spaces*: X is positively curved (PC) in the Aleksandrov sense if and only if X is K-SC with $K = 1$.
- *Aleksandrov spaces*: if X is an Aleksandrov space whose curvature is bounded from below by a negative constant $-k$ and $D = \mathrm{diam}(X) < +\infty$, then X

is a K-SC space with $K = \frac{D\sqrt{k}}{\tanh(D\sqrt{k})}$. This class includes all Riemannian manifolds whose sectional curvature is bounded from below.

- *Product and L^2-spaces*: if $\{(X_i, d_i)\}_{i\in\mathbb{N}}$ is a countable collection of K-SC spaces, then the product $\prod_{i\in\mathbb{N}} X_i$ with the usual product distance is a K-SC space. If μ is a finite measure on some separable measure space Ω, then $\mathscr{X} := L^2_\mu(\Omega; X) = \{f : \Omega \to X : \int_\Omega d^2(f(\omega), x_0)\,d\mu(\omega) < +\infty$ for some $x_0 \in X\}$ endowed with the distance $d^2_{\mathscr{X}}(f, g) = \int_\Omega d^2(f(\omega), g(\omega))\,d\mu(\omega)$ is K-SC whenever X is K-SC.
- *Wasserstein space*: $(\mathscr{P}_2(X), W_2)$ is K-SC if and only if X is K-SC (see the next section).

We will also assume that the (upper) angle between a couple of geodesics emanating from the same point satisfies a suitable condition.

Definition 12 (upper angles). Let x^1, x^2 be two geodesics emanating from the same initial point $x_0 := x_0^1 = x_0^2$. Their *upper angle* $\sphericalangle_u(x^1, x^2) \in [0, \pi]$ is defined by

$$\cos(\sphericalangle_u(x^1, x^2)) := \liminf_{s,t\downarrow 0} \frac{d^2(x_0, x_s^1) + d^2(x_0, x_t^2) - d^2(x_s^1, x_t^2)}{2d(x_0, x_s^1)d(x_0, x_t^2)}.$$

Definition 13 (local angle condition (LAC)). We say that $D(\phi) \subset X$ satisfies the *local angle condition* (LAC) if for any triple of geodesics $x^i : [0, 1] \to D(\phi)$, $i = 1, 2, 3$, emanating from the same initial point x_0 the corresponding angles $\theta^{ij} := \sphericalangle_u(x^i, x^j)$ satisfy one of the following equivalent conditions:

1. $\theta^{12} + \theta^{23} + \theta^{31} \le 2\pi$.
2. There exist a Hilbert space H and vectors $w^i \in H$ such that $\langle w^i, w^j \rangle_H = \cos(\theta^{ij})$ for $1 \le i, j \le 3$.
3. For any choice of $\xi^1, \xi^2, \xi^3 \ge 0$ one has that $\sum_{i,j=1}^3 \cos(\theta^{ij})\xi^i\xi^j \ge 0$.

6.4.3.2 Examples of (LAC) spaces

- *A Banach space X* satisfies (LAC) if and only if X is a Hilbert space.
- *Riemannian manifolds and Aleksandrov spaces* with curvature bounded from below satisfy (LAC). In particular, if (6.85) holds with $K = 1$ then X satisfies (LAC).
- *Product and L^2-spaces*: $\prod_{i\in\mathbb{N}} X_i$ satisfies (LAC) if and only if each (X_i, d_i) does; $L^2_\mu(\Omega; X)$ satisfies (LAC) if and only if X satisfies it.
- *Wasserstein space*: $\mathscr{P}_2(X)$ satisfies (LAC) if and only if X does.

Theorem 6.18 (generation theorem for geodesically λ-convex functionals in K-SC and (LAC) spaces). *Let (X, d) be a complete metric space and let $\phi : (-\infty, +\infty]$ be a proper, l.s.c. and λ-geodesically convex functional.*

(1) For every $\tau, \eta > 0$ with $1 + \tau\lambda > 0$ and $U^0_{\tau,\eta} = u_0 \in D(\phi)$ the relaxed minimizing movement scheme (6.61a,b) admits at least one solution $(U^n_{\tau,\eta})_{n \in \mathbb{N}}$.

(2) If $D(\phi)$ is a K-SC space and satisfies the (LAC) then the discrete solution $\bar{U}_{\tau,\eta}$ converges to u as $\tau \downarrow 0$ uniformly in each compact interval.

(3) The limit u is the unique solution of (EVI$_\lambda$). In particular, ϕ admits a λ-gradient flow according to Definition 4, thus satisfying all the properties stated in Theorem 6.9.

6.5 Wasserstein spaces and diffusion equations

6.5.1 The Wasserstein space

Here, just to set the notation, we collect some basic definitions and properties of the Wasserstein space which will be used in the following. For a more detailed overview on this topic we refer to [67, 68], [3] and [4].

6.5.1.1 Transport maps and couplings

We denote by X_i, for some $i \in \mathbb{N}$, a separable and complete metric space. $\mathscr{P}(X)$ is the space of Borel probability measures on X.

If $\mu \in \mathscr{P}(X_1)$ and $t : X_1 \to X_2$ is a Borel map, we denote by $t_\#\mu \in \mathscr{P}(X_2)$ the *push-forward of μ through t*, defined by

$$t_\#\mu(B) := \mu(t^{-1}(B)) \quad \text{for every } B \in \mathscr{B}(X_2). \tag{6.86}$$

We denote by π^i, for $i = 1, \dots, n$, the canonical projection operator from a product space $X_1 \times \cdots \times X_n$ into X_i, defined by

$$\pi^i(x_1, \dots, x_n) := x_i.$$

Given $\mu_1 \in \mathscr{P}(X_1)$ and $\mu_2 \in \mathscr{P}(X_2)$, the class $\Gamma(\mu_1, \mu_2)$ of *transport plans* or *couplings* between μ_1 and μ_2 is defined by

$$\Gamma(\mu_1, \mu_2) := \{\gamma \in \mathscr{P}(X_1 \times X_2) : \pi^1_\#\gamma = \mu_1, \ \pi^2_\#\gamma = \mu_2\}.$$

To each couple of measures $\mu_1 \in \mathscr{P}(X_1)$, $\mu_2 = t_{\#}\mu_1 \in \mathscr{P}(X_2)$ linked by a Borel map $t : X_1 \rightarrow X_2$ we can associate the coupling

$$\mu := (i_{X_1} \times t)_{\#}\mu_1 \in \mathscr{P}(X_1 \times X_2), \quad i_{X_1} \text{ being the identity map on } X_1.$$
(6.87)

If μ is representable as in (6.87) we say that μ is induced by t and t is a *transport map* between μ_1 and μ_2. Each coupling $\mu \in \Gamma(\mu_1, \mu_2)$ concentrated on a μ-measurable graph in $X_1 \times X_2$ admits the representation (6.87) for some μ_1-measurable map t, which therefore transports μ_1 into μ_2.

6.5.1.2 Wasserstein distance

Given a complete and separable metric space (X, d) we denote by $\mathscr{P}_2(X)$ the space of Borel probability measures with finite quadratic moment: $\mu \in \mathscr{P}(X)$ belongs to $\mathscr{P}_2(X)$ iff

$$\int_X d^2(x, x_o) \, d\mu(x) < +\infty \quad \text{for some (and thus any) point } x_o \in X. \quad (6.88)$$

For every couple of measures $\mu, \nu \in \mathscr{P}_2(X)$ we consider the Kantorovich problem for the cost d^2:

$$W_2^2(\mu, \nu) := \min\left\{ \int_{X \times X} d^2(x, y) \, d\gamma(x, y) : \gamma \in \Gamma(\mu, \nu) \right\}. \quad (6.89)$$

It is not difficult to check, by the direct method of calculus of variations, that the minimum problem (6.89) admits at least a solution. The subset of $\Gamma(\mu, \nu)$ given by the *optimal transport plans* for (6.89) will be denoted by $\Gamma_{\text{opt}}(\mu, \nu)$. Notice that if there exists $\gamma = (i_{X_1} \times t)_{\#}\mu \in \Gamma(\mu, \nu)$, we have

$$\int_{X \times X} d^2(x, y) \, d\gamma(x, y) = \int_X d^2(x, t(x)) \, d\mu(x).$$

The quantity $W_2(\mu, \nu)$ defined by (6.89) is a distance between the measures $\mu, \nu \in \mathscr{P}_2(X)$ which enjoys remarkable properties.

Theorem 6.19. *Let (X, d) be a complete and separable metric space. Then, W_2 defines a distance on $\mathscr{P}_2(X)$ and $(\mathscr{P}_2(X), W_2)$ is a complete and separable metric space. Moreover, for a given sequence $\{\mu_k\}_{k \in \mathbb{N}} \subset \mathscr{P}_2(X)$ we have*

$$\lim_{k \to +\infty} W_2(\mu_k, \mu) = 0 \Leftrightarrow \begin{cases} \int_X f \, d\mu_k \to \int_X f \, d\mu & \text{for every } f \in C_b^0(X) \\ \lim_{R \uparrow +\infty} \int_{X \setminus B_R(x_0)} d^2(x, x_0) \, d\mu_k(x) = 0 & \text{uniformly w.r.t. } k \in \mathbb{N}. \end{cases}$$

The metric space $(\mathscr{P}_2(X), W_2)$ is called the $(L^2$-) *Wasserstein space* on X. When $X = \mathbb{R}^d$ we denote by $\mathscr{P}_2^a(\mathbb{R}^d)$ the subset of $\mathscr{P}_2(\mathbb{R}^d)$ defined by

$$\mathscr{P}_2^a(\mathbb{R}^d) := \{\mu \in \mathscr{P}_2(\mathbb{R}^d) : \mu \ll \mathscr{L}^d\}. \tag{6.90}$$

Here we recall the following basic result on the existence and uniqueness of optimal transport plans induced by maps (which are then called *optimal transport maps*) in the case in which the initial measure μ belongs to $\mathscr{P}_2^a(\mathbb{R}^d)$.

Theorem 6.20 (existence and uniqueness of optimal transport maps [12, 38]). *For any $\mu \in \mathscr{P}_2^a(\mathbb{R}^d)$ and $v \in \mathscr{P}_2(\mathbb{R}^d)$, Kantorovich's optimal transport problem (6.89) has a unique solution γ, which is concentrated on the graph of a transport map t. t is the unique minimizer of Monge's optimal transport problem on $\mathbb{R}^d$ for the Euclidean distance*

$$\min\left\{\int_{\mathbb{R}^d} |x - r(x)|^2 \, d\mu(x) : r_\# \mu = v\right\}.$$

The map t is cyclically monotone and there exists a convex open set $\Omega \subset \mathbb{R}^d$ with $\mu(\mathbb{R}^d \setminus \Omega) = 0$ and a convex function $\phi : \Omega \to \mathbb{R}$ such that $t(x) = \nabla\phi(x)$ for μ-a.e. $x \in \Omega$.

6.5.1.3 Geodesics and curvature properties of $(\mathscr{P}_2(\mathbb{R}^d), W_2)$

Theorem 6.21 (geodesics in the Wasserstein space). *Given $\mu, v \in \mathscr{P}_2(\mathbb{R}^d)$ and $\gamma \in \Gamma_{\mathrm{opt}}(\mu, v)$, the curve*

$$[0, 1] \ni s \mapsto \mu_s = \left((1 - s)\pi^1 + s\pi^2\right)_\# \gamma$$

is a constant-speed geodesic between μ and v, i.e. it satisfies

$$W_2(\mu_s, \mu_t) = |s - t| W_2(\mu_0, \mu_1) \quad \text{for every } s, t \in [0, 1].$$

Vice versa, any constant-speed geodesic between μ and v can be built in this way.

If $\gamma = (i \times t)_\# \mu$, then

$$\mu_s = \left((1 - s)i + st\right)_\# \mu, \quad s \in [0, 1].$$

In particular, $(\mathscr{P}_2(\mathbb{R}^d), W_2)$ is a geodesic space.

In view of the application to the Wasserstein framework of the theory of gradient flows in metric spaces developed in the previous section, we recall the following theorem (see Theorem 7.3.2 and Example 7.3.3 of [3]):

Theorem 6.22 $((\mathscr{P}_2(\mathbb{R}^d), W_2)$ **is a PC-space).** *For any* $\mu_0, \mu_1, \mu_2 \in \mathscr{P}_2(\mathbb{R}^d)$ *we have*

$$W_2^2(\mu_s, \mu_2) \geq (1-s)W_2^2(\mu_0, \mu_2) + s W_2^2(\mu_1, \mu_2) - s(1-s)W_2^2(\mu_0, \mu_1)$$

$$\text{for every } s \in [0, 1], \quad (6.91)$$

where μ_s *is any constant-speed geodesic between* μ_0 *and* μ_1.

Moreover, when $d \geq 2$ *there is no constant* $\lambda \in \mathbb{R}$ *such that* $W_2^2(\cdot, \mu_2)$ *is* λ-*convex along geodesics.*

According to Aleksandrov's notion of curvature for metric spaces, (6.91) can be interpreted by saying that the Wasserstein space is a *positively curved metric space* (or PC-space). Then, the square of the Wasserstein distance along geodesics does not satisfy the 1-convexity assumption (6.79), which would be the most natural to prove the generation Theorem 6.17 for the gradient flows of λ-convex functionals.

However, the theory developed in the previous section allows for a great flexibility in the choice of the connecting curves. In particular, for the Wasserstein space on $\mathbb{R}^d$ the 1-convexity property (6.79) is satisfied along the following class of curves:

Definition 14 (generalized geodesics). A generalized geodesic joining μ_2 to μ_3 (with base point μ_1) is a curve of the type

$$[0, 1] \ni s \mapsto \mu_s^{2 \to 3} := ((1-s)\pi^2 + s\pi^3)_{\#}\mu, \quad (6.92)$$

where

$$\mu \in \Gamma(\mu_1, \mu_2, \mu_3) \quad \text{and} \quad \pi^{1,2}{}_{\#}\mu \in \Gamma_{\mathrm{opt}}(\mu_1, \mu_2), \quad \pi^{1,3}{}_{\#}\mu \in \Gamma_{\mathrm{opt}}(\mu_1, \mu_3). \quad (6.93)$$

Here, $\Gamma(\mu_1, \mu_2, \mu_3) := \{\gamma \in \mathscr{P}(\mathbb{R}^d \times \mathbb{R}^d \times \mathbb{R}^d) : \pi_{\#}^i\gamma = \mu^i, i = 1, 2, 3\}$ and $\pi^{i,j} : \mathbb{R}^d \times \mathbb{R}^d \times \mathbb{R}^d \to \mathbb{R}^d \times \mathbb{R}^d$ is the projection on the ith and jth coordinate.

Proposition 6.23 (1-convexity of the Wasserstein distance along generalized geodesics). *Let* $\mu_1, \mu_2, \mu_3 \in \mathscr{P}_2(\mathbb{R}^d)$ *and let* $\mu \in \Gamma(\mu_1, \mu_2, \mu_3)$ *such that* $\pi_{\#}^{1,i}\mu \in \Gamma_{\mathrm{opt}}(\mu_1, \mu_i)$, *for* $i = 2, 3$. *Then,*

$$W_2^2(\mu_s^{2 \to 3}, \mu_1) \leq (1-s)W_2^2(\mu_1, \mu_2) + s W_2^2(\mu_1, \mu_3) - s(1-s)W_2^2(\mu_2, \mu_3)$$

$$\text{for every } s \in [0, 1].$$

In particular, the function $\frac{1}{2}W_2^2(\mu_1, \cdot)$ *is 1-convex along generalized geodesics.*

6.5.2 Absolutely continuous curves in $(\mathscr{P}_2(\mathbb{R}^d), W_2)$

We recall here some basic properties of absolutely continuous curves in the Wasserstein space, which are related to the "dynamic interpretation" by Benamou and Brenier [9]. The main result is the following [3, Theorem 8.3.1]:

Theorem 6.24 (absolutely continuous curves and the continuity equation). *Let $\mu_t : (0, +\infty) \to \mathscr{P}_2(\mathbb{R}^d)$ be an absolutely continuous curve and let $|\mu'| \in L^1(0, +\infty)$ be its metric derivative. Then there exists a Borel vector field $\boldsymbol{v} : (x, t) \mapsto \boldsymbol{v}_t(x)$ such that*

$$\boldsymbol{v}_t \in L^2(\mu_t; \mathbb{R}^d), \quad ||\boldsymbol{v}_t||_{L^2(\mu_t; \mathbb{R}^d)} \leq |\mu'|(t) \quad \text{for } \mathscr{L}^1\text{-a.e. } t \in (0, +\infty) \quad (6.94)$$

and the continuity equation

$$\frac{\partial}{\partial t}\mu_t + \nabla \cdot (\boldsymbol{v}_t \mu_t) = 0 \quad \text{in } \mathbb{R}^d \times (0, +\infty) \quad (6.95)$$

holds in the sense of distributions.
  Moreover,

$$\boldsymbol{v}_t \in \overline{\{\nabla\varphi : \varphi \in C_c^\infty(\mathbb{R}^d)\}}^{L^2(\mu_t; \mathbb{R}^d)} \quad \text{for } \mathscr{L}^1\text{-a.e. } t \in (0, +\infty). \quad (6.96)$$

Conversely, if a curve $(0, +\infty) \ni t \mapsto \mu_t \in \mathscr{P}_2(\mathbb{R}^d)$ is continuous w.r.t. the weak topology on $\mathscr{P}(\mathbb{R}^d)$ and it satisfies the continuity equation (6.95) for some Borel vector field $\boldsymbol{v}_t$ with

$$\int_0^{+\infty} ||\boldsymbol{v}_t||_{L^2(\mu_t; \mathbb{R}^d)} < +\infty,$$

then μ_t is an absolutely continuous curve and $|\mu'|(t) \leq ||\boldsymbol{v}_t||_{L^2(\mu_t; \mathbb{R}^d)}$ for $\mathscr{L}^1$-a.e. $t \in (0, +\infty)$.

   Then, the minimal norm for the vector fields $\boldsymbol{v}_t$ satisfying (6.95) for an absolutely continuous curve μ_t is given by its metric derivative. Furthermore, such "minimal" vector fields satisfy (6.96). This fact suggests the following definition (we refer to [3, Chapter 8.4]).

Definition 15 (tangent space). Let $\mu \in \mathscr{P}_2(\mathbb{R}^d)$. We define the *tangent space* to $\mathscr{P}_2(\mathbb{R}^d)$ at the point μ as

$$\text{Tan}_\mu \mathscr{P}_2(\mathbb{R}^d) := \overline{\{\nabla\varphi : \varphi \in C_c^\infty(\mathbb{R}^d)\}}^{L^2(\mu_t; \mathbb{R}^d)}. \quad (6.97)$$

Proposition 6.25 (tangent vectors to absolutely continuous curves). *Let $\mu_t : (0, +\infty) \to \mathscr{P}_2(\mathbb{R}^d)$ be an absolutely continuous curve and let $\boldsymbol{v}_t \in L^2(\mu_t; \mathbb{R}^d)$ be a Borel vector field such that (6.95) holds. Then $\boldsymbol{v}_t$ satisfies (6.94) if and* *only if* $\boldsymbol{v}_t \in \mathrm{Tan}_{\mu_t} \mathscr{P}_2(\mathbb{R}^d)$ *for* $\mathscr{L}^1$*-a.e.* $t \in (0, +\infty)$*. The vector* $\boldsymbol{v}_t$ *is uniquely determined* $\mathscr{L}^1$*-a.e. by (6.94) and (6.95).*

Tangent vector fields are also strictly related to the first-order infinitesimal behavior of the Wasserstein distance along absolutely continuous curves.

Proposition 6.26. *Let* $\mu_t : (0, +\infty) \to \mathscr{P}_2(\mathbb{R}^d)$ *be an absolutely continuous curve and let* $\boldsymbol{v}_t \in \mathrm{Tan}_{\mu_t} \mathscr{P}_2(\mathbb{R}^d)$ *be the tangent vector characterized by Proposition 6.25. Then, for* $\mathscr{L}^1$*-a.e.* $t \in (0, +\infty)$ *the following property holds:*

$$\lim_{h \to 0} \frac{W_2(\mu_{t+h}, (\boldsymbol{i} + h\boldsymbol{v}_t)_{\#}\mu_t)}{|h|} = 0. \qquad (6.98)$$

Then, if μ_t *and* μ_{t+h} *are linked by an optimal transport map* $\boldsymbol{t}_{\mu_t}^{\mu_{t+h}}$*, we have*

$$\lim_{h \to 0} \frac{\boldsymbol{t}_{\mu_t}^{\mu_{t+h}} - \boldsymbol{i}}{h} = \boldsymbol{v}_t \quad \text{in } L^2(\mu_t; \mathbb{R}^d).$$

As an application of (6.98) we are able to show the $\mathscr{L}^1$-a.e. differentiability of $t \mapsto W_2(\mu_t, \sigma)$ along absolutely continuous curves μ_t in terms of tangent vectors and optimal transport plans; this provides a useful formula for the left-hand side of the (EVI$_\lambda$):

$$\frac{1}{2} \frac{\mathrm{d}}{\mathrm{d}t} W_2^2(\mu_t, \sigma) \le \phi(\sigma) - \phi(\mu_t) \quad \text{for every } \sigma \in D(\phi).$$

Theorem 6.27. *Let* $\mu_t : (0, +\infty) \to \mathscr{P}_2(\mathbb{R}^d)$ *be an absolutely continuous curve, let* $\boldsymbol{v}_t \in \mathrm{Tan}_{\mu_t} \mathscr{P}_2(\mathbb{R}^d)$ *be the tangent vector characterized by Proposition 6.25 and let* $\sigma \in \mathscr{P}_2(X)$*. Then*

$$\frac{1}{2} \frac{\mathrm{d}}{\mathrm{d}t} W_2^2(\mu_t, \sigma) = \int_{\mathbb{R}^d \times \mathbb{R}^d} \langle x - y, \boldsymbol{v}_t(x) \rangle \, \mathrm{d}\boldsymbol{\gamma}(x, y) \quad \text{for every } \boldsymbol{\gamma} \in \Gamma_{\mathrm{opt}}(\mu_t, \sigma),$$

$$(6.99)$$

for $\mathscr{L}^1$*-a.e* $t \in (0, +\infty)$*. In particular, if* $\mu_t \in \mathscr{P}_2^a(\mathbb{R}^d)$*,*

$$\frac{1}{2} \frac{\mathrm{d}}{\mathrm{d}t} W_2^2(\mu_t, \sigma) = \int_{\mathbb{R}^d} \langle x - \boldsymbol{t}_{\mu_t}^\sigma(x), \boldsymbol{v}_t(x) \rangle \, \mathrm{d}\mu_t(x), \qquad (6.100)$$

where $\boldsymbol{t}_{\mu_t}^\sigma$ *is the unique optimal transport map between* μ_t *and* σ*.*

6.5.3 Geodesically λ-convex functionals in $\mathscr{P}_2(\mathbb{R}^d)$

In this section we introduce the three main classes of λ-geodesically convex functionals on the Wasserstein space $(\mathscr{P}_2(\mathbb{R}^d), W_2)$ introduced by McCann [47] (for the proofs of the main results we refer to Chapter 9 of [3]).

Example 1 (potential energy). Let $V : \mathbb{R}^d \to \mathbb{R}$ be a λ_V-convex function for some $\lambda_V \in \mathbb{R}$ and let us define the potential energy

$$\mathscr{V}(\mu) := \int_{\mathbb{R}^d} V(x) \, d\mu(x) \quad \text{for every } \mu \in \mathscr{P}_2(\mathbb{R}^d). \tag{6.101}$$

For every $\mu_1, \mu_2 \in D(\mathscr{V})$ and $\boldsymbol{\mu} \in \Gamma(\mu_1, \mu_2)$ we have

$$\mathscr{V}\left(\left[(1-s)\pi^1 + s\pi^2\right]_\# \boldsymbol{\mu}\right) \leq (1-s)\mathscr{V}(\mu_1) + s\mathscr{V}(\mu_2) - \frac{\lambda_V}{2}s(1-s)$$

$$\times \int_{\mathbb{R}^d \times \mathbb{R}^d} |x-y|^2 \, d\boldsymbol{\mu}(x,y). \tag{6.102}$$

In particular, $\mathscr{V}$ is geodesically λ_V-convex on $\mathscr{P}_2(\mathbb{R}^d)$.

Example 2 (interaction energy). Let $\lambda_W \leq 0$ and let $W : \mathbb{R}^d \to \mathbb{R}$ be a λ_W-convex function with $W(-x) = W(x)$ for every $x \in \mathbb{R}^d$, and let us set

$$\mathscr{W}(\mu) := \frac{1}{2} \iint_{\mathbb{R}^d \times \mathbb{R}^d} W(x-y) \, d\mu(x) \, d\mu(y) \quad \text{for every } \mu \in \mathscr{P}_2(\mathbb{R}^d). \tag{6.103}$$

For every $\mu_1, \mu_2 \in D(\mathscr{W})$ and $\boldsymbol{\mu} \in \Gamma(\mu_1, \mu_2)$ we have

$$\mathscr{W}\left(\left[(1-s)\pi^1 + s\pi^2\right]_\# \boldsymbol{\mu}\right) \leq (1-s)\mathscr{W}(\mu_1) + s\mathscr{W}(\mu_2) - \frac{\lambda_W}{2}s(1-s)$$

$$\times \int_{\mathbb{R}^d \times \mathbb{R}^d} |x-y|^2 \, d\boldsymbol{\mu}(x,y). \tag{6.104}$$

In particular, $\mathscr{W}$ is geodesically λ_W-convex on $\mathscr{P}_2(\mathbb{R}^d)$.

Example 3 (internal energy). Let $U : [0, +\infty) \to \mathbb{R}$ be a convex function such that

$$U(0) = 0, \quad \liminf_{s \downarrow 0} \frac{U(s)}{s^\alpha} > -\infty \text{ for some } \alpha > \frac{d}{d+2}, \quad \lim_{s \to +\infty} \frac{U(s)}{s} = +\infty \tag{6.105}$$

the map $s \mapsto s^d U(s^{-d})$ is convex and non-increasing on $(0, +\infty)$. $\tag{6.106}$

The internal energy functional

$$\mathscr{U}(\mu) := \begin{cases} \int_{\mathbb{R}^d} U(\rho(x)) \, d\mathscr{L}^d(x) & \text{if } \mu \ll \mathscr{L}^d, \ \rho = \frac{d\mu}{d\mathscr{L}^d}, \\ +\infty & \text{otherwise} \end{cases} \tag{6.107}$$

is geodesically convex and lower semicontinuous in $\mathscr{P}_2(\mathbb{R}^d)$. Among the functionals $\mathscr{U}$ with U satisfying (6.105) and (6.106) we have

$$\text{the entropy functional } U(s) = s \log s, \tag{6.108}$$

$$\text{the power functional } U(s) = \frac{s^m}{m-1}, \quad m > 1. \tag{6.109}$$

Property (6.106) is also satisfied in the range $1 - \frac{1}{d} < m < 1$; however, since in this case U is not superlinear at infinity (the third condition of (6.105)), we have to consider the relaxed lower semicontinuous functional

$$\mathscr{U}^*(\mu) = \int_{\mathbb{R}^d} \frac{\rho^m}{m-1} \, d\mathscr{L}^d, \quad \text{if } \mu = \rho\mathscr{L}^d + \mu^\perp, \ \mu^\perp \perp \mathscr{L}^d.$$

Example 4 (relative entropy). Let μ, γ be two measures in $\mathscr{P}(\mathbb{R}^d)$. Then, the relative entropy of μ w.r.t. γ is the functional defined by

$$\text{Ent}_\gamma(\mu) := \begin{cases} \int \frac{d\mu}{d\gamma} \log \frac{d\mu}{d\gamma} \, d\gamma, & \text{if } \mu \ll \gamma, \\ +\infty & \text{otherwise.} \end{cases} \tag{6.110}$$

We note that (6.110) corresponds to the internal energy associated with the function (6.108) when $\gamma = \mathscr{L}^d$ (which nevertheless is not in $\mathscr{P}(\mathbb{R}^d)$).

Proposition 6.28. *The relative entropy* Ent_γ *is geodesically convex in* $\mathscr{P}_2(\mathbb{R}^d)$ *if and only if one of the following conditions holds:*

1. $\gamma = e^{-V}\mathscr{L}^d$ *for some convex function* $V : \mathbb{R}^d \to \mathbb{R}$; $\tag{6.111}$

2. γ *is log-concave, i.e. for every couple of open sets* $A, B \subset \mathbb{R}^d$, $t \in [0, 1]$
$$\log \gamma((1-t)A + tB) \geq (1-t)\log \gamma(A) + t \log \gamma(B) \tag{6.112}$$

Now we introduce the notion of convexity, which will be crucial to apply the metric theory of gradient flows developed in Section 6.4 to the main Examples 1–4 of geodesically λ-convex functionals in the Wasserstein space.

Definition 16 (convexity along generalized geodesics). Given $\lambda \in \mathbb{R}$, we say that $\phi : \mathscr{P}_2(\mathbb{R}^d) \to (-\infty, +\infty]$ is λ-convex along generalized geodesics if for any $\mu_1, \mu_2, \mu_3 \in D(\phi)$ there exists a generalized geodesic $[0, 1] \ni s \mapsto \mu_s^{2 \to 3}$ joining μ_2 to μ_3 induced by a plan $\boldsymbol{\mu}$ satisfying (6.93) such that

$$\phi(\mu_s^{2 \to 3}) \leq (1-s)\phi(\mu_2) + s\phi(\mu_3) - \frac{\lambda}{2}s(1-s) \int |x_2 - x_3|^2 \, d\boldsymbol{\mu}(x_1, x_2, x_3)$$
$$\text{for every } s \in [0, 1]. \tag{6.113}$$

Lemma 6.29 $((1/\tau + \lambda)$-*convexity of* $\Phi(\tau, \mu_1; \cdot))$. *Let* $\phi : \mathscr{P}_2(\mathbb{R}^d) \to (-\infty, +\infty]$ *be a proper functional which is* λ-*convex along generalized geodesics* *for some* $\lambda \in \mathbb{R}$. *Then, for each* $\mu_1 \in D(\phi)$ *and for each* $0 < \tau < \frac{1}{\lambda^-}$ *the functional*

$$\Phi(\tau, \mu_1; \mu) := \frac{1}{2\tau} W_2^2(\mu_1, \mu) + \phi(\mu) \text{ satisfies the convexity assumption (6.82).}$$

By Lemma 6.29, whenever ϕ is proper, l.s.c., and λ-convex along generalized geodesics in $\mathscr{P}_2(\mathbb{R}^d)$, we can apply Theorem 6.17 and get the existence, uniqueness, and regularizing estimates for the solutions of the EVI_λ.

The examples of geodesically convex functionals in $\mathscr{P}_2(\mathbb{R}^d)$ which have been introduced in this section are also convex along the generalized geodesics.

Theorem 6.30. *The functionals on* $\mathscr{P}_2(\mathbb{R}^d)$ *considered by Examples 1, 2 (with* $\lambda \leq 0$*), 3 (under condition (6.106)), and 4 (under condition (6.112)) are* λ*-convex along generalized geodesics.*

6.5.4 Gradient flows in $(\mathscr{P}_2(\mathbb{R}^d), W_2)$ and evolutionary PDEs

In this section we show some applications of the generation result 6.17 to the existence, well-posedness, and asymptotic behavior of nonnegative solutions $\rho : \mathbb{R}^d \times (0, +\infty) \to \mathbb{R}$ of evolutionary PDEs of the type

$$\frac{\partial}{\partial t}\rho - \nabla \cdot \left(\rho \nabla \frac{\delta\phi}{\delta\rho}\right) = 0 \quad \text{in } \mathbb{R}^d \times (0, +\infty), \tag{6.114}$$

where $\frac{\delta\phi(\rho)}{\delta\rho}$ is the first variation of a suitable integral functional; here, we consider the case of functionals which are a positive linear combination of the three kinds of contributions considered in Examples 1, 2, and 3, i.e. $\phi(\rho) := \alpha_1 \mathscr{U}(\rho) + \alpha_2 \mathscr{V}(\rho) + \alpha_3 \mathscr{W}(\rho)$ where $\alpha_i \geq 0$ and

$$\mathscr{U}(\rho) := \int_{\mathbb{R}^d} U(\rho(x)) \, dx,$$

$$\mathscr{V}(\rho) := \int_{\mathbb{R}^d} V(x)\rho(x) \, dx, \tag{6.115}$$

$$\mathscr{W}(\rho) := \frac{1}{2} \int_{\mathbb{R}^d \times \mathbb{R}^d} W(x - y)\rho(x)\rho(y) \, dx dy,$$

so that

$$\frac{\delta\phi(\rho)}{\delta\rho} = \alpha_1 U'(\rho) + \alpha_2 V + \alpha_3 W * \rho. \tag{6.116}$$

In the particular cases of the Fokker–Planck equation ($\phi = \mathscr{U} + \mathscr{V}$ and $U(r) = r \log r$)

$$\frac{\partial}{\partial t}\rho - \nabla \cdot (\nabla\rho + \rho\nabla V) = 0 \quad \text{in } \mathbb{R}^d \times (0, +\infty), \tag{6.117}$$

and of the nonlinear diffusion equations ($\phi = \mathcal{U}$, $U(r) = \frac{1}{m-1}r^m$)

$$\frac{\partial}{\partial t}\rho - \Delta\rho^m = 0, \quad m \geq 1 - \frac{1}{d}, \tag{6.118}$$

the Wasserstein approach has been introduced by the remarkable papers of Jordan *et al.* [36] and Otto [56] and then extended in many interesting directions, covering a wide range of applications; e.g. see [1, 5, 6, 11, 17–22, 33, 34, 45, 49, 50, 54, 57].

The results presented here are just examples of the transport approach.

Theorem 6.31. *Let V, W, U be as in Examples 1, 2, and 3, let $\mathcal{V}, \mathcal{W}, \mathcal{U}$ be defined as in (6.115), and let $\phi := \alpha_1\mathcal{U} + \alpha_2\mathcal{V} + \alpha_3\mathcal{W}$. For every $\mu_0 \in \mathcal{P}_2(\mathbb{R}^d)$ there exists a unique solution $\mu_t \in \mathrm{Lip}_{\mathrm{loc}}(0, +\infty; \mathcal{P}_2(\mathbb{R}^d))$ satisfying $\mathrm{EVI}_\lambda(\mathcal{P}_2(\mathbb{R}^d), W_2, \phi), \lambda := \alpha_2\lambda_V + \alpha_3\lambda_W$,*

$$\frac{1}{2}\frac{d}{dt}W_2^2(\mu_t, \sigma) \leq \phi(\sigma) - \phi(\mu_t) - \frac{\lambda}{2}W_2^2(\mu_t, \sigma) \quad \text{for every } \sigma \in D(\phi) \tag{6.119}$$

with $\lim_{t\downarrow 0}\mu_t = \mu_0$ in $\mathcal{P}_2(\mathbb{R}^d)$; the curve μ satisfies all the properties stated in Theorem 6.9, the continuity equation

$$\frac{\partial}{\partial t}\mu_t + \nabla \cdot (\mu_t\,v_t) = 0 \quad \text{in } \mathbb{R}^d \times (0, +\infty), \quad \text{with} \tag{6.120}$$

$$v_t \in \mathrm{Tan}_{\mu_t}\mathcal{P}_2(\mathbb{R}^d) \quad \mathcal{L}^1\text{-a.e. in } (0, +\infty) \quad \text{and}$$

$$t \mapsto \int_{\mathbb{R}^d}|v_t|^2\,d\mu_t = |\mu_t'|^2 \in L^\infty_{\mathrm{loc}}(0, +\infty), \tag{6.121}$$

and for $\mathcal{L}^1$-a.e. $t \in (0, +\infty)$ the velocity vector $v_t \in \mathrm{Tan}_{\mu_t}\mathcal{P}_2(\mathbb{R}^d)$ satisfies the "subdifferential inequality"

$$\int_{\mathbb{R}^d}\langle v_t(x), x - y\rangle + \frac{\lambda}{2}|y - x|^2\,d\gamma_t(x, y) \leq \phi(\sigma) - \phi(\mu_t)$$

$$\text{for every } \gamma_t \in \Gamma_{\mathrm{opt}}(\mu_t, \sigma). \tag{6.122}$$

We can give an explicit characterization of the system (6.120), (6.122). Here, we consider the simpler case when U, V, W are differentiable and satisfy a *doubling condition*: for a function $f : \mathbb{R}^h \to \mathbb{R}$ it means that there exists a constant $C > 0$ such that

$$f(x + y) \leq C(1 + f(x) + f(y)) \quad \text{for every } x, y \in \mathbb{R}^h. \tag{6.123}$$

We also set

$$L_U(r) := rU'(r) - U(r) \quad \text{if } r > 0, \quad L_U(0) = 0. \tag{6.124}$$

Theorem 6.32. *Under the same assumptions of the previous theorem, let us also suppose that U, V, W are differentiable and satisfy the doubling condition (6.123). The locally Lipschitz curve μ characterized by (6.119) (or by (6.120), (6.122)) solves the following evolutionary PDEs in $\mathbb{R}^d \times (0, +\infty)$:*
Transport equation, $\phi = \mathscr{V}$, $v_t = -\nabla V$:

$$\frac{\partial}{\partial_t} \mu_t - \nabla \cdot (\mu_t \nabla V) = 0. \qquad (6.125)$$

Nonlocal interaction equation, $\phi = \mathscr{W}$, $v_t = -(\nabla W) * \mu_t$

$$\frac{\partial}{\partial_t} \mu_t - \nabla \cdot (\mu_t (\nabla W * \mu_t)) = 0. \qquad (6.126)$$

Fokker–Planck equation, $\phi = \mathscr{U} + \mathscr{V}$, $U(r) = r \log r$, $-\mu_t v_t = \nabla \mu_t + \mu_t \nabla V$

$$\frac{\partial}{\partial_t} \mu_t - \nabla \cdot (\nabla \mu_t + \mu_t \nabla V) = 0. \qquad (6.127)$$

In this case, $\mu_t = \rho_t \mathscr{L}^d$ with $\rho_t \in W^{1,1}_{\text{loc}}(\mathbb{R}^d)$ for $\mathscr{L}^1$-a.e. $t \in (0, +\infty)$.
Nonlinear diffusion equation, $\phi = \mathscr{U}$, $\mu_t v_t = -\nabla L_U(\rho_t)$ where $\mu_t = \rho_t \mathscr{L}^d \ll \mathscr{L}^d$,

$$\frac{\partial}{\partial_t} \mu_t - \Delta(L_U(\rho_t)) = 0, \qquad (6.128)$$

with $L_U(\rho_t) \in W^{1,1}_{\text{loc}}(\mathbb{R}^d)$ for $\mathscr{L}^1$-a.e. $t \in (0, +\infty)$.
Drift-diffusion with nonlocal interactions, $\phi = \mathscr{U} + \mathscr{V} + \mathscr{W}$, $-\mu_t v_t = \nabla L_U(\rho_t) + \mu_t \nabla V + \mu_t((\nabla W) * \mu_t)$, $\mu_t = \rho_t \mathscr{L}^d \ll \mathscr{L}^d$,

$$\frac{\partial}{\partial_t} \mu_t - \nabla \cdot \left(\nabla L_U(\rho_t) + \mu_t \nabla V + \mu_t((\nabla W) * \mu_t)\right) = 0. \qquad (6.129)$$

We refer to [3, Chapter 11] for the proofs and for more general and detailed results; here, we just give a sketch of the argument showing that (6.120), (6.122) yield (6.127) when $\phi = \mathscr{U} + \mathscr{V}$ in the case of $U(r) = r \log r$.

Let us fix a time $t > 0$ where (6.122) holds, a smooth test function $\zeta \in C^\infty_c(\mathbb{R}^d)$, and $t_\varepsilon := i + \varepsilon \nabla \zeta$. If $|\varepsilon| \max_{\mathbb{R}^d} \|D^2 \zeta\| < 1$ the coupling $\gamma_\varepsilon := (i, t_\varepsilon)_\# \mu_t$ is optimal between μ_t and $(t_\varepsilon)_\# \mu_t$ so that (6.122) yields

$$-\varepsilon \int_{\mathbb{R}^d} \langle v_t(x), \nabla \zeta(x) \rangle \, d\mu_t(x) \le \phi((t_\varepsilon)_\# \mu_t) - \phi(\mu_t).$$

Setting

$$\rho_t := \frac{d\mu_t}{d\mathscr{L}^d}, \qquad \rho_t^\varepsilon := \frac{d(t_\varepsilon)_\# \mu_t}{d\mathscr{L}^d}$$

we get

$$-\varepsilon \int_{\mathbb{R}^d} \langle \boldsymbol{v}_t, \nabla \zeta \rangle \, \mathrm{d}\mu_t \leq \int_{\mathbb{R}^d} \rho_t^\varepsilon \log \rho_t^\varepsilon \, \mathrm{d}\mathscr{L}^d - \int_{\mathbb{R}^d} \rho_t \log \rho_t \, \mathrm{d}\mathscr{L}^d$$
$$+ \int_{\mathbb{R}^d} \big(V(t_\varepsilon(x)) - V(x)\big) \, \mathrm{d}\mu_t(x).$$

Applying the change of variables formula

$$\rho_t^\varepsilon(t_\varepsilon(x)) \det[\boldsymbol{i} + \varepsilon \mathrm{D}^2 \zeta(x)] = \rho_t(x),$$

we obtain

$$-\varepsilon \int_{\mathbb{R}^d} \langle \boldsymbol{v}_t, \nabla \zeta \rangle \, \mathrm{d}\mu_t \leq - \int_{\mathbb{R}^d} \rho(x) \log\big(\det[\boldsymbol{i} + \varepsilon \mathrm{D}^2 \zeta(x)]\big) \, \mathrm{d}\mathscr{L}^d$$
$$+ \int_{\mathbb{R}^d} \big(V(t_\varepsilon(x)) - V(x)\big) \, \mathrm{d}\mu_t(x). \qquad (6.130)$$

Finally, dividing by ε and taking the limit of (6.130) as ε tends to 0 we get

$$- \int_{\mathbb{R}^d} \langle \boldsymbol{v}_t, \nabla \zeta \rangle \, \mathrm{d}\mu_t = \int_{\mathbb{R}^d} \Big(- \Delta \zeta(x) + \nabla V(x) \cdot \nabla \zeta\Big) \, \mathrm{d}\mu_t$$
$$\text{for every } \zeta \in C_c^\infty(\mathbb{R}^d),$$

so that μ satisfies the distributional formulation of (6.127).

6.5.5 The heat flow on Riemannian manifolds and metric-measure spaces

We conclude these notes by giving a short account of possible applications of the Wasserstein setting to the generation of the heat flow in Riemannian manifolds and metric-measure spaces.

Let us start with a compact and smooth Riemannian manifold (M, g); we denote by d_g its Riemannian distance and by $\gamma = \mathrm{Vol}_g \in \mathscr{P}(M)$ its (normalized) volume measure.

In $\mathscr{P}_2(M)$ we consider the relative entropy functional Ent_γ as in (6.110). Sturm and von Renesse [66] proved the following:

Theorem 6.33. *The relative entropy functional* Ent_γ *is geodesically* λ-*convex in* $\mathscr{P}_2(M)$ *if and only if* M *satisfies the lower Ricci curvature bound*

$$\mathrm{Ric}(M) \geq \lambda, \quad \textit{i.e.} \quad \mathrm{Ric}_x(v, v) \geq \lambda |v|_g^2 \quad \textit{for all } x \in M \textit{ and } v \in \mathrm{Tan}_x(M).$$
$$(6.131)$$

In this case, it is possible to show (see [27,32,53,55,57,68]) that the relative entropy functional Ent_γ generates a λ-gradient flow $S_t : \mathscr{P}_2(M) \to \mathscr{P}_2(M)$ according to Definition 4, which coincides with the classical heat flow on M.

Theorem 6.34. *The relative entropy functional* Ent_γ *generates a* λ-*gradient flow* S_t *in* $\mathscr{P}_2(M)$ *according to Definition 4 (and thus satisfying all the properties stated in Theorems 6.9 and 6.13). A curve* $\mu_t \in \mathscr{P}_2(M)$ *is a solution of* $\text{EVI}_\lambda(M, d_g, \text{Ent}_\gamma)$ *if and only if its density* $\rho_t = d\mu_t/d\gamma$ *solves the heat equation*

$$\frac{\partial}{\partial t}\rho_t - \Delta_g \rho_t = 0 \quad in \ M \times (0, +\infty),$$

where Δ_g *is the Laplace–Beltrami operator on* M.

The adimensionality of the form of the entropy functional (6.110) and the purely metric character of the EVI suggest that one can use them to define a heat flow on more general *measure-metric* spaces (X, d, γ), where (X, d) is a complete and separable metric space and $\gamma \in \mathscr{P}(X)$. Indeed, as it has been often pointed out in the previous sections, the EVI formulation gives nice regularity, stability, and asymptotic properties for the related flow. We briefly sketch two possible approaches.

6.5.5.1 Approximation by measured Gromov–Hausdorff convergence

We consider a sequence of smooth and compact Riemannian manifolds (M^h, d^h, Vol^h) converging to a limit measure-metric space (X, d, γ) in the measured Gromov–Hausdorff convergence: it means [64] that a sequence $\{\hat{d}^h\}_{k\in\mathbb{N}}$ of (complete, separable) *coupling semidistances* on the disjoint union $M^h \sqcup X$ exists such that the restriction of $\hat{d}^h$ on M^h (resp. X) coincides with d^h (resp. d) and

$$\lim_{k\uparrow\infty} \hat{W}_2^h(\text{Vol}^h, \gamma) = 0, \quad \hat{W}_2^h \text{ is the Wasserstein distance on } \mathscr{P}_2(M^h \sqcup X)$$

$$\text{induced by } \hat{d}^h. \quad (6.132)$$

A sequence $\mu^h \in \mathscr{P}_2(M^h)$ converges to $\mu \in \mathscr{P}_2(X)$ if $\lim_{k\uparrow+\infty} \hat{W}_2^h(\mu^h, \mu) = 0$. Adapting the arguments of Theorem 6.14 it is possible to prove the following asymptotic result:

Theorem 6.35 [63]. *Let us assume that the compact Riemannian manifolds* M^h *satisfy the uniform lower bound on the Ricci curvature* $\text{Ric}(M^h) \geq \lambda$ *for some* $\lambda \in \mathbb{R}$ *independent of* k *and converge to* (X, d, γ) *in the measured Gromov–Hausdorff sense. Then the relative entropy functional* Ent_γ *admits a* λ-*gradient flow* S_t *on* $\mathscr{P}_2(X)$ *and for every sequence of initial measures*

$\mu_0^h \in \mathscr{P}_2(M^h)$ *converging to* $\mu_0 \in \mathscr{P}_2(X)$ *the corresponding solution* μ_t^h *of the heat flow on* M^h *converges to* $\mathsf{S}_t(\mu_0)$ *in* $\mathscr{P}_2(X)$ *for every* $t > 0$.

Applying Theorem 6.10 one finds in particular that the limit entropy functional Ent_γ is *strongly* geodesically λ-convex (at least when the support of γ is X), a stability result that has been proved by [42, 65].

6.5.5.2 Intrinsic costruction

Starting from Theorem 6.33, Sturm [65] and Lott and Villani [42] introduced the concept of metric-measure spaces (X, d, γ) satisfying a lower Ricci curvature bound, by requiring that the relative entropy functional Ent_γ is geodesically λ-convex in $\mathscr{P}_2(X)$.

Definition 17 (lower Ricci curvature bounds for metric-measure spaces). We say that a metric-measure space (X, d, γ) has Ricci curvature bounded from below by a certain $\lambda \in \mathbb{R}$ (and we write $\mathrm{Ric}(X) \geq \lambda$) if the relative entropy Ent_γ is λ-geodesically convex on X.

It is then natural to look for other intrinsic properties of X which are sufficient to deduce the existence of the associated EVI semigroup. It is interesting to notice that if the relative entropy functional generates a λ-gradient flow S_t then S_t is a semigroup of *linear* operators [63]. In the case of compact PC Alexandrov spaces the existence of a λ-contracting gradient flow can be deduced by a general unpublished result of [58] and has recently been proved by Ohta [52].

In more general cases, we can apply Theorem 6.18:

Theorem 6.36. *Let us suppose that* (X, d, γ) *is a complete and separable metric-measure space with Ricci curvature bounded from below, according to Definition 17, and measure* γ *with full support* $\mathrm{supp}(\gamma) = X$. *If* X *satisfies the local angle condition Definition 13 and it is* K-*semiconcave as in Definition 11, then the relative entropy functional* Ent_γ *generates a* λ-*gradient flow on* $\mathscr{P}_2(X)$ *which can be uniquely extended to a Markov semigroup (i.e. linear, order preserving, strongly continuous, contractive) in every space* $L^p(\gamma)$, $p \in [1, +\infty)$.

References

[1] M. Agueh. Existence of solutions to degenerate parabolic equations via the Monge–Kantorovich theory. *Adv. Differential Equations*, 10(3):309–360, 2005.

[2] L. Ambrosio. Minimizing movements. *Rend. Accad. Naz. Sci. XL Mem. Mat. Appl. (5)*, 19:191–246, 1995.

[3] L. Ambrosio, N. Gigli, and G. Savaré. *Gradient flows in metric spaces and in the space of probability measures*. Lectures in Mathematics ETH Zürich. Birkhäuser Verlag, Basel, 2005.

[4] L. Ambrosio and G. Savaré. Gradient flows of probability measures. In *Handbook of Evolution Equations (III)*. Elsevier, 2006.

[5] L. Ambrosio, G. Savaré, and L. Zambotti. Existence and stability for Fokker–Planck equations with log-concave reference measure. *Probab. Theory Relat. Fields*, 145(3–4):517–564, 2009.

[6] L. Ambrosio and S. Serfaty. A gradient flow approach to an evolution problem arising in superconductivity. *Comm. Pure Appl. Math.*, 61(11):1495–1539, 2008.

[7] C. Baiocchi. Discretization of evolution variational inequalities. In F. Colombini, A. Marino, L. Modica, and S. Spagnolo, editors, *Partial Differential Equations and the Calculus of Variations*, Vol. I, pages 59–92. Birkhäuser Boston, Boston, MA, 1989.

[8] V. Barbu. *Nonlinear Semigroups and Differential Equations in Banach Spaces*. Editura Academiei Republicii Socialiste România, Bucharest, 1976. Translated from the Romanian.

[9] J.-D. Benamou and Y. Brenier. A computational fluid mechanics solution to the Monge–Kantorovich mass transfer problem. *Numer. Math.*, 84(3):375–393, 2000.

[10] P. Bénilan. Solutions intégrales d'équations d'évolution dans un espace de Banach. *C. R. Acad. Sci. Paris Sér. A-B*, 274:A47–A50, 1972.

[11] A. Blanchet, V. Calvez, and J.A. Carrillo. Convergence of the mass-transport steepest descent scheme for the subcritical Patlak–Keller–Segel model. *SIAM J. Numer. Anal.*, 46:691–721, 2008.

[12] Y. Brenier. Polar factorization and monotone rearrangement of vector-valued functions. *Comm. Pure Appl. Math.*, 44(4):375–417, 1991.

[13] H. Brézis. Monotonicity methods in Hilbert spaces and some applications to nonlinear partial differential equations. In *Contribution to Nonlinear Functional Analysis, Proc. Symposium Math. Res. Center, Univ. Wisconsin, Madison, 1971*, pages 101–156. Academic Press, New York, 1971.

[14] H. Brézis. *Opérateurs Maximaux Monotones et Semi-groupes de Contractions dans les Espaces de Hilbert*. North-Holland Publishing Co., Amsterdam, 1973. North-Holland Mathematics Studies, No. 5. Notas de Matemática (50).

[15] D. Burago, Y. Burago, and S. Ivanov. *A Course in Metric Geometry*, volume 33 of *Graduate Studies in Mathematics*. American Mathematical Society, Providence, RI, 2001.

[16] T. Cardinali, G. Colombo, F. Papalini, and M. Tosques. On a class of evolution equations without convexity. *Nonlinear Anal.*, 28(2):217–234, 1997.

[17] E.A. Carlen and W. Gangbo. Constrained steepest descent in the 2-Wasserstein metric. *Ann. Math. (2)*, 157(3):807–846, 2003.

[18] E.A. Carlen and W. Gangbo. Solution of a model Boltzmann equation via steepest descent in the 2-Wasserstein metric. *Arch. Ration. Mech. Anal.*, 172(1):21–64, 2004.

[19] J.A. Carrillo, S. Lisini, G. Savaré, and D. Slepcev. Nonlinear mobility continuity equations and generalized displacement convexity. *J. Funct. Anal.*, 258(4):1273–1309, 2010.

[20] J.A. Carrillo, M. Di Francesco, and C. Lattanzio. Contractivity of Wasserstein metrics and asymptotic profiles for scalar conservation laws. *J. Differential Equations*, 231(2):425–458, 2006.

[21] J.A. Carrillo, R.J. McCann, and C. Villani. Kinetic equilibration rates for granular media and related equations: entropy dissipation and mass transportation estimates. *Rev. Mat. Iberoamericana*, 19(3):971–1018, 2003.

[22] J.A. Carrillo, R.J. McCann, and C. Villani. Contractions in the 2-Wasserstein length space and thermalization of granular media. *Arch. Ration. Mech. Anal.*, 179(2):217–263, 2006.

[23] P. Clément. Introduction to gradient flows in metric spaces. Lecture Notes, University of Bielefeld, 2009. Available online at https://igk.math.uni-bielefeld.de/study-materials/notes-clement-part2.pdf.

[24] M.G. Crandall and T.M. Liggett. Generation of semi-groups of nonlinear transformations on general Banach spaces. *Am. J. Math.*, 93:265–298, 1971.

[25] M.G. Crandall and A. Pazy. Semi-groups of nonlinear contractions and dissipative sets. *J. Functional Analysis*, 3:376–418, 1969.

[26] G. Dal Maso. *An Introduction to Γ-Convergence*, volume 8 of *Progress in Nonlinear Differential Equations and Their Applications*. Birkhäuser, Boston, 1993.

[27] S. Daneri and G. Savaré. Eulerian calculus for the displacement convexity in the Wasserstein distance. *SIAM J. Math. Anal.*, 40(3):1104–1122, 2008.

[28] E. De Giorgi. New problems on minimizing movements. In C. Baiocchi and J.L. Lions, editors, *Boundary Value Problems for PDE and Applications*, pages 81–98. Masson, 1993.

[29] E. De Giorgi, M. Degiovanni, A. Marino, and M. Tosques. Evolution equations for a class of nonlinear operators. *Atti Accad. Naz. Lincei Rend. Cl. Sci. Fis. Mat. Natur. (8)*, 75(1–2):1–8 (1984), 1983.

[30] E. De Giorgi, A. Marino, and M. Tosques. Problems of evolution in metric spaces and maximal decreasing curve. *Atti Accad. Naz. Lincei Rend. Cl. Sci. Fis. Mat. Natur. (8)*, 68(3):180–187, 1980.

[31] M. Degiovanni, A. Marino, and M. Tosques. Evolution equations with lack of convexity. *Nonlinear Anal.*, 9(12):1401–1443, 1985.

[32] M. Erbar. The heat equation on manifolds as a gradient flow in the Wasserstein space. *Annales de l'Institut Henri Poincaré – Probabilités et Statistiques*, 46(1):1–23, 2010.

[33] L.C. Evans, O. Savin, and W. Gangbo. Diffeomorphisms and nonlinear heat flows. *SIAM J. Math. Anal.*, 37(3):737–751 (electronic), 2005.

[34] S. Fang, J. Shao, and T.K. Sturm. Wasserstein space over the wiener space. webdoc.sub.gwdg.de, Jan 2008.

[35] U. Gianazza and G. Savaré. Abstract evolution equations on variable domains: an approach by minimizing movements. *Ann. Sc. Norm. Sup. Pisa Cl. Sci. (4)*, 23:149–178, 1996.

[36] R. Jordan, D. Kinderlehrer, and F. Otto. The variational formulation of the Fokker–Planck equation. *SIAM J. Math. Anal.*, 29(1):1–17 (electronic), 1998.

[37] J. Jost. *Nonpositive Curvature: Geometric and Analytic Aspects*. Lectures in Mathematics ETH Zürich. Birkhäuser Verlag, Basel, 1997.

[38] M. Knott and C.S. Smith. On the optimal mapping of distributions. *J. Optim. Theory Appl.*, 43(1):39–49, 1984.

[39] Y. Kōmura. Nonlinear semi-groups in Hilbert space. *J. Math. Soc. Japan*, 19:493–507, 1967.

[40] J.-L. Lions. *Quelques Méthodes de Résolution des Problèmes aux Limites non Linéaires*. Dunod, Gauthier-Villars, Paris, 1969.

[41] J.-L. Lions and G. Stampacchia. Variational inequalities. *Comm. Pure Appl. Math.*, 20:493–519, 1967.

[42] J. Lott and C. Villani. Ricci curvature for metric-measure spaces via optimal transport. *Ann. Math. (2)*, 169(3):903–991, 2009.

[43] S. Luckhaus. Solutions for the two-phase Stefan problem with the Gibbs–Thomson law for the melting temperature. *Eur. Jnl. Appl. Math.*, 1:101–111, 1990.

[44] A. Marino, C. Saccon, and M. Tosques. Curves of maximal slope and parabolic variational inequalities on nonconvex constraints. *Ann. Scuola Norm. Sup. Pisa Cl. Sci. (4)*, 16(2):281–330, 1989.

[45] D. Matthes, R.J. McCann, and G. Savaré. A family of nonlinear fourth order equations of gradient flow type. *Comm. Partial Differential Equations*, 34(10–12):1352–1397, 2009.

[46] U.F. Mayer. Gradient flows on nonpositively curved metric spaces and harmonic maps. *Comm. Anal. Geom.*, 6(2):199–253, 1998.

[47] R.J. McCann. A convexity principle for interacting gases. *Adv. Math.*, 128(1):153–179, 1997.

[48] A. Mielke, F. Theil, and V.I. Levitas. A variational formulation of rate-independent phase transformations using an extremum principle. *Arch. Ration. Mech. Anal.*, 162(2):137–177, 2002.

[49] L. Natile, M.A. Peletier, and G. Savaré. Contraction of general transportation costs along solutions to Fokker–Planck equations with monotone drifts. *arXiv:1002.0088v1*, 2010.

[50] L. Natile and G. Savaré. A Wasserstein approach to the one-dimensional sticky particle system. *arxiv:0902.4373v2*, 2009.

[51] R.H. Nochetto, G. Savaré, and C. Verdi. A posteriori error estimates for variable time-step discretizations of nonlinear evolution equations. *Comm. Pure Appl. Math.*, 53(5):525–589, 2000.

[52] S.-i. Ohta. Gradient flows on wasserstein spaces over compact alexandrov spaces. Technical report, Universität Bonn, 2007.

[53] S.-i. Ohta. Gradient flows on Wasserstein spaces over compact Alexandrov spaces. *Am. J. Math.*, 131(2):475–516, 2009.

[54] F. Otto and C. Villani. Generalization of an inequality by Talagrand and links with the logarithmic Sobolev inequality. *J. Funct. Anal.*, 173(2):361–400, 2000.

[55] F. Otto and C. Villani. Generalization of an inequality by Talagrand and links with the logarithmic Sobolev inequality. *J. Funct. Anal.*, 173(2):361–400, 2000.

[56] F. Otto. The geometry of dissipative evolution equations: the porous medium equation. *Comm. Partial Differential Equations*, 26(1-2):101–174, 2001.

[57] F. Otto and M. Westdickenberg. Eulerian calculus for the contraction in the Wasserstein distance. *SIAM J. Math. Anal.*, 37(4):1227–1255 (electronic), 2005.

[58] G. Perelman and A. Petrunin. Quasigeodesics and gradient curves in alexandrov spaces. Unpublished preprint, available online at www.math.psu.edu/petrunin/papers/papers.html.

[59] J. Rulla. Error analysis for implicit approximations to solutions to Cauchy problems. *SIAM J. Numer. Anal.*, 33:68–87, 1996.

[60] G. Savaré. Gradient flows and evolution variational inequalities in metric spaces. *In preparation*, 2014.

[61] G. Savaré. Weak solutions and maximal regularity for abstract evolution inequalities. *Adv. Math. Sci. Appl.*, 6(2):377–418, 1996.

[62] G. Savaré. Error estimates for dissipative evolution problems. In *Free Boundary Problems (Trento, 2002)*, volume 147 of *Internat. Ser. Numer. Math.*, pages 281–291. Birkhäuser, Basel, 2004.

[63] G. Savaré. Gradient flows and diffusion semigroups in metric spaces under lower curvature bounds. *C. R. Math. Acad. Sci. Paris*, 345(3):151–154, 2007.

[64] K.-T. Sturm. Convex functionals of probability measures and nonlinear diffusions on manifolds. *J. Math. Pures Appl. (9)*, 84(2):149–168, 2005.

[65] K.-T. Sturm. On the geometry of metric measure spaces. I. *Acta Math.*, 196(1):65–131, 2006.

[66] K.-T. Sturm and M.-K. von Renesse. Transport inequalities, gradient estimates, entropy, and Ricci curvature. *Comm. Pure Appl. Math.*, 58(7):923–940, 2005.

[67] C. Villani. *Topics in Optimal Transportation*, volume 58 of *Graduate Studies in Mathematics*. American Mathematical Society, Providence, RI, 2003.

[68] C. Villani. *Optimal Transport. Old and New*, volume 338 of *Grundlehren der Mathematischen Wissenschaften*. Springer-Verlag, Berlin, 2009.

7

Ricci curvature, entropy, and optimal transport

SHIN-ICHI OHTA

Kyoto University, Japan

Abstract

This chapter comprises the lecture notes on the interplay between optimal transport and Riemannian geometry. On a Riemannian manifold, the convexity of entropy along optimal transport in the space of probability measures characterizes lower bounds of the Ricci curvature. We then discuss geometric properties of general metric measure spaces satisfying this convexity condition.

7.1 Introduction

This chapter is extended notes based on the author's lecture series at the summer school at Université Joseph Fourier, Grenoble: "Optimal Transportation: Theory and Applications." The aim of these five lectures (corresponding to Sections 7.3–7.7) was to review the recent impressive development on the interplay between optimal transport theory and Riemannian geometry. Ricci curvature and entropy are the key ingredients. See [Lo2] for a survey in the same spirit with a slightly different selection of topics.

Optimal transport theory is concerned with the behavior of transport between two probability measures in a metric space. We say that such transport is optimal if it minimizes a certain cost function typically defined from the distance of the metric space. Optimal transport naturally inherits the geometric structure of the underlying space; in particular Ricci curvature plays a crucial role for describing optimal transport in Riemannian manifolds. In fact, optimal transport

Mathematics Subject Classification (2000): 53C21, 53C23, 53C60, 28A33, 28D20
Keywords: Ricci curvature, entropy, optimal transport, curvature-dimension condition

is always performed along geodesics, and we obtain Jacobi fields as their variational vector fields. The behavior of these Jacobi fields is controlled by the Ricci curvature as is usual in comparison geometry. In this way, a lower Ricci curvature bound turns out to be equivalent to a certain convexity property of entropy in the space of probability measures. The latter convexity condition is called the curvature-dimension condition, and it can be formulated without using the differentiable structure. Therefore, the curvature-dimension condition can be regarded as a "definition" of a lower Ricci curvature bound for general metric measure spaces, and implies many analogous properties in an interesting way.

A prerequisite is the basic knowledge of optimal transport theory and Wasserstein geometry. Riemannian geometry is also necessary in Sections 7.3, 7.4, and is helpful for better understanding of the other sections. We refer to [AGS], [Vi1], [Vi2] and other articles in this proceeding for optimal transport theory, [CE], [Ch] and [Sak] for the basics of (comparison) Riemannian geometry. We discuss Finsler geometry in Section 7.7, for which we refer to [BCS], [Sh2], and [Oh5]. Besides them, the main references are [CMS1], [CMS2], [vRS], [St3], [St4], [LV2], [LV1], and [Vi2, Chapter III].

The organization of this chapter is as follows. After summarizing some notation we use, Section 7.3 is devoted to the definition of the Ricci curvature of Riemannian manifolds and to the classical Bishop–Gromov volume comparison theorem. In Section 7.4, we start with the Brunn–Minkowski inequalities in (unweighted or weighted) Euclidean spaces, and explain the equivalence between a lower (weighted) Ricci curvature bound for a (weighted) Riemannian manifold and the curvature-dimension condition. In Section 7.5, we give the precise definition of the curvature-dimension condition for metric measure spaces, and see that it is stable under the measured Gromov–Hausdorff convergence. Section 7.6 is concerned with several geometric applications of the curvature-dimension condition followed by related open questions. In Section 7.7, we verify that this kind of machinery is useful also in Finsler geometry. We finally discuss three related topics in Section 7.8. Interested readers can find more references in the Further Reading sections at the end of each section (except the last section).

Some subjects in this article are more comprehensively discussed in [Vi2, Part III]. Despite these inevitable overlaps with Villani's massive book, we try to argue in a more geometric way, and mention recent development. Analytic applications of the curvature-dimension condition are not dealt with in these notes, for which we refer to [LV1], [LV2], and [Vi2, Chapter III], among others.

7.2 Notation

Throughout the article except Section 7.7, (M, g) is an n-dimensional, connected, complete C^∞-Riemannian manifold without boundary such that $n \geq 2$, vol_g stands for the Riemannian volume measure of g. A *weighted Riemannian manifold* (M, g, m) will mean a Riemannian manifold (M, g) endowed with a conformal deformation $m = e^{-\psi} \mathrm{vol}_g$ of vol_g with $\psi \in C^\infty(M)$. Similarly, a weighted Euclidean space $(\mathbb{R}^n, \|\cdot\|, m)$ will be a Euclidean space with a measure $m = e^{-\psi} \mathrm{vol}_n$, where vol_n stands for the n-dimensional Lebesgue measure.

A metric space is called a *geodesic space* if any two points $x, y \in X$ can be connected by a rectifiable curve $\gamma : [0, 1] \longrightarrow X$ of length $d(x, y)$ with $\gamma(0) = x$ and $\gamma(1) = y$. Such minimizing curves parameterized proportionally to arc length are called *minimal geodesics*. The open ball of center x and radius r will be denoted by $B(x, r)$. We remark that, thanks to the Hopf–Rinow theorem (cf. [Bal, Theorem 2.4]), a complete, locally compact geodesic space is proper; i.e., every bounded closed set is compact.

In this chapter, we mean by a *metric measure space* a triple (X, d, m) consisting of a complete, separable geodesic space (X, d) and a Borel measure m on it. Our definition of the curvature-dimension condition will include the additional (but natural) condition that $0 < m(B(x, r)) < \infty$ holds for all $x \in X$ and $0 < r < \infty$. We extend m to an outer measure in the Brunn–Minkowski inequalities (Theorems 7.4, 7.5, 7.10; see Remark 2 for more details).

For a complete, separable metric space (X, d), $\mathcal{P}(X)$ stands for the set of Borel probability measures on X. Define $\mathcal{P}_2(X) \subset \mathcal{P}(X)$ as the set of measures of finite second moment (i.e., $\int_X d(x, y)^2 \, d\mu(y) < \infty$ for some (and hence all) $x \in X$). We denote by $\mathcal{P}_b(X) \subset \mathcal{P}_2(X)$, $\mathcal{P}_c(X) \subset \mathcal{P}_b(X)$ the sets of measures of bounded or compact support, respectively. Given a measure m on X, denote by $\mathcal{P}^{\mathrm{ac}}(X, m) \subset \mathcal{P}(X)$ the set of absolutely continuous measures with respect to m. Then d_2^W stands for the L^2-(*Kantorovich–Rubinstein–*) *Wasserstein distance* of $\mathcal{P}_2(X)$. The push-forward of a measure μ by a map $\mathcal{F}$ will be written as $\mathcal{F}_\sharp \mu$.

As usual in comparison geometry, the following functions will frequently appear in our discussions. For $K \in \mathbb{R}$, $N \in (1, \infty)$ and $0 < r$ ($< \pi\sqrt{(N-1)/K}$ if $K > 0$), we set

$$
\mathbf{s}_{K,N}(r) := \begin{cases} \sqrt{(N-1)/K} \, \sin(r\sqrt{K/(N-1)}) & \text{if } K > 0, \\ r & \text{if } K = 0, \\ \sqrt{-(N-1)/K} \, \sinh(r\sqrt{-K/(N-1)}) & \text{if } K < 0. \end{cases} \tag{7.1}
$$

This is the solution to the differential equation

$$s''_{K,N} + \frac{K}{N-1} s_{K,N} = 0 \qquad (7.2)$$

with the initial conditions $s_{K,N}(0) = 0$ and $s'_{K,N}(0) = 1$. For $n \in \mathbb{N}$ with $n \geq 2$, $s_{K,n}(r)^{n-1}$ is proportional to the area of the sphere of radius r in the n-dimensional space form of constant sectional curvature $K/(n-1)$ (see Theorem 7.1 and the paragraph after it). In addition, using $s_{K,N}$, we define

$$\beta^t_{K,N}(r) := \left(\frac{s_{K,N}(tr)}{t s_{K,N}(r)} \right)^{N-1}, \qquad \beta^t_{K,\infty}(r) := e^{K(1-t^2)r^2/6} \qquad (7.3)$$

for K, N, r as above and $t \in (0, 1)$. This function plays a vital role in the key infinitesimal inequality (7.23) of the curvature-dimension condition.

7.3 Ricci curvature and comparison theorems

We begin with the basic concepts of curvature in Riemannian geometry and several comparison theorems involving lower bounds of the Ricci curvature. Instead of giving the detailed definition, we intend to explain the geometric intuition of the sectional and Ricci curvatures through comparison geometry.

Curvature is one of the most important quantities in Riemannian geometry. By putting some conditions on the value of the (sectional or Ricci) curvature, we obtain various quantitative and qualitative controls of distance, measure, geodesics and so forth. Comparison geometry is specifically interested in spaces whose curvature is bounded by a constant from above or below. In other words, we consider a space which is more positively or negatively curved than a space form of constant curvature, and compare these spaces from various viewpoints.

The n-dimensional (simply connected) *space form* $\mathbb{M}^n(k)$ of constant sectional curvature $k \in \mathbb{R}$ is the unit sphere $\mathbb{S}^n$ for $k = 1$; the Euclidean space $\mathbb{R}^n$ for $k = 0$; and the hyperbolic space $\mathbb{H}^n$ for $k = -1$. Scaling gives general space forms for all $k \in \mathbb{R}$; e.g., $\mathbb{M}^n(k)$ for $k > 0$ is the sphere of radius $1/\sqrt{k}$ in $\mathbb{R}^{n+1}$ with the induced Riemannian metric.

7.3.1 Sectional curvature

Given linearly independent tangent vectors $v, w \in T_x M$, the *sectional curvature* $\mathcal{K}(v, w) \in \mathbb{R}$ reflects the asymptotic behavior of the distance function $d(\gamma(t), \eta(t))$ near $t = 0$ between geodesics $\gamma(t) = \exp_x(tv)$ and

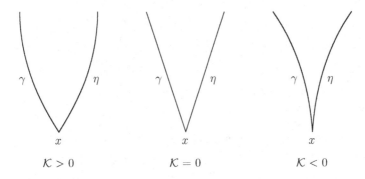

$$K > 0 \qquad\qquad K = 0 \qquad\qquad K < 0$$

Figure 7.1

$\eta(t) = \exp_x(tw)$. That is to say, the asymptotic behavior of $d(t) := d(\gamma(t), \eta(t))$ near $t = 0$ is the same as the distance between geodesics, with the same speed and angle between them, in the space form of curvature $k = \mathcal{K}(v, w)$. (See Figure 7.1 which represents isometric embeddings of γ and η into $\mathbb{R}^2$ such that $d(\gamma(t), \eta(t))$ coincides with the Euclidean distance.)

Assuming $\|v\| = \|w\| = 1$ for simplicity, we can compute $d(t)$ in the space form $\mathbb{M}^n(k)$ by using the spherical/Euclidean/hyperbolic law of cosines as (cf. [Sak, Section IV.1])

$$
\begin{aligned}
\cos\left(\sqrt{k}d(t)\right) &= \cos^2(\sqrt{k}t) + \sin^2(\sqrt{k}t)\cos\angle(v, w) & \text{for } k > 0, \\
d(t)^2 &= 2t^2 - 2t^2\cos\angle(v, w) & \text{for } k = 0, \\
\cosh\left(\sqrt{-k}d(t)\right) &= \cosh^2(\sqrt{-k}t) - \sinh^2(\sqrt{-k}t)\cos\angle(v, w) & \text{for } k < 0.
\end{aligned}
$$

Observe that the dimension n does not appear in these formulas. The sectional curvature $\mathcal{K}(v, w)$ depends only on the 2-plane (in T_xM) spanned by v and w, and coincides with the Gaussian curvature at x if $n = 2$.

A more precise relation between curvature and geodesics can be described through Jacobi fields. A C^∞-vector field J along a geodesic $\gamma : [0, l] \longrightarrow M$ is called a *Jacobi field* if it solves the *Jacobi equation*

$$D_{\dot\gamma}D_{\dot\gamma}J(t) + R\big(J(t), \dot\gamma(t)\big)\dot\gamma(t) = 0 \qquad (7.4)$$

for all $t \in [0, l]$. Here, $D_{\dot\gamma}$ denotes the covariant derivative along γ, and $R : T_xM \otimes T_xM \longrightarrow T_x^*M \otimes T_xM$ is the *curvature tensor* determined by the Riemannian metric g. Another equivalent way of introducing a Jacobi field is to define it as the variational vector field $J(t) = (\partial\sigma/\partial s)(0, t)$ of some C^∞-variation $\sigma : (-\varepsilon, \varepsilon) \times [0, l] \longrightarrow M$ such that $\sigma(0, t) = \gamma(t)$ and that every

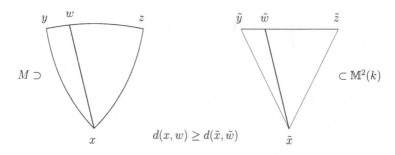

$$M \supset \qquad\qquad d(x, w) \geq d(\tilde{x}, \tilde{w}) \qquad\qquad \subset \mathbb{M}^2(k)$$

Figure 7.2

$\sigma_s := \sigma(s, \cdot)$ is geodesic. (This characterization of Jacobi fields needs only the class of geodesics, and then it is possible to regard (7.4) as the definition of R.) For linearly independent vectors $v, w \in T_x M$, the precise definition of the sectional curvature is

$$\mathcal{K}(v, w) := \frac{\langle R(w, v)v, w \rangle}{\|v\|^2 \|w\|^2 - \langle v, w \rangle^2}.$$

It might be helpful to compare (7.4) with (7.2).

Remark 1 (Alexandrov spaces). Although it is not our main subject, we briefly comment on comparison geometry involving lower bounds of the sectional curvature. As the sectional curvature is defined for each two-dimensional subspace in tangent spaces, it controls the behavior of two-dimensional subsets in M, in particular triangles. The classical *Alexandrov–Toponogov comparison theorem* asserts that $\mathcal{K} \geq k$ holds for some $k \in \mathbb{R}$ if and only if every geodesic triangle in M is thicker than the triangle with the same side lengths in $\mathbb{M}^2(k)$. See Figure 7.2 for more details, where M is of $\mathcal{K} \geq k$, and then $d(x, w) \geq d(\tilde{x}, \tilde{w})$ holds between geodesic triangles with the same side lengths ($d(x, y) = d(\tilde{x}, \tilde{y})$, $d(y, z) = d(\tilde{y}, \tilde{z})$, $d(z, x) = d(\tilde{z}, \tilde{x})$) as well as $d(y, w) = d(\tilde{y}, \tilde{w})$.

The point is that we can forget about the dimension of M because the sectional curvature cares only about two-dimensional subsets. The above triangle comparison property is written by using only distance and geodesics, so that it can be formulated for metric spaces having enough geodesics (i.e., geodesic spaces). Such spaces are called *Alexandrov spaces*, and there are deep geometric and analytic theories on them (see [BGP], [OtS], [BBI, Chapters 4, 10]). We discuss optimal transport and Wasserstein geometry on Alexandrov spaces in Section 7.8.2.

Figure 7.3

7.3.2 Ricci curvature

Given a unit vector $v \in T_x M$, we define the *Ricci curvature* of v as the trace of the sectional curvature $\mathcal{K}(v, \cdot)$,

$$\mathrm{Ric}(v) := \sum_{i=1}^{n-1} \mathcal{K}(v, e_i),$$

where $\{e_i\}_{i=1}^{n-1} \cup \{v\}$ is an orthonormal basis of $T_x M$. We will mean by $\mathrm{Ric} \geq K$ for $K \in \mathbb{R}$ that $\mathrm{Ric}(v) \geq K$ holds for all unit vectors $v \in TM$. As we discussed in the previous subsection, sectional curvature controls geodesics and distance. Ricci curvature has less information since we take the trace, and naturally controls the behavior of the measure vol_g.

The following is one of the most important theorems in comparison Riemannian geometry, which asserts that a lower bound of the Ricci curvature implies an upper bound of the volume growth. The proof is done via calculations involving Jacobi fields. Recall (7.1) for the definition of the function $s_{K,n}$.

Theorem 7.1 (Bishop–Gromov volume comparison). *Assume that* $\mathrm{Ric} \geq K$ *holds for some* $K \in \mathbb{R}$. *Then we have, for any* $x \in M$ *and* $0 < r < R$ ($\leq$ $\pi \sqrt{(n-1)/K}$ *if* $K > 0$),

$$\frac{\mathrm{vol}_g(B(x,R))}{\mathrm{vol}_g(B(x,r))} \leq \frac{\int_0^R s_{K,n}(t)^{n-1}\, dt}{\int_0^r s_{K,n}(t)^{n-1}\, dt}. \tag{7.5}$$

Proof. Given a unit vector $v \in T_x M$, we fix a unit speed minimal geodesic $\gamma : [0, l] \longrightarrow M$ with $\dot{\gamma}(0) = v$ and an orthonormal basis $\{e_i\}_{i=1}^{n-1} \cup \{v\}$ of $T_x M$. Then we consider the variation $\sigma_i : (-\varepsilon, \varepsilon) \times [0, l] \longrightarrow M$ defined by $\sigma_i(s, t) := \exp_x(tv + ste_i)$ for $i = 1, \ldots, n-1$, and introduce the Jacobi fields $\{J_i\}_{i=1}^{n-1}$ along γ given by

$$J_i(t) := \frac{\partial \sigma_i}{\partial s}(0, t) = D(\exp_x)_{tv}(te_i) \in T_{\gamma(t)} M$$

(see Figure 7.3, where $s > 0$).

Note that $J_i(0) = 0$, $D_{\dot\gamma} J_i(0) = e_i$, $\langle J_i, \dot\gamma \rangle \equiv 0$ (by the Gauss lemma) and $\langle D_{\dot\gamma} J_i, \dot\gamma \rangle \equiv 0$ (by (7.4) and $\langle R(J_i, \dot\gamma)\dot\gamma, \dot\gamma \rangle \equiv 0$). We also remark that $\gamma(t)$ is not conjugate to x for all $t \in (0, l)$ (and hence $\{J_i(t)\}_{i=1}^{n-1} \cup \{\dot\gamma(t)\}$ is a basis of $T_{\gamma(t)}M$) since γ is minimal. Hence we find an $(n-1) \times (n-1)$ matrix $\mathcal{U}(t) = (u_{ij}(t))_{i,j=1}^{n-1}$ such that $D_{\dot\gamma} J_i(t) = \sum_{j=1}^{n-1} u_{ij}(t) J_j(t)$ for $t \in (0, l)$. We define two more $(n-1) \times (n-1)$ matrices

$$\mathcal{A}(t) := \left(\langle J_i(t), J_j(t) \rangle \right)_{i,j=1}^{n-1}, \quad \mathcal{R}(t) := \left(\langle R(J_i(t), \dot\gamma(t))\dot\gamma(t), J_j(t) \rangle \right)_{i,j=1}^{n-1}.$$

Note that $\mathcal{A}$ and $\mathcal{R}$ are symmetric matrices. Moreover, we have $\mathrm{tr}(\mathcal{R}(t)\mathcal{A}(t)^{-1}) = \mathrm{Ric}(\dot\gamma(t))$ as $\mathcal{A}(t)$ is the matrix representation of the metric g in the basis $\{J_i(t)\}_{i=1}^{n-1}$ of the orthogonal complement $\dot\gamma(t)^\perp$ of $\dot\gamma(t)$. To be precise, choosing an $(n-1) \times (n-1)$ matrix $C = (c_{ij})_{i,j=1}^{n-1}$ such that $\{\sum_{j=1}^{n-1} c_{ij} J_j(t)\}_{i=1}^{n-1}$ is orthonormal, we observe $I_n = C\mathcal{A}(t)C^t$ (C^t is the transpose of C) and

$$\mathrm{Ric}\left(\dot\gamma(t)\right) = \sum_{i,j,k=1}^{n-1} \langle R(c_{ij} J_j(t), \dot\gamma(t))\dot\gamma(t), c_{ik} J_k(t) \rangle = \mathrm{tr}\left(C(t)\mathcal{R}(t)C(t)^t\right)$$

$$= \mathrm{tr}\left(\mathcal{R}(t)\mathcal{A}(t)^{-1}\right).$$

Claim 7.2. (a) *It holds that* $\mathcal{U}\mathcal{A} = \mathcal{A}\mathcal{U}^t$. *In particular, we have* $2\mathcal{U} = \mathcal{A}'\mathcal{A}^{-1}$. (b) *The matrix* $\mathcal{U}$ *is symmetric and we have* $\mathrm{tr}(\mathcal{U}^2) \geq (\mathrm{tr}\,\mathcal{U})^2/(n-1)$.

Proof. (a) The first assertion easily follows from the Jacobi equation (7.4) and the symmetry of $\mathcal{R}$; indeed,

$$\frac{d}{dt}\{\langle D_{\dot\gamma} J_i, J_j \rangle - \langle J_i, D_{\dot\gamma} J_j \rangle\} = \langle D_{\dot\gamma} D_{\dot\gamma} J_i, J_j \rangle - \langle J_i, D_{\dot\gamma} D_{\dot\gamma} J_j \rangle$$

$$= -\langle R(J_i, \dot\gamma)\dot\gamma, J_j \rangle + \langle J_i, R(J_j, \dot\gamma)\dot\gamma \rangle = 0.$$

Thus we have $\mathcal{A}' = \mathcal{U}\mathcal{A} + \mathcal{A}\mathcal{U}^t = 2\mathcal{U}\mathcal{A}$, which shows the second assertion.
   (b) Recall that

$$\frac{\partial \sigma_i}{\partial t}(0, t) = \dot\gamma(t), \quad \frac{\partial \sigma_i}{\partial s}(0, t) = J_i(t)$$

hold for $t \in (0, l)$. As $[\partial/\partial s, \partial/\partial t] = 0$, we have

$$D_{\dot\gamma} J_i(t) = D_t\left(\frac{\partial \sigma_i}{\partial s}\right)(0, t) = D_s\left(\frac{\partial \sigma_i}{\partial t}\right)(0, t).$$

Now, we introduce the function

$$f : \exp_x\left(\left\{tv + \sum_{i=1}^{n-1} s_i te_i \,\middle|\, t \in [0, l], \, |s_i| < \varepsilon\right\}\right) \longrightarrow \mathbb{R}$$

so that $f(\exp_x(tv + \sum_{i=1}^{n-1} s_i t e_i)) = t$. We derive from $\nabla f(\sigma_i(s, t)) = (\partial \sigma_i / \partial t)(s, t)$ that

$$D_s\left(\frac{\partial \sigma_i}{\partial t}\right)(0, t) = D_{J_i}(\nabla f)(\gamma(t)) = \nabla^2 f(J_i(t)),$$

where $\langle \nabla^2 f(w), w'\rangle = \operatorname{Hess} f(w, w')$. This means that $\mathcal{U}$ is the matrix presentation of the symmetric form $\nabla^2 f$ (restricted in $\dot{\gamma}^{\perp}$) with respect to the basis $\{J_i\}_{i=1}^{n-1}$. Therefore, $\mathcal{U}$ is symmetric. By denoting the eigenvalues of $\mathcal{U}$ by $\lambda_1, \ldots, \lambda_{n-1}$, the Cauchy–Schwarz inequality shows that

$$(\operatorname{tr}\mathcal{U})^2 = \left(\sum_{i=1}^{n-1} \lambda_i\right)^2 \le (n-1)\sum_{i=1}^{n-1} \lambda_i^2 = (n-1)\operatorname{tr}(\mathcal{U}^2).$$

$\square$

We calculate, by using Claim 7.2,

$$\left[(\det A)^{1/2(n-1)}\right]' = \frac{1}{2(n-1)}(\det A)^{1/2(n-1)-1} \cdot \det A \operatorname{tr}(A'A^{-1})$$

$$= \frac{1}{n-1}(\det A)^{1/2(n-1)} \operatorname{tr}\mathcal{U}.$$

Then Claim 7.2(b) yields

$$\left[(\det A)^{1/2(n-1)}\right]'' = \frac{1}{(n-1)^2}(\det A)^{1/2(n-1)}(\operatorname{tr}\mathcal{U})^2 + \frac{1}{n-1}(\det A)^{1/2(n-1)} \operatorname{tr}(\mathcal{U}')$$

$$\le \frac{1}{n-1}(\det A)^{1/2(n-1)}\{\operatorname{tr}(\mathcal{U}^2) + \operatorname{tr}(\mathcal{U}')\}.$$

We also deduce from Claim 7.2(a) and (7.4) that

$$\mathcal{U}' = \frac{1}{2}A''A^{-1} - \frac{1}{2}(A'A^{-1})^2$$

$$= \frac{1}{2}(-2\mathcal{R} + 2\mathcal{U}A\mathcal{U})A^{-1} - 2\mathcal{U}^2 = -\mathcal{R}A^{-1} - \mathcal{U}^2.$$

This implies the (matrix) *Riccati equation*

$$\mathcal{U}' + \mathcal{U}^2 + \mathcal{R}A^{-1} = 0.$$

Taking the trace gives

$$(\operatorname{tr}\mathcal{U})' + \operatorname{tr}(\mathcal{U}^2) + \operatorname{Ric}(\dot{\gamma}) = 0.$$

Thus we obtain from our hypothesis $\operatorname{Ric} \ge K$ the differential inequality

$$\left[(\det A)^{1/2(n-1)}\right]'' \le -\frac{K}{n-1}(\det A)^{1/2(n-1)}. \tag{7.6}$$

This is a version of the fundamental *Bishop comparison theorem* which plays a prominent role in comparison geometry. Comparing (7.6) with (7.2), we have

$$\frac{d}{dt}\left\{\left[(\det\mathcal{A})^{1/2(n-1)}\right]'\mathbf{s}_{K,n} - (\det\mathcal{A})^{1/2(n-1)}\mathbf{s}'_{K,n}\right\}$$

$$= \left[(\det\mathcal{A})^{1/2(n-1)}\right]''\mathbf{s}_{K,n} - (\det\mathcal{A})^{1/2(n-1)}\mathbf{s}''_{K,n} \le 0,$$

and hence $(\det\mathcal{A})^{1/2(n-1)}/\mathbf{s}_{K,n}$ is nonincreasing. Then integrating $\sqrt{\det\mathcal{A}}$ in unit vectors $v \in T_x M$ implies the *area comparison theorem*

$$\frac{\mathrm{area}_g(S(x,R))}{\mathrm{area}_g(S(x,r))} \le \frac{\mathbf{s}_{K,n}(R)^{n-1}}{\mathbf{s}_{K,n}(r)^{n-1}}, \tag{7.7}$$

where $S(x,r) := \{y \in M \mid d(x,y) = r\}$ and area_g stands for the $(n-1)$-dimensional Hausdorff measure associated with g (in other words, the volume measure of the $(n-1)$-dimensional Riemannian metric of $S(x,r)$ induced from g).

Now, we integrate (7.7) in the radial direction. Set $\mathbf{A}(t) := \mathrm{area}_g(S(x,t))$ and $\mathbf{S}(t) := \mathbf{s}_{K,n}(t)^{n-1}$, and recall that $\mathbf{A}/\mathbf{S}$ is nonincreasing. Hence we obtain the key inequality

$$\int_0^r \mathbf{A}\,dt \int_r^R \mathbf{S}\,dt \ge \frac{\mathbf{A}(r)}{\mathbf{S}(r)} \int_0^r \mathbf{S}\,dt \int_r^R \mathbf{S}\,dt \ge \int_0^r \mathbf{S}\,dt \int_r^R \mathbf{A}\,dt. \tag{7.8}$$

From here to the desired estimate (7.5) is an easy calculation as follows:

$$\mathrm{vol}_g\left(B(x,r)\right) \int_0^R \mathbf{s}_{K,n}(t)^{n-1}\,dt = \int_0^r \mathbf{A}\,dt \int_r^R \mathbf{S}\,dt + \int_0^r \mathbf{A}\,dt \int_0^r \mathbf{S}\,dt$$

$$\ge \int_0^r \mathbf{S}\,dt \int_r^R \mathbf{A}\,dt + \int_0^r \mathbf{A}\,dt \int_0^r \mathbf{S}\,dt$$

$$= \mathrm{vol}_g\left(B(x,R)\right) \int_0^r \mathbf{s}_{K,n}(t)^{n-1}\,dt.$$

$$\square$$

The sphere of radius r in the space form $\mathbb{M}^n(k)$ has area $a_n\mathbf{s}_{(n-1)k,n}(r)^{n-1}$, where a_n is the area of $\mathbb{S}^{n-1}$, and the ball of radius r has volume $a_n \int_0^r \mathbf{s}_{(n-1)k,n}(t)^{n-1}\,dt$. Thus the right-hand side of (7.5) ((7.7), respectively) coincides with the ratio of the volume of balls (the area of spheres, respectively) of radius R and r in $\mathbb{M}^n(K/(n-1))$.

Theorem 7.1 for $K > 0$ immediately implies a diameter bound. This ensures that the condition $R \le \pi\sqrt{(n-1)/K}$ in Theorem 7.1 is natural.

Corollary 7.3 (Bonnet–Myers diameter bound). *If* Ric $\geq K > 0$, *then we have*

$$\text{diam } M \leq \pi \sqrt{\frac{n-1}{K}}. \tag{7.9}$$

Proof. Put $R := \pi \sqrt{(n-1)/K}$ and assume diam $M \geq R$. Given $x \in M$, Theorem 7.1 implies that

$$\limsup_{\varepsilon \downarrow 0} \frac{\text{vol}_g(B(x,R) \setminus B(x,R-\varepsilon))}{\text{vol}_g(B(x,R))} = \limsup_{\varepsilon \downarrow 0} \left\{ 1 - \frac{\text{vol}_g(B(x,R-\varepsilon))}{\text{vol}_g(B(x,R))} \right\}$$

$$\leq \limsup_{\varepsilon \downarrow 0} \frac{\int_{R-\varepsilon}^{R} \mathbf{s}_{K,n}(t)^{n-1} \, dt}{\int_0^R \mathbf{s}_{K,n}(t)^{n-1} \, dt} = 0.$$

This shows $\text{area}_g(S(x,R)) = 0$ and hence diam $M \leq R$. To be precise, it follows from $\text{area}_g(S(x,R)) = 0$ that every point in $S(x,R)$ must be a conjugate point of x. Therefore any geodesic emanating from x is not minimal after passing through $S(x,R)$, and hence diam $M = R$. (A more direct proof in terms of metric geometry can be found in Theorem 7.14(i).)   □

The bound (7.9) is sharp, and equality is achieved only by the sphere $\mathbb{M}^n(K/(n-1))$ of radius $\sqrt{(n-1)/K}$ (compare this with Theorem 7.15).

As we mentioned in Remark 1, lower sectional curvature bounds are characterized by simple triangle comparison properties involving only distance, and there is a successful theory of metric spaces satisfying them. Then it is natural to ask the following question.

Question 1. How to characterize lower Ricci curvature bounds without using differentiable structure?

This had been a long-standing important question, and we will see an answer in Section 7.4 (Theorem 7.6). Such a condition naturally involves measure and dimension besides distance, and should be preserved under the convergence of metric measure spaces (see Section 7.5).

7.3.3 Further reading

See, for instances, [CE], [Ch], and [Sak] for the fundamentals of Riemannian geometry and comparison theorems. A property corresponding to the Bishop comparison theorem (7.6) was proposed as a lower Ricci curvature bound for metric measure spaces by Cheeger and Colding [CC] (as well as

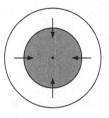
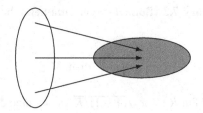

Bishop–Gromov   Brunn–Minkowski/curvature-dimension

Figure 7.4

Gromov [Gr]), and used to study the limit spaces of Riemannian manifolds
with uniform lower Ricci curvature bounds. The deep theory of such limit
spaces is one of the main motivations for asking Question 1, so that the
stability deserves a particular interest (see Section 7.5 for more details). The
systematic investigation of (7.6) in metric measure spaces has not been done
until [Oh1] and [St4], where we call this property the measure contraction
property. We will revisit this in Section 7.8.3. Here, we only remark that the
measure contraction property is strictly weaker than the curvature-dimension
condition.

7.4 A characterization of lower Ricci curvature bound
via optimal transport

The Bishop–Gromov volume comparison theorem (Theorem 7.1) can be
regarded as a concavity estimate of $\mathrm{vol}_g^{1/n}$ along the contraction of the ball
$B(x, R)$ to its center x. This is generalized to optimal transport between pairs of
uniform distributions (the Brunn–Minkowski inequality) and, moreover, pairs
of probability measures (the curvature-dimension condition). Figure 7.4 represents the difference between contraction and transport (see also Figures 7.5,
and 7.8).

    The main theorem in this section is Theorem 7.6, which asserts that, on a
weighted Riemannian manifold, the curvature-dimension condition is equivalent to a lower bound of the corresponding weighted Ricci curvature. In order
to avoid lengthy calculations, we begin with Euclidean spaces with or without
weight, and see the relation between the Brunn–Minkowski inequality and the
weighted Ricci curvature. Then the general Riemannian situation is only briefly
explained. We hope that our simplified argument will help the readers to catch
the idea of the curvature-dimension condition.

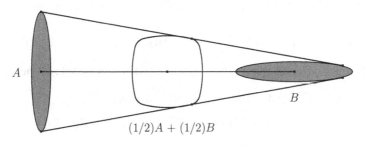

Figure 7.5

7.4.1 Brunn–Minkowski inequalities in Euclidean spaces

For later convenience, we explain fundamental facts of optimal transport theory on Euclidean spaces. Given $\mu_0, \mu_1 \in \mathcal{P}_c(\mathbb{R}^n)$ with $\mu_0 \in \mathcal{P}^{\mathrm{ac}}(\mathbb{R}^n, \mathrm{vol}_n)$, there is a convex function $f : \mathbb{R}^n \longrightarrow \mathbb{R}$ such that the map

$$\mathcal{F}_t(x) := (1 - t)x + t\nabla f(x), \quad t \in [0, 1],$$

gives the unique optimal transport from μ_0 to μ_1 (Brenier's theorem, [Br]). Precisely, $t \longmapsto \mu_t := (\mathcal{F}_t)_\sharp \mu_0$ is the unique minimal geodesic from μ_0 to μ_1 with respect to the L^2-Wasserstein distance. We remark that the convex function f is twice differentiable a.e. (Alexandrov's theorem, cf. [Vi2, Chapter 14]). Thus, ∇f makes sense and $\mathcal{F}_t$ is differentiable a.e. Moreover, the *Monge–Ampère equation*

$$\rho_1\big(\mathcal{F}_1(x)\big)\det\big(D\mathcal{F}_1(x)\big) = \rho_0(x) \tag{7.10}$$

holds for μ_0-a.e. x.

Now, we are ready for proving the classical Brunn–Minkowski inequality in the (unweighted) Euclidean space $(\mathbb{R}^n, \|\cdot\|, \mathrm{vol}_n)$. Briefly speaking, it asserts that $\mathrm{vol}_n^{1/n}$ is concave. We shall give a proof based on optimal transport theory. Given two (nonempty) sets $A, B \subset \mathbb{R}^n$ and $t \in [0, 1]$, we set

$$(1 - t)A + tB := \{(1 - t)x + ty \,|\, x \in A, \ y \in B\}.$$

(See Figure 7.5, where $(1/2)A + (1/2)B$ has much more measure than A and B.)

Theorem 7.4 (Brunn–Minkowski inequality). *For any measurable sets $A, B \subset \mathbb{R}^n$ and $t \in [0, 1]$, we have*

$$\mathrm{vol}_n\big((1 - t)A + tB\big)^{1/n} \geq (1 - t)\,\mathrm{vol}_n(A)^{1/n} + t\,\mathrm{vol}_n(B)^{1/n}. \tag{7.11}$$

Proof. We can assume that both A and B are bounded and of positive measure. The case of $\mathrm{vol}_n(A) = 0$ is easily checked by choosing a point $x \in A$, as we have

$$\mathrm{vol}_n\left((1-t)A+tB\right)^{1/n} \geq \mathrm{vol}_n\left((1-t)\{x\}+tB\right)^{1/n} = t\,\mathrm{vol}_n(B)^{1/n}.$$

If either A or B is unbounded, then applying (7.11) to bounded sets yields

$$\mathrm{vol}_n\left((1-t)\{A \cap B(0, R)\}+t\{B \cap B(0, R)\}\right)^{1/n}$$
$$\geq (1-t)\,\mathrm{vol}_n\left(A \cap B(0, R)\right)^{1/n}+t\,\mathrm{vol}_n\left(B \cap B(0, R)\right)^{1/n}.$$

We take the limit as R goes to infinity and obtain

$$\mathrm{vol}_n\left((1-t)A+tB\right)^{1/n} \geq (1-t)\,\mathrm{vol}_n(A)^{1/n}+t\,\mathrm{vol}_n(B)^{1/n}.$$

Consider the uniform distributions on A and B,

$$\mu_0 = \rho_0\,\mathrm{vol}_n := \frac{\chi_A}{\mathrm{vol}_n(A)}\,\mathrm{vol}_n, \qquad \mu_1 = \rho_1\,\mathrm{vol}_n := \frac{\chi_B}{\mathrm{vol}_n(B)}\,\mathrm{vol}_n,$$

where χ_A stands for the characteristic function of A. As μ_0 is absolutely continuous, there is a convex function $f : \mathbb{R}^n \longrightarrow \mathbb{R}$ such that the map $\mathcal{F}_1 = (1-t)\,\mathrm{Id}_{\mathbb{R}^n}+t\nabla f$, $t \in [0, 1]$, is the unique optimal transport from μ_0 to μ_1. Between the uniform distributions μ_0 and μ_1, the Monge–Ampère equation (7.10) simply means that

$$\det(D\mathcal{F}_1) = \frac{\mathrm{vol}_n(B)}{\mathrm{vol}_n(A)}$$

μ_0-a.e.

Note that $D\mathcal{F}_1 = \mathrm{Hess}\,f$ is symmetric and positive definite μ_0-a.e., since f is convex and $\det(D\mathcal{F}_1) > 0$. We shall estimate $\det(D\mathcal{F}_t) = \det((1-t)I_n + tD\mathcal{F}_1)$ from above and below. To do so, we denote the eigenvalues of $D\mathcal{F}_1$ by $\lambda_1, \ldots, \lambda_n > 0$ and apply the inequality of arithmetic and geometric means to see

$$\left\{\frac{(1-t)^n}{\det((1-t)I_n+tD\mathcal{F}_1)}\right\}^{1/n} + \left\{\frac{t^n\det(D\mathcal{F}_1)}{\det((1-t)I_n+tD\mathcal{F}_1)}\right\}^{1/n}$$
$$= \left\{\prod_{i=1}^{n}\frac{1-t}{(1-t)+t\lambda_i}\right\}^{1/n} + \left\{\prod_{i=1}^{n}\frac{t\lambda_i}{(1-t)+t\lambda_i}\right\}^{1/n}$$
$$\leq \frac{1}{n}\sum_{i=1}^{n}\left\{\frac{1-t}{(1-t)+t\lambda_i}+\frac{t\lambda_i}{(1-t)+t\lambda_i}\right\} = 1.$$

Thus, on the one hand, we have

$$\det(D\mathcal{F}_t)^{1/n} \geq (1-t)+t\det(D\mathcal{F}_1)^{1/n} = (1-t)+t\left\{\frac{\mathrm{vol}_n(B)}{\mathrm{vol}_n(A)}\right\}^{1/n}. \qquad (7.12)$$

On the other hand, the Hölder inequality and the change of variables formula yield

$$\int_A \det(D\mathcal{F}_t)^{1/n}\, d\mu_0 \leq \left(\int_A \det(D\mathcal{F}_t)\, d\mu_0\right)^{1/n} = \left(\frac{1}{\mathrm{vol}_n(A)} \int_{\mathcal{F}_t(A)} d\,\mathrm{vol}_n\right)^{1/n}.$$

Therefore, we obtain

$$\int_A \det(D\mathcal{F}_t)^{1/n}\, d\mu_0 \leq \left\{\frac{\mathrm{vol}_n(\mathcal{F}_t(A))}{\mathrm{vol}_n(A)}\right\}^{1/n} \leq \left\{\frac{\mathrm{vol}_n((1-t)A + tB)}{\mathrm{vol}_n(A)}\right\}^{1/n}.$$

Combining these, we complete the proof of (7.11). $\qquad\qquad\square$

Remark 2. We remark that the set $(1 - t)A + tB$ is not necessarily measurable (regardless of the measurability of A and B). Hence, to be precise, $\mathrm{vol}_n((1 - t)A + tB)$ is considered as an outer measure given by $\inf_W \mathrm{vol}_n(W)$, where $W \subset \mathbb{R}^n$ runs over all measurable sets containing $(1 - t)A + tB$. The same remark is applied to Theorems 7.5 and 7.10.

Next we treat the weighted case $(\mathbb{R}^n, \|\cdot\|, m)$, where $m = e^{-\psi}\,\mathrm{vol}_n$ with $\psi \in C^\infty(\mathbb{R}^n)$. Then we need to replace $1/n$ in (7.11) with $1/N$ for some $N \in (n, \infty)$, and the analog of (7.11) leads us to an important condition on ψ.

Theorem 7.5 (Brunn–Minkowski inequality with weight). *Take $N \in (n, \infty)$. A weighted Euclidean space $(\mathbb{R}^n, \|\cdot\|, m)$ with $m = e^{-\psi}\,\mathrm{vol}_n$, $\psi \in C^\infty(\mathbb{R}^n)$, satisfies*

$$m\big((1-t)A + tB\big)^{1/N} \geq (1-t)m(A)^{1/N} + tm(B)^{1/N} \tag{7.13}$$

for all measurable sets $A, B \subset \mathbb{R}^n$ and all $t \in [0, 1]$ if and only if

$$\mathrm{Hess}\,\psi(v, v) - \frac{\langle \nabla\psi(x), v\rangle^2}{N - n} \geq 0 \tag{7.14}$$

holds for all unit vectors $v \in T_x\mathbb{R}^n$.

Proof. We first prove that (7.14) implies (7.13). Similar to Theorem 7.4, we assume that A and B are bounded and of positive measure, and set

$$\mu_0 := \frac{\chi_A}{m(A)}m, \qquad \mu_1 := \frac{\chi_B}{m(B)}m.$$

We again find a convex function $f : \mathbb{R}^n \longrightarrow \mathbb{R}$ such that $\mu_t := (\mathcal{F}_t)_\sharp\mu_0$ is the minimal geodesic from μ_0 to μ_1, where $\mathcal{F}_t := (1 - t)\,\mathrm{Id}_{\mathbb{R}^n} + t\nabla f$. Instead of $\det(D\mathcal{F}_t)$, we consider

$$\det_m\big(D\mathcal{F}_t(x)\big) := e^{\psi(x) - \psi(\mathcal{F}_t(x))}\det\big(D\mathcal{F}_t(x)\big).$$

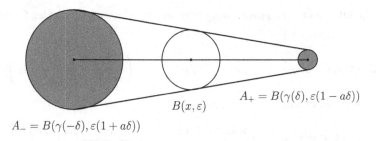

$$A_- = B(\gamma(-\delta), \varepsilon(1 + a\delta))$$

$$B(x, \varepsilon)$$

$$A_+ = B(\gamma(\delta), \varepsilon(1 - a\delta))$$

Figure 7.6

The coefficient $e^{\psi(x)-\psi(\mathcal{F}_t(x))}$ represents the ratio of the weights at x and $\mathcal{F}_t(x)$. As in Theorem 7.4 (see, especially, (7.12)), it is sufficient to show the concavity of $\det_m(D\mathcal{F}_t(x))^{1/N}$ to derive the desired inequality (7.13). Fix $x \in A$ and put

$$\gamma(t) := \mathcal{F}_t(x), \quad \Phi_m(t) := \det_m\big(D\mathcal{F}_t(x)\big)^{1/N}, \quad \Phi(t) := \det\big(D\mathcal{F}_t(x)\big)^{1/n}.$$

On the one hand, it is proved in (7.12) that $\Phi(t) \geq (1-t)\Phi(0) + t\Phi(1)$. On the other hand, the assumption (7.14) implies that $e^{-\psi(\mathcal{F}_t(x))/(N-n)}$ is a concave function in t. These together imply (7.13) via the Hölder inequality. To be precise, we have

$$\Phi_m(t) = e^{\{\psi(x)-\psi(\mathcal{F}_t(x))\}/N}\Phi(t)^{n/N}$$
$$\geq e^{\psi(x)/N}\{(1-t)e^{-\psi(x)/(N-n)} + te^{-\psi(\mathcal{F}_1(x))/(N-n)}\}^{(N-n)/N}$$
$$\times \{(1-t)\Phi(0) + t\Phi(1)\}^{n/N},$$

and then the Hölder inequality yields

$$\Phi_m(t) \geq e^{\psi(x)/N}\{(1-t)e^{-\psi(x)/N}\Phi(0)^{n/N} + te^{-\psi(\mathcal{F}_1(x))/N}\Phi(1)^{n/N}\}$$
$$= (1-t)\Phi_m(0) + t\Phi_m(1).$$

To see the converse, we fix an arbitrary unit vector $v \in T_x\mathbb{R}^n$ and set $\gamma(t) := x + tv$ for $t \in \mathbb{R}$ and $a := \langle\nabla\psi(x), v\rangle/(N-n)$. Given $\varepsilon > 0$ and $\delta \in \mathbb{R}$ with $\varepsilon, |\delta| \ll 1$, we consider two open balls (see Figure 7.6, where $a\delta > 0$)

$$A_+ := B\big(\gamma(\delta), \varepsilon(1 - a\delta)\big), \qquad A_- := B\big(\gamma(-\delta), \varepsilon(1 + a\delta)\big).$$

Note that $A_+ = A_- = B(x, \varepsilon)$ for $\delta = 0$ and that $(1/2)A_- + (1/2)A_+ = B(x, \varepsilon)$. We also observe that

$$m(A_\pm) = e^{-\psi(\gamma(\pm\delta))}c_n\varepsilon^n(1 \mp a\delta)^n + O(\varepsilon^{n+1}),$$

where $c_n = \mathrm{vol}_n(B(0,1))$ and $O(\varepsilon^{n+1})$ is independent of δ. Applying (7.13) to $A_\pm$ with $t = 1/2$, we obtain

$$m\big(B(x,\varepsilon)\big) \geq \frac{1}{2^N}\{m(A_-)^{1/N} + m(A_+)^{1/N}\}^N. \qquad (7.15)$$

We know that $m(B(x,\varepsilon)) = e^{-\psi(x)}c_n\varepsilon^n + O(\varepsilon^{n+1})$. In order to estimate the right-hand side, we calculate

$$\frac{\partial^2}{\partial \delta^2}\Big[e^{-\psi(\gamma(\delta))/N}(1 - a\delta)^{n/N}\Big]\Big|_{\delta=0}$$

$$= \left\{-\frac{\mathrm{Hess}\,\psi(v,v)}{N} + \frac{\langle\nabla\psi,v\rangle^2}{N^2} + 2\frac{\langle\nabla\psi,v\rangle}{N}\frac{n}{N}a + \frac{n}{N}\Big(\frac{n}{N} - 1\Big)a^2\right\}e^{-\psi(x)/N}$$

$$= \left\{-\mathrm{Hess}\,\psi(v,v) + \frac{\langle\nabla\psi,v\rangle^2}{N-n} - \frac{n}{N(N-n)}\big((N-n)a - \langle\nabla\psi,v\rangle\big)^2\right\}\frac{e^{-\psi(x)/N}}{N}.$$

Owing to the choice of $a = \langle\nabla\psi(x),v\rangle/(N - n)$ (as the maximizer), we have

$$\frac{\partial^2}{\partial \delta^2}\Big[e^{-\psi(\gamma(\delta))/N}(1 - a\delta)^{n/N}\Big]\Big|_{\delta=0} = \left\{\frac{\langle\nabla\psi(x),v\rangle^2}{N-n} - \mathrm{Hess}\,\psi(v,v)\right\}\frac{e^{-\psi(x)/N}}{N}.$$

Thus, we find, by the Taylor expansion of $e^{-\psi(\gamma(\delta))/N}(1 - a\delta)^{n/N}$ at $\delta = 0$,

$$\frac{m(A_-)^{1/N} + m(A_+)^{1/N}}{(c_n\varepsilon^n)^{1/N}}$$

$$= 2e^{-\psi(x)/N} - \left\{\mathrm{Hess}\,\psi(v,v) - \frac{\langle\nabla\psi,v\rangle^2}{N-n}\right\}\frac{e^{-\psi(x)/N}}{N}\delta^2 + O(\delta^4) + O(\varepsilon).$$

Hence, we obtain by letting ε go to zero in (7.15) that

$$e^{-\psi(x)} \geq \frac{1}{2^N}\left[2e^{-\psi(x)/N} - \left\{\mathrm{Hess}\,\psi(v,v) - \frac{\langle\nabla\psi,v\rangle^2}{N-n}\right\}\frac{e^{-\psi(x)/N}}{N}\delta^2 + O(\delta^4)\right]^N$$

$$= e^{-\psi(x)}\left[1 - \frac{1}{2}\left\{\mathrm{Hess}\,\psi(v,v) - \frac{\langle\nabla\psi,v\rangle^2}{N-n}\right\}\delta^2\right] + O(\delta^4).$$

Therefore, we conclude

$$\mathrm{Hess}\,\psi(v,v) - \frac{\langle\nabla\psi(x),v\rangle^2}{N-n} \geq 0.$$

$$\square$$

Applying (7.13) to $A = \{x\}$, $B = B(x,R)$, and $t = r/R$ implies

$$\frac{m(B(x,R))}{m(B(x,r))} \leq \Big(\frac{R}{r}\Big)^N \qquad (7.16)$$

for all $x \in \mathbb{R}^n$ and $0 < r < R$. Thus, compared with Theorem 7.1, $(\mathbb{R}^n, \| \cdot \|, m)$ satisfying (7.14) behaves like an "N-dimensional" space of nonnegative Ricci curvature (see Theorem 7.12 for a more general theorem in terms of the curvature-dimension condition).

7.4.2 Characterizing lower Ricci curvature bounds

Now we switch to the weighted Riemannian situation (M, g, m), where $m = e^{-\psi} \text{vol}_g$ with $\psi \in C^\infty(M)$. Ricci curvature controls vol_g as we saw in Section 7.3, and Theorem 7.5 suggests that the quantity

$$\text{Hess}\, \psi(v, v) - \frac{\langle \nabla \psi, v \rangle^2}{N - n}$$

has an essential information in controlling the effect of the weight. Their combination indeed gives the weighted Ricci curvature as follows (cf. [BE], [Qi], [Lo1]).

Definition 1 (weighted Ricci curvature). Given a unit tangent vector $v \in T_x M$ and $N \in [n, \infty]$, the *weighted Ricci curvature* $\text{Ric}_N(v)$ is defined by

(1) $\text{Ric}_n(v) := \begin{cases} \text{Ric}(v) + \text{Hess}\, \psi(v, v) & \text{if } \langle \nabla \psi(x), v \rangle = 0, \\ -\infty & \text{otherwise}; \end{cases}$

(2) $\text{Ric}_N(v) := \text{Ric}(v) + \text{Hess}\, \psi(v, v) - \dfrac{\langle \nabla \psi(x), v \rangle^2}{N - n}$ for $N \in (n, \infty)$;

(3) $\text{Ric}_\infty(v) := \text{Ric}(v) + \text{Hess}\, \psi(v, v)$.

We say that $\text{Ric}_N \geq K$ holds for $K \in \mathbb{R}$ if $\text{Ric}_N(v) \geq K$ holds for all unit vectors $v \in TM$.

Note that $\text{Ric}_N \leq \text{Ric}_{N'}$ holds for $n \leq N \leq N' < \infty$. Ric_∞ is also called the *Bakry–Émery tensor*. If the weight is trivial in the sense that ψ is constant, then Ric_N coincides with Ric for all $N \in [n, \infty]$. One of the most important examples possessing nontrivial weight is the following.

Example 1 (Euclidean spaces with log-concave measures). Consider a weighted Euclidean space $(\mathbb{R}^n, \| \cdot \|, m)$ with $m = e^{-\psi} \text{vol}_n$, $\psi \in C^\infty(\mathbb{R}^n)$. Then clearly $\text{Ric}_\infty(v) = \text{Hess}\, \psi(v, v)$, thus $\text{Ric}_\infty \geq 0$ if ψ is convex. The most typical and important example satisfying $\text{Ric}_\infty \geq K > 0$ is the Gaussian measure

$$m = \left(\frac{K}{2\pi}\right)^{n/2} e^{-K\|x\|^2/2}\, \text{vol}_n, \qquad \psi(x) = \frac{K}{2}\|x\|^2 + \frac{n}{2}\log\left(\frac{2\pi}{K}\right).$$

Note that $\text{Hess}\, \psi \geq K$ holds independently of the dimension n.

Before stating the main theorem of the section, we mention that optimal transport in a Riemannian manifold is described in the same manner as the Euclidean spaces (due to [Mc2], [CMS1]). Given $\mu_0, \mu_1 \in \mathcal{P}_c(M)$ with $\mu_0 = \rho_0 \operatorname{vol}_g \in \mathcal{P}^{\mathrm{ac}}(M, \operatorname{vol}_g)$, there is a $(d^2/2)$-*convex function* $f : M \longrightarrow \mathbb{R}$ such that $\mu_t := (\mathcal{F}_t)_\sharp \mu_0$ with $\mathcal{F}_t(x) := \exp_x[t\nabla f(x)]$, $t \in [0, 1]$, gives the unique minimal geodesic from μ_0 to μ_1. We do not give the definition of $(d^2/2)$-convex functions, but only remark that they are twice differentiable a.e. Furthermore, the absolute continuity of μ_0 implies that μ_t is absolutely continuous for all $t \in [0, 1)$, so that we can set $\mu_t = \rho_t \operatorname{vol}_g$. Since $\mathcal{F}_t$ is differentiable μ_0-a.e., we can consider the Jacobian $\|(D\mathcal{F}_t)_x\|$ (with respect to vol_n) which satisfies the Monge–Ampère equation

$$\rho_0(x) = \rho_t\big(\mathcal{F}_t(x)\big)\|(D\mathcal{F}_t)_x\| \tag{7.17}$$

for μ_0-a.e. x.

We next introduce two entropy functionals. Given $N \in [n, \infty)$ and an absolutely continuous probability measure $\mu = \rho m \in \mathcal{P}^{\mathrm{ac}}(M, m)$, we define the *Rényi entropy* as

$$S_N(\mu) := -\int_M \rho^{1-1/N}\, dm. \tag{7.18}$$

We also define the *relative entropy* with respect to the reference measure m by

$$\operatorname{Ent}_m(\mu) := \int_M \rho \log \rho\, dm. \tag{7.19}$$

Note that Ent_m has the opposite sign to the Boltzmann entropy. The domain of these functionals will be extended in Section 7.5 ((7.25), (7.26)) to probability measures possibly with nontrivial singular part. In this section, however, we consider only absolutely continuous measures for the sake of simplicity. As any two points in $\mathcal{P}_c^{\mathrm{ac}}(M, m)$ are connected by a unique minimal geodesic contained in $\mathcal{P}_c^{\mathrm{ac}}(M, m)$, the convexity of S_N and Ent_m in $\mathcal{P}_c^{\mathrm{ac}}(M, m)$ makes sense.

Recall (7.3) for the definition of the function $\beta_{K,N}^t$. The following theorem is due to von Renesse, Sturm and many others, see this section's Further reading for more details.

Theorem 7.6 (a characterization of Ricci curvature bound). *For a weighted Riemannian manifold (M, g, m) with $m = e^{-\psi} \operatorname{vol}_g$, $\psi \in C^\infty(M)$, we have $\operatorname{Ric}_N \geq K$ for some $K \in \mathbb{R}$ and $N \in [n, \infty)$ if and only if any pair of measures $\mu_0 = \rho_0 m$, $\mu_1 = \rho_1 m \in \mathcal{P}_c^{\mathrm{ac}}(M, m)$ satisfies*

$$S_N(\mu_t) \leq -(1 - t) \int_{M \times M} \beta_{K,N}^{1-t}\big(d(x, y)\big)^{1/N} \rho_0(x)^{-1/N}\, d\pi(x, y)$$

$$- t \int_{M \times M} \beta_{K,N}^{t}\big(d(x, y)\big)^{1/N} \rho_1(y)^{-1/N}\, d\pi(x, y) \tag{7.20}$$

for all $t \in (0, 1)$, where $(\mu_t)_{t \in [0,1]} \subset \mathcal{P}_c^{ac}(M, m)$ is the unique minimal geodesic from μ_0 to μ_1 in the L^2-Wasserstein space $(\mathcal{P}_2(M), d_2^W)$, and π is the unique optimal coupling of μ_0 and μ_1.

Similarly, $\mathrm{Ric}_\infty \geq K$ is equivalent to

$$\mathrm{Ent}_m(\mu_t) \leq (1 - t)\mathrm{Ent}_m(\mu_0) + t\,\mathrm{Ent}_m(\mu_1) - \frac{K}{2}(1 - t)t d_2^W(\mu_0, \mu_1)^2.$$
(7.21)

Outline of proof. We give a sketch of the proof for $N < \infty$ along the lines of [St4] and [LV1]. The case of $N = \infty$ goes along the essentially same line.

First, we assume $\mathrm{Ric}_N \geq K$. Fix $\mu_0 = \rho_0 m$, $\mu_1 = \rho_1 m \in \mathcal{P}_c^{ac}(M, m)$ and take a $(d^2/2)$-convex function $f : M \longrightarrow \mathbb{R}$ such that $\mathcal{F}_t(x) := \exp_x[t\nabla f(x)]$, $t \in [0, 1]$, provides the unique minimal geodesic $\mu_t = \rho_t m = (\mathcal{F}_t)_\sharp \mu_0$ from μ_0 to μ_1. Taking the weight $e^{-\psi}$ into account, we introduce the Jacobian $\mathbf{J}_t^\psi(x) := e^{\psi(x) - \psi(\mathcal{F}_t(x))} \|(D\mathcal{F}_t)_x\|$ with respect to m (like $\det_m$ in Theorem 7.5). Then it follows from the Monge–Ampère equation (7.17) with respect to vol_g that

$$\rho_0(x) = \rho_t(\mathcal{F}_t(x))\mathbf{J}_t^\psi(x)$$
(7.22)

for μ_0-a.e. x (i.e., the Monge–Ampère equation with respect to m).

Now, the essential point is that optimal transport is performed along geodesics $t \longmapsto \exp_x[t\nabla f(x)] = \mathcal{F}_t(x)$. Therefore, its variational vector fields are Jacobi fields (recall (7.4)), and controlled by Ricci curvature. Together with the weight control as in Theorem 7.5, calculations somewhat similar to (but more involved than) Theorem 7.1 shows our key inequality

$$\mathbf{J}_t^\psi(x)^{1/N} \geq (1 - t)\beta_{K,N}^{1-t}\big(d(x, \mathcal{F}_1(x))\big)^{1/N} + t\beta_{K,N}^t\big(d(x, \mathcal{F}_1(x))\big)^{1/N}\mathbf{J}_1^\psi(x)^{1/N}.$$
(7.23)

This inequality can be thought of as an infinitesimal version of the Brunn–Minkowski inequality (see (7.13) and Theorem 7.10(i) as well). As the change of variables formula and the Monge–Ampère equation (7.22) yield

$$S_N(\mu_t) = -\int_M \rho_t(\mathcal{F}_t)^{1-1/N}\mathbf{J}_t^\psi\, dm = -\int_M \left(\frac{\mathbf{J}_t^\psi}{\rho_0}\right)^{1/N} d\mu_0,$$

we obtain from (7.23) (and (7.22) again) that

$$S_N(\mu_t) \leq -(1 - t)\int_M \frac{\beta_{K,N}^{1-t}(d(x, \mathcal{F}_1(x)))^{1/N}}{\rho_0(x)^{1/N}}\, d\mu_0(x)$$
$$- t\int_M \frac{\beta_{K,N}^t(d(x, \mathcal{F}_1(x)))^{1/N}}{\rho_1(\mathcal{F}_1(x))^{1/N}}\, d\mu_0(x).$$

This is the desired inequality (7.20), for $\pi = (\mathrm{Id}_M \times \mathcal{F}_1)_\sharp \mu_0$.

Second, we assume (7.20). Then applying it to uniform distributions on balls (as in the proof of Theorem 7.5) shows $\mathrm{Ric}_N \geq K$. More precisely, we use the generalized Brunn–Minkowski inequality (Theorem 7.10) instead of (7.13). $\square$

As $\beta^t_{0,N} \equiv 1$, the inequality (7.20) is simplified into the convexity of S_N

$$S_N(\mu_t) \leq (1-t)S_N(\mu_0) + tS_N(\mu_1)$$

when $K = 0$. For $K \neq 0$, however, the K-convexity of S_N

$$S_N(\mu_t) \leq (1-t)S_N(\mu_0) + tS_N(\mu_1) - \frac{K}{2}(1-t)t d_2^W(\mu_0, \mu_1)^2$$

turns out uninteresting (see [St2, Theorem 1.3]). This is a reason why we need to consider a more subtle inequality like (7.20).

Theorem 7.6 gives an answer to Question 1, for the conditions (7.20), (7.21) are written in terms of only distance and measure, without using the differentiable structure. Then it is interesting to consider these conditions for general metric measure spaces as "synthetic Ricci curvature bounds", and we should verify the stability. We discuss them in the next section.

7.4.3 Further reading

We refer to [AGS], [Vi1], and [Vi2, Part I] for the basics of optimal transport theory and Wasserstein geometry. McCann's [Mc2] fundamental result on the shape of optimal transport maps is generalized to not necessarily compactly supported measures in [FF] and [FG] (see also [Vi2, Chapter 10]).

See [Ga] and [Le, Section 2.2] for the Brunn–Minkowski inequality and related topics. The Bakry–Émery tensor Ric_∞ was introduced in [BE], and its generalization Ric_N is due to Qian [Qi]. See also [Lo1] for geometric and topological applications, [Mo, Chapter 18] and the references therein for minimal surface theory in weighted manifolds (which are called *manifolds with density* there).

After McCann's [Mc1] pinoneering work on the convexity of the relative entropy along geodesics in the Wasserstein space (called the *displacement convexity*) over Euclidean spaces, Cordero–Erausquin et al. [CMS1] first showed that $\mathrm{Ric} \geq 0$ implies (7.21) with $K = 0$ in unweighted Riemannian manifolds. They [CMS2] further proved that $\mathrm{Ric}_\infty \geq K$ implies (7.21) in the weighted situation. Then Theorem 7.6 is due to von Renesse and Sturm [vRS], [St2] for $N = \infty$, and independently to Sturm [St3], [St4] and Lott and Villani [LV2], [LV1] for $N < \infty$.

We comment on recent work on a variant of (7.20). Studied in [BaS1] is the following inequality (called the *reduced curvature-dimension condition*):

$$S_N(\mu_t) \leq -(1-t) \int_{M \times M} \beta_{K,N+1}^{1-t} \big(d(x,y)\big)^{1/N} \rho_0(x)^{-1/N} \, d\pi(x,y)$$

$$- t \int_{M \times M} \beta_{K,N+1}^{t} \big(d(x,y)\big)^{1/N} \rho_1(y)^{-1/N} \, d\pi(x,y). \qquad (7.24)$$

Note the difference between

$$t\beta_{K,N}^{t}(r)^{1/N} = t^{1/N} \left(\frac{\mathsf{s}_{K,N}(tr)}{\mathsf{s}_{K,N}(r)} \right)^{1-1/N}, \qquad t\beta_{K,N+1}^{t}(r)^{1/N} = \frac{\mathsf{s}_{K,N+1}(tr)}{\mathsf{s}_{K,N+1}(r)}.$$

We remark that (7.24) coincides with (7.20) when $K = 0$, and is weaker than (7.20) for general $K \neq 0$. The condition (7.24) is also equivalent to $\mathrm{Ric}_N \geq K$ for Riemannian manifolds. In the setting of metric measure spaces, (7.24) has some advantages such as the tensorization and the localization properties (see Section 7.8.3 for more details). One drawback is that, as it is weaker than (7.20), (7.24) derives slightly worse estimates than (7.20) (in the Bishop–Gromov volume comparison (Theorem 7.12), the Bonnet–Myers diameter bound (Theorem 7.14), the Lichnerowicz inequality (Theorem 7.16), etc.). Nevertheless, such weaker estimates are sufficient for several topological applications.

See also [St2] and [OT] for the K-convexity of generalized entropies (or free energies) and its characterization and applications. It is discussed in [St2, Theorem 1.7] that there is a class of functionals whose K-convexity is equivalent to $\mathrm{Ric} \geq K$ and $\dim \leq N$ for unweighted Riemannian manifolds. The choice of a functional is by no means unique, and it is unclear how this observation relates to the curvature-dimension condition.

7.5 The curvature-dimension condition and stability

Motivated by Theorem 7.6, we introduce the curvature-dimension condition for metric measure spaces and show that it is stable under the measured Gromov–Hausdorff convergence. In this and the next sections, (X, d, m) will always be a metric measure space in the sense of Section 7.2.

7.5.1 The curvature-dimension condition

We can regard the conditions (7.20) and (7.21) as convexity estimates of the functionals S_N and Ent_m. For the sake of consistency with the monotonicity of Ric_N in N ($\mathrm{Ric}_N \leq \mathrm{Ric}_{N'}$ for $N \leq N'$), we introduce important classes of functionals (due to McCann [Mc1]) including S_N and Ent_m.

For $N \in [1, \infty)$, denote by $\mathcal{DC}_N$ (*displacement convexity class*) the set of continuous convex functions $U : [0, \infty) \longrightarrow \mathbb{R}$ such that $U(0) = 0$ and that the function $\varphi(s) = s^N U(s^{-N})$ is convex on $(0, \infty)$. Similarly, define $\mathcal{DC}_\infty$ as the set of continuous convex functions $U : [0, \infty) \longrightarrow \mathbb{R}$ such that $U(0) = 0$ and that $\varphi(s) = e^s U(e^{-s})$ is convex on $\mathbb{R}$. In both cases, the convexity of U implies that φ is nonincreasing. Observe the monotonicity, $\mathcal{DC}_{N'} \subset \mathcal{DC}_N$ holds for $1 \leq N \leq N' \leq \infty$.

For $\mu \in \mathcal{P}(X)$, using its Lebesgue decomposition $\mu = \rho m + \mu^s$ into absolutely continuous and singular parts, we set

$$U_m(\mu) := \int_X U(\rho) \, dm + U'(\infty)\mu^s(X), \qquad U'(\infty) := \lim_{r \to \infty} \frac{U(r)}{r}.$$

Note that $U'(\infty)$ indeed exists as $U(r)/r$ is nondecreasing. In the case where $U'(\infty) = \infty$, we set $\infty \cdot 0 := 0$ by convention. The most important element of $\mathcal{DC}_N$ is the function $U(r) = Nr(1 - r^{-1/N})$, which induces the Rényi entropy (7.18) in a slightly deformed form as

$$U_m(\mu) = N \int_X \rho(1 - \rho^{-1/N}) \, dm + N\mu^s(X) = N\left(1 - \int_X \rho^{1-1/N} \, dm\right).$$

$$(7.25)$$

This extends (7.18) to whole $\mathcal{P}(X)$. Letting N go to infinity, we have $U(r) = r \log r \in \mathcal{DC}_\infty$ as well as the relative entropy (extending (7.19))

$$U_m(\mu) = \int_X \rho \log \rho \, dm + \infty \cdot \mu^s(X). \qquad (7.26)$$

Let us denote by $\Gamma(X)$ the set of minimal geodesics $\gamma : [0, 1] \longrightarrow X$ endowed with the distance

$$d_{\Gamma(X)}(\gamma_1, \gamma_2) := \sup_{t \in [0,1]} d_X(\gamma_1(t), \gamma_2(t)).$$

Define the evaluation map $e_t : \Gamma(X) \longrightarrow X$ for $t \in [0, 1]$ as $e_t(\gamma) := \gamma(t)$, and note that this is 1-Lipschitz. A probability measure $\Pi \in \mathcal{P}(\Gamma(X))$ is called a *dynamical optimal transference plan* if $\pi := (e_0 \times e_n)_\sharp \Pi$ is an optimal coupling of $\alpha(0)$ and $\alpha(n)$. Then, the curve $\alpha(t) := (e_t)_\sharp \Pi$, $t \in [0, 1]$, is a minimal geodesic in $(\mathcal{P}_2(X), d_2^W)$. We remark that Π is not uniquely determined by α and π; that is to say, different plans Π and Π' could generate the same minimal geodesic α and optimal coupling π. If (X, d) is locally compact (and hence proper), then any minimal geodesic in $\mathcal{P}_2(X)$ is associated with a (not necessarily unique) dynamical optimal transference plan ([LV2, Proposition 2.10], [Vi2, Corollary 7.22]).

Now we are ready to present the precise definition of the curvature-dimension condition in the form due to Lott and Villani (after Sturm and others, see Further reading of this and the previous sections).

Definition 2 (the curvature-dimension condition). Suppose that $m(B(x, r)) \in (0, \infty)$ holds for all $x \in X$ and $r \in (0, \infty)$. For $K \in \mathbb{R}$ and $N \in (1, \infty]$, we say that a metric measure space (X, d, m) satisfies the *curvature-dimension condition* CD(K, N) if, for any $\mu_0 = \rho_0 m + \mu_0^s$, $\mu_1 = \rho_1 m + \mu_1^s \in \mathcal{P}_b(X)$, there exists a dynamical optimal transference plan $\Pi \in \mathcal{P}(\Gamma(X))$ associated with a minimal geodesic $\alpha(t) = (e_t)_\sharp \Pi$, $t \in [0, 1]$, from μ_0 to μ_1 and an optimal coupling $\pi = (e_0 \times e_1)_\sharp \Pi$ of μ_0 and μ_1 such that we have

$$
U_m\big(\alpha(t)\big) \leq (1-t) \int_{X \times X} \beta_{K,N}^{1-t}(d(x, y)) U\left(\frac{\rho_0(x)}{\beta_{K,N}^{1-t}(d(x, y))}\right) d\pi_x(y) dm(x)
$$
$$
+ t \int_{X \times X} \beta_{K,N}^{t}(d(x, y)) U\left(\frac{\rho_1(y)}{\beta_{K,N}^{t}(d(x, y))}\right) d\pi_y(x) dm(y)
$$
$$
+ U'(\infty)\{(1-t)\mu_0^s(X) + t\mu_1^s(X)\} \tag{7.27}
$$

for all $U \in \mathcal{DC}_N$ and $t \in (0, 1)$, where π_x and π_y denote disintegrations of π by μ_0 and μ_1; i.e., $d\pi(x, y) = d\pi_x(y) d\mu_0(x) = d\pi_y(x) d\mu_1(y)$.

In the special case of $K = 0$, as $\beta_{0,N}^t \equiv 1$, the inequality (7.27) means the convexity of U_m

$$
U_m\big(\alpha(t)\big) \leq (1-t) U_m(\mu_0) + t U_m(\mu_1),
$$

without referring to the optimal coupling π. In the case where both μ_0 and μ_1 are absolutely continuous, we have $d\pi(x, y) = \rho_0(x) d\pi_x(y) dm(x) = \rho_1(y) d\pi_y(x) dm(y)$ and hence (7.27) is rewritten in a more symmetric form as

$$
U_m\big(\alpha(t)\big) \leq (1-t) \int_{X \times X} \frac{\beta_{K,N}^{1-t}(d(x, y))}{\rho_0(x)} U\left(\frac{\rho_0(x)}{\beta_{K,N}^{1-t}(d(x, y))}\right) d\pi(x, y)
$$
$$
+ t \int_{X \times X} \frac{\beta_{K,N}^{t}(d(x, y))}{\rho_1(y)} U\left(\frac{\rho_1(y)}{\beta_{K,N}^{t}(d(x, y))}\right) d\pi(x, y). \tag{7.28}
$$

Note that choosing $U(r) = Nr(1 - r^{-1/N})$ and $U(r) = r \log r$ in (7.28) reduce to (7.20) and (7.21), respectively.

Remark 3. (a) It is easily checked that, if (X, d, m) satisfies CD(K, N), then the scaled metric measure space $(X, cd, c'm)$ for $c, c' > 0$ satisfies CD(K/c^2, N).

(b) In Definition 2, to be precise, we need to impose the condition

$$
m\big(X \setminus B(x, \pi \sqrt{(N-1)/K})\big) = 0
$$

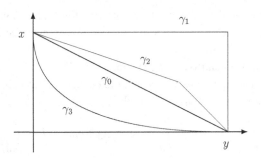

Figure 7.7

for all $x \in X$ if $K > 0$ and $N < \infty$, in order to stay inside the domain of $\beta_{K,N}^t$. Nevertheless, this is always the case by virtue of the generalized Bonnet–Myers theorem (Theorem 7.14) below.

(c) Recall that $U(r)/r$ is nondecreasing, and observe that $\beta_{K,N}^t(r)$ is increasing in K and decreasing in N. Combining this with the monotonicity $\mathcal{DC}_{N'} \subset \mathcal{DC}_N$ for $N \leq N'$ (and (b) above), we see that $\mathrm{CD}(K, N)$ implies $\mathrm{CD}(K', N')$ for all $K' \leq K$ and $N' \geq N$. Therefore, in the condition $\mathrm{CD}(K, N)$, K represents a lower bound of the Ricci curvature and N represents an upper bound of the dimension.

(d) The validity of (7.27) along only "some" geodesics is essential to establish the stability. In fact, if we impose it for all geodesics between μ_0 and μ_1, then it is not stable under convergence (in the sense of Theorem 7.9). This is because, when a sequence $\{(X_i, d_i)\}_{i \in \mathbb{N}}$ converges to the limit space (X, d), there may be a geodesic in X which cannot be represented as the limit of a sequence of geodesics in X_i. Therefore, the convexity along geodesics in X_i does not necessarily imply the convexity along all geodesics in X. One typical example is a sequence of ℓ_p^n-spaces as p goes to 1 or ∞. Only straight lines are geodesics in ℓ_p^n with $1 < p < \infty$, while ℓ_1^n and ℓ_∞^n have many more geodesics. (See Figure 7.7, where γ_i for all $i = 0, \ldots, 3$ are geodesics for ℓ_1^2, while only the straight-line segment γ_0 is a geodesic for ℓ_p^2 with $1 < p < \infty$.)

(e) In Riemannian manifolds or, more generally, nonbranching proper metric measure spaces, we can reduce (7.27) to a special case from two aspects as follows. If (7.28) holds for $U(r) = Nr(1 - r^{-1/N})$ (or $U(r) = r \log r$ if $N = \infty$) and all measures in $\mathcal{P}_b^{ac}(X, m)$ (and hence in $\mathcal{P}_c^{ac}(X, m)$) with continuous densities, then (7.27) holds for all $U \in \mathcal{DC}_N$ and all measures in $\mathcal{P}_b(X)$ ([St4, Proposition 4.2], [LV2, Proposition 3.21, Lemma 3.24]). In this sense, (7.20) and (7.21) are essential among the class of inequalities (7.27). A geodesic space is said to be nonbranching if geodesics in it do not branch (see Section 7.8.1 for the precise definition).

(f) In Riemannian manifolds, (7.23) implies (7.27) for all $U \in \mathcal{DC}_N$. Indeed, $\alpha(t) = \rho_t m$ is absolutely continuous and the change of variables formula and the Monge–Ampère equation (7.22) imply

$$U_m\big(\alpha(t)\big) = \int_M U(\rho_t)\, dm = \int_M U\big(\rho_t(\mathcal{F}_t)\big)\mathbf{J}_t^{\psi}\, dm = \int_M U\left(\frac{\rho_0}{\mathbf{J}_t^{\psi}}\right)\frac{\mathbf{J}_t^{\psi}}{\rho_0}\, d\mu_0.$$

For $N < \infty$, as $\varphi(s) = s^N U(s^{-N})$ is nonincreasing and convex, (7.23) yields

$$
\begin{aligned}
U_m\big(\alpha(t)\big) &\le \int_M \varphi\Bigg((1-t)\frac{\beta_{K,N}^{1-t}(d(x,\mathcal{F}_1(x)))^{1/N}}{\rho_0(x)^{1/N}} \\
&\qquad + t\frac{\beta_{K,N}^{t}(d(x,\mathcal{F}_1(x)))^{1/N}}{\rho_1(\mathcal{F}_1(x))^{1/N}}\Bigg)\, d\mu_0(x) \\
&\le \int_M \Bigg\{(1-t)\varphi\left(\frac{\beta_{K,N}^{1-t}(d(x,\mathcal{F}_1(x)))^{1/N}}{\rho_0(x)^{1/N}}\right) \\
&\qquad + t\varphi\left(\frac{\beta_{K,N}^{t}(d(x,\mathcal{F}_1(x)))^{1/N}}{\rho_1(\mathcal{F}_1(x))^{1/N}}\right)\Bigg\}\, d\mu_0(x).
\end{aligned}
$$

The case of $N = \infty$ is similar. This means that the infinitesimal expression of $CD(K, N)$ is always (7.23) whatever U is, and various ways of integration give rise to the definition of $CD(K, N)$ involving $\mathcal{DC}_N$.

Although $\mathbf{J}_t^{\psi}$ relies on the differentiable structure of M, it is possible to rewrite (7.23) through (7.22) as

$$
\begin{aligned}
\rho_t\big(\mathcal{F}_t(x)\big)^{-1/N} &\ge (1-t)\beta_{K,N}^{1-t}\big(d(x,\mathcal{F}_1(x))\big)^{1/N}\rho_0(x)^{-1/N} \\
&\qquad + t\beta_{K,N}^{t}\big(d(x,\mathcal{F}_1(x))\big)^{1/N}\rho_1\big(\mathcal{F}_1(x)\big)^{-1/N}
\end{aligned}
$$

(see [St4, Proposition 4.2(iv)]). This makes sense in metric measure spaces; however, the integrated inequalities (i.e., (7.27), (7.28)) are more convenient for verifying the stability.

(g) The role of the dynamical optimal transference plan Π may seem unclear in Definition 2, as only α and π appear in (7.27). We use only α and π also in applications in Section 7.6. As we shall see in Theorem 7.9, it is the stability in which Π plays a crucial role.

7.5.2 Stability and geometric background

As we mentioned in Remark 1, one geometric motivation behind the curvature-dimension condition is the theory of Alexandrov spaces. That is to say, we would like to find a way of formulating and investigating singular spaces of Ricci curvature bounded below in some sense (recall Question 1). Then, what kind of singular spaces should we consider? Here comes into play another deep theory of the precompactness with respect to the convergence of metric (measure) spaces. Briefly speaking, the precompactness ensures that a sequence of Riemannian manifolds with a uniform lower Ricci curvature bound contains a convergent subsequence. Such a limit space is not a manifold any more, but should inherit some properties. In order to make use of the curvature-dimension condition in the limit, we need to establish that it is preserved under the convergence (actually, the limit of Alexandrov spaces is again an Alexandrov space).

We say that a map $\varphi : Y \longrightarrow X$ between metric spaces is ε-*approximating* for $\varepsilon > 0$ if

$$|d_X(\varphi(y), \varphi(z)) - d_Y(y, z)| \le \varepsilon$$

holds for all $y, z \in Y$ and if $\overline{B(\varphi(Y), \varepsilon)} = X$.

Definition 3 (measured Gromov–Hausdorff convergence). Consider a sequence of metric measure spaces $\{(X_i, d_i, m_i)\}_{i \in \mathbb{N}}$ and another metric measure space (X, d, m).

(1) (Compact case) Assume that (X_i, d_i) for all $i \in \mathbb{N}$ and (X, d) are compact. We say that $\{(X_i, d_i, m_i)\}_{i \in \mathbb{N}}$ converges to (X, d, m) in the sense of the *measured Gromov–Hausdorff convergence* if there are sequences of positive numbers $\{\varepsilon_i\}_{i \in \mathbb{N}}$ and Borel maps $\{\varphi_i : X_i \longrightarrow X\}_{i \in \mathbb{N}}$ such that $\lim_{i \to \infty} \varepsilon_i = 0$, φ_i is an ε_i-approximating map, and that $(\varphi_i)_\sharp m_i$ weakly converges to m.

(2) (Noncompact case) Assume that (X_i, d_i) for all $i \in \mathbb{N}$ and (X, d) are proper, and fix base points $x_i \in X_i$ and $x \in X$. We say that $\{(X_i, d_i, m_i, x_i)\}_{i \in \mathbb{N}}$ converges to (X, d, m, x) in the sense of the *pointed measured Gromov–Hausdorff convergence* if, for all $R > 0$, $\{(\overline{B(x_i, R)}, d_i, m_i)\}_{i \in \mathbb{N}}$ converges to $(\overline{B(x, R)}, d, m)$ in the sense of the measured Gromov–Hausdorff convergence (as in (1) above).

If we consider only distance structures (X_i, d_i) and (X, d) and remove the weak convergence condition on φ_i, then it is the *(pointed) Gromov–Hausdorff convergence* under which the lower sectional curvature bound in the sense of Alexandrov is known to be preserved. The following observation ([LV2, Proposition 4.1]) says that the Gromov–Hausdorff convergence of a sequence of metric spaces is propagated to the Wasserstein spaces over them.

Proposition 7.7. *If a sequence of compact metric spaces $\{(X_i, d_i)\}_{i \in \mathbb{N}}$ converges to a compact metric space (X, d) in the sense of the Gromov–Hausdorff convergence, then so does the sequence of Wasserstein spaces $\{(\mathcal{P}(X_i), d_2^W)\}_{i \in \mathbb{N}}$ to $(\mathcal{P}(X), d_2^W)$.*

More precisely, ε_i-approximating maps $\varphi_i : X_i \longrightarrow X$ give rise to $\theta(\varepsilon_i)$-approximating maps $(\varphi_i)_\sharp : \mathcal{P}(X_i) \longrightarrow \mathcal{P}(X)$ such that θ is a universal function satisfying $\lim_{\varepsilon \downarrow 0} \theta(\varepsilon) = 0$.

In the noncompact case, the pointed Gromov–Hausdorff convergence of $\{(X_i, d_i, x_i)\}_{i \in \mathbb{N}}$ to (X, d, x) similarly implies the Gromov–Hausdorff convergence of $\{(\mathcal{P}(\overline{B(x_i, R)}), d_2^W)\}_{i \in \mathbb{N}}$ to $(\mathcal{P}(\overline{B(x, R)}), d_2^W)$ for all $R > 0$ (instead of the convergence of $(\overline{B(\delta_{x_i}, R)}, d_2^W)$ to $(\overline{B(\delta_x, R)}, d_2^W)$).

The following inspiring precompactness theorem is established by Gromov [Gr, Section 5.A] for the Gromov–Hausdorff convergence, and extended by Fukaya [Fu] to the measured case.

Theorem 7.8 (Gromov–Fukaya precompactness). *Let $\{(M_i, g_i, \operatorname{vol}_{g_i}, x_i)\}_{i \in \mathbb{N}}$ be a sequence of pointed Riemannian manifolds such that*

$$\operatorname{Ric}_{g_i} \geq K, \qquad \dim M_i \leq N$$

uniformly hold for some $K \in \mathbb{R}$ and $N \in \mathbb{N}$. Then it contains a subsequence which is convergent to some pointed proper metric measure space (X, d, m, x) in the sense of the pointed measured Gromov–Hausdorff convergence.

To be more precise, we choose a complete space as the limit, and then the properness follows from our hypotheses $\operatorname{Ric} \geq K$ and $\dim \leq N$. The key ingredient of the proof is the Bishop–Gromov volume comparison (Theorem 7.1) from which we derive an upper bound of the doubling constant $\sup_{x \in M, r \leq R} \operatorname{vol}_g(B(x, 2r)) / \operatorname{vol}_g(B(x, r))$ for each $R \in (0, \infty)$.

By virtue of Theorem 7.8, starting from a sequence of Riemannian manifolds with a uniform lower Ricci curvature bound, we find the limit space of some subsequence. Such a limit space is not a manifold any more, but should have some inherited properties. The stability of the curvature-dimension condition ensures that we can use it for the investigation of these limit spaces.

We remark that the measures of balls $\operatorname{vol}_{g_i}(B(x_i, R))$ for $i \in \mathbb{N}$ have a uniform upper bound depending only on K, N, and R thanks to Theorem 7.1. However, it could tend to zero, and then we cannot obtain any information on (X, d, m). Therefore, we should take a scaling $c_i \operatorname{vol}_{g_i}$ with some appropriate constant $c_i > 1$. It does not change anything because the weighted Ricci curvature is invariant under scalings of the measure (the weight function of $\tilde{m} = cm$

is $\psi_{\tilde{m}} = \psi_m - \log c$). By the same reasoning, it is natural to assume that any bounded open ball has a finite positive measure in the next theorem (see also Remark 3(a)).

Theorem 7.9 (stability). *Assume that a sequence of pointed proper metric measure spaces $\{(X_i, d_i, m_i, x_i)\}_{i\in\mathbb{N}}$ uniformly satisfies* $\mathsf{CD}(K, N)$ *for some $K \in \mathbb{R}$ and $N \in (1, \infty]$ and that it converges to a pointed proper metric measure space (X, d, m, x) in the sense of the pointed measured Gromov–Hausdorff convergence. If, moreover, $0 < m(B(x, r)) < \infty$ holds for all $x \in X$ and $r \in (0, \infty)$, then (X, d, m) satisfies* $\mathsf{CD}(K, N)$.

Outline of proof. The proof of stability goes as follows (along the lines of [LV2], [LV1]). First of all, as we consider only measures with bounded (and hence compact) support in Definition 2, we can restrict ourselves to measures with continuous density and compact support. Indeed, it implies by approximation the general case ([LV2, Proposition 3.21, Lemma 3.24], see also Remark 3(e)). Given continuous measures $\mu = \rho m$, $\nu = \sigma m \in \mathcal{P}_c^{\mathrm{ac}}(X, m)$ and ε_i-approximating maps $\varphi_i : X_i \longrightarrow X$ as in Definition 3, we consider

$$\mu_i = \frac{\rho \circ \varphi_i}{\int_{X_i} \rho \circ \varphi_i \, dm_i} \cdot m_i, \quad \nu_i = \frac{\sigma \circ \varphi_i}{\int_{X_i} \sigma \circ \varphi_i \, dm_i} \cdot m_i \in \mathcal{P}_c^{\mathrm{ac}}(X_i, m_i)$$

and take a dynamical optimal transference plan $\Pi_i \in \mathcal{P}(\Gamma(X_i))$ from μ_i to ν_i satisfying (7.28). Note that $(\varphi_i)_\sharp \mu_i$ and $(\varphi_i)_\sharp \nu_i$ weakly converge to μ and ν, respectively, thanks to the continuity of ρ and σ.

By a compactness argument ([LV1, Theorem A.45]), extracting a subsequence if necessary, Π_i converges to some dynamical transference plan $\Pi \in \mathcal{P}(\Gamma(X))$ from μ to ν such that, setting

$$\alpha_i(t) := (e_t)_\sharp \Pi_i, \quad \alpha(t) := (e_t)_\sharp \Pi, \quad \pi_i := (e_0 \times e_1)_\sharp \Pi_i, \quad \pi := (e_0 \times e_1)_\sharp \Pi,$$

$(\varphi_i)_\sharp \alpha_i$ and $(\varphi_i \times \varphi_i)_\sharp \pi_i$ weakly converge to α and π, respectively. Then it follows from Proposition 7.7 that α is a minimal geodesic from μ to ν and that π is an optimal coupling of μ and ν.

On the one hand, the right-hand side of (7.28) for π_i converges to that for π by virtue of the continuous densities. On the other hand, the monotonicity

$$U_{(\varphi_i)_\sharp m_i}\big((\varphi_i)_\sharp [\alpha_i(t)]\big) \leq U_{m_i}\big(\alpha_i(t)\big)$$

and the lower semi-continuity

$$U_m(\alpha(t)) \leq \liminf_{i\to\infty} U_{(\varphi_i)_\sharp m_i}\big((\varphi_i)_\sharp [\alpha_i(t)]\big)$$

hold true in general ([LV2, Theorem B.33]). Therefore, we obtain (7.28) for
Π and complete the proof.   □

7.5.3 Further reading

The definition of the curvature-dimension condition is much indebted to
McCann's influential work [Mc1] introducing the important class of functions
$\mathcal{DC}_N$ as well as the displacement convexity along geodesics in the Wasserstein
space (see Further reading in Section 7.4.3). Otto and Villani's work [OV]
on the relation between such convexity of the entropy and several functional
inequalities was also inspiring.

The term "curvature-dimension condition" is used by Sturm [St3], [St4] (and
also in [Vi2]) following Bakry and Émery's celebrated work [BE]. Sturm's condition requires that (7.27) is satisfied for all absolutely continuous measures and
$U = S_{N'}$ for all $N' \in [N, \infty]$. Lott and Villani [LV2], [LV1], independently of
Sturm, introduced the condition as in Theorem 7.6 and call it N-*Ricci curvature*
bounded from below by K. These conditions are equivalent in nonbranching
spaces (see Remark 3(e) and Section 7.8.1). In locally compact nonbranching
spaces, it is also possible to extend (7.27) from compactly supported measures
to not necessarily compactly supported measures (see [FV]).

See [Fu], [Gr, Chapter 3, Section 5.A] and [BBI, Chapters 7 and 8] for the
basics of (measured) Gromov–Hausdorff convergence and for precompactness
theorems. The stability under the measured Gromov–Hausdorff convergence
we presented above is due to Lott and Villani [LV2], [LV1]. Sturm [St3],
[St4] also proved the stability with respect to a different (his own) notion
of convergence induced from his **D**-*distance* between metric measure spaces.
Roughly speaking, the **D**-distance takes couplings not only for measures, but
also for distances (see [St3] for more details).

We also refer to the celebrated work of Cheeger and Colding [CC] (mentioned in Further reading in Section 7.3.3) for related geometric approach
toward the investigation of limit spaces of Riemannian manifolds of Ricci
curvature bounded below. Their strategy is to fully use the fact that it is the
limit of Riemannian manifolds. They reveal the detailed local structure of such
limit spaces; however, it also turns out that the limit spaces can have highly
wild structures (see the survey [We] and the references therein). We can not
directly extend Cheeger and Colding's theory to metric measure spaces with
the curvature-dimension condition. Their key tool is the Cheeger–Gromoll-
type splitting theorem, but Banach spaces prevent us applying it under the
curvature-dimension condition (see Section 7.6.3.3 for more details).

7.6 Geometric applications

Metric measure spaces satisfying the curvature-dimension condition $\mathrm{CD}(K, N)$ enjoy many properties common to "spaces of dimension $\leq N$ and Ricci curvature $\geq K$." To be more precise, though $N \in (1, \infty]$ is not necessarily an integer, we will obtain estimates numerically extended to noninteger N. Proofs based on optimal transport theory themselves are interesting and inspiring. Although we concentrate on geometric applications in this chapter, there are also many analytic applications, including the Talagrand inequality, logarithmic Sobolev inequality (and hence the normal concentration of measures), global Poincaré inequality, and so forth (see [LV2], [LV1]).

7.6.1 Generalized Brunn–Minkowski inequality and applications

Our first application is a generalization of the Brunn–Minkowski inequality (7.11) and (7.13) to curved spaces. This follows from the curvature-dimension condition (7.28) applied to S_N and Ent_m (i.e., (7.20) and (7.21)) between uniform distributions on two measurable sets. In the particular case of $K = 0$, we obtain the concavity of $m^{1/N}$ or $\log m$ as in (7.11) and (7.13). Given two sets $A, B \subset X$ and $t \in (0, 1)$, we denote by $Z_t(A, B)$ the set of points $\gamma(t)$ such that $\gamma : [0, 1] \longrightarrow X$ is a minimal geodesic with $\gamma(0) \in A$ and $\gamma(1) \in B$. We remark that $Z_t(A, B)$ is not necessarily measurable regardless of the measurability of A and B; however, it is not a problem because m is regular (see Remark 2).

The following theorem is essentially contained in von Renesse and Sturm [vRS] for $N = \infty$, and due to Sturm [St4] for $N < \infty$. Again, we will be implicitly indebted to Theorem 7.14 that guarantees that $\mathrm{diam}\, X \leq \pi\sqrt{(N-1)/K}$ if $K > 0$ and $N < \infty$ (see Remark 3(b)). Figure 7.8 represents a rough image of the theorem: $Z_{1/2}(A, B)$ has more measure in a positively curved space and less measure in a negatively curved space (compare this with Figure 7.1).

Theorem 7.10 (generalized Brunn–Minkowski inequality). *Take a metric measure space* (X, d, m) *satisfying* $\mathrm{CD}(K, N)$ *and two measurable sets* A, $B \subset X$.

(i) If $N \in (1, \infty)$*, then we have*

$$m\big(Z_t(A, B)\big)^{1/N} \geq (1-t) \inf_{x \in A,\, y \in B} \beta_{K,N}^{1-t}\big(d(x, y)\big)^{1/N} \cdot m(A)^{1/N}$$
$$+ t \inf_{x \in A,\, y \in B} \beta_{K,N}^{t}\big(d(x, y)\big)^{1/N} \cdot m(B)^{1/N}$$

for all $t \in (0, 1)$.

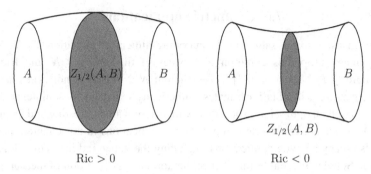

$$\mathrm{Ric} > 0 \qquad\qquad\qquad \mathrm{Ric} < 0$$

Figure 7.8

(ii) If $N = \infty$ and $0 < m(A), m(B) < \infty$, then we have

$$\log m\big(Z_t(A, B)\big)$$

$$\geq (1 - t)\log m(A) + t\log m(B) + \frac{K}{2}(1 - t)t d_2^W\left(\frac{\chi_A}{m(A)}m, \frac{\chi_B}{m(B)}m\right)^2$$

for all $t \in (0, 1)$.

Proof. Similar to Theorem 7.4, we can assume that A and B are bounded and of positive measure. Set

$$\mu_0 := \frac{\chi_A}{m(A)} \cdot m, \quad \mu_1 := \frac{\chi_B}{m(B)} \cdot m, \quad \hat{\beta}^t := \inf_{x \in A,\, y \in B} \beta_{K,N}^t\big(d(x, y)\big).$$

(i) We consider $U(r) = Nr(1 - r^{-1/N})$ and recall from (7.25) that, for $\mu = \rho m + \mu^s$,

$$U_m(\mu) = N\left(1 - \int_X \rho^{1-1/N}\, dm\right).$$

Hence, it follows from $\mathrm{CD}(K, N)$ that there is a minimal geodesic $\alpha : [0, 1] \longrightarrow \mathcal{P}(X)$ from μ_0 to μ_1 as well as an optimal coupling π such that, for all $t \in (0, 1)$,

$$-\int_X \rho_t^{1-1/N}\, dm \leq -(1 - t)\int_{X \times X} \{m(A)\beta_{K,N}^{1-t}\big(d(x, y)\big)\}^{1/N}\, d\pi(x, y)$$

$$-t\int_{X \times X} \{m(B)\beta_{K,N}^t\big(d(x, y)\big)\}^{1/N}\, d\pi(x, y)$$

$$\leq -(1 - t)(\hat{\beta}^{1-t})^{1/N}m(A)^{1/N} - t(\hat{\beta}^t)^{1/N}m(B)^{1/N},$$

where we set $\alpha(t) = \rho_t m + \mu_t^s$. Then the Hölder inequality yields

$$\int_X \rho_t^{-1/N} \cdot \rho_t \, dm \leq \left(\int_{\text{supp}\,\rho_t} \rho_t^{-1} \cdot \rho_t \, dm \right)^{1/N}$$

$$= m(\text{supp}\,\rho_t)^{1/N} \leq m\big(Z_t(A, B)\big)^{1/N}.$$

This completes the proof for $N < \infty$.

(ii) We argue similarly and obtain from (7.21) that

$$\text{Ent}_m\big(\alpha(t)\big) \leq -(1-t)\log m(A) - t\log m(B) - \frac{K}{2}(1-t)t d_2^W(\mu_0, \mu_1)^2.$$

Note that, since $\text{Ent}_m(\alpha(t)) < \infty$, $\alpha(t)$ is absolutely continuous and written as $\alpha(t) = \rho_t m$. Furthermore, Jensen's inequality applied to the convex function $s \longmapsto s\log s$ shows

$$\text{Ent}_m\big(\alpha(t)\big) = m(\text{supp}\,\rho_t) \int_{\text{supp}\,\rho_t} \rho_t \log \rho_t \, \frac{dm}{m(\text{supp}\,\rho_t)}$$

$$\geq m(\text{supp}\,\rho_t) \int_{\text{supp}\,\rho_t} \rho_t \frac{dm}{m(\text{supp}\,\rho_t)} \cdot \log \left(\int_{\text{supp}\,\rho_t} \rho_t \frac{dm}{m(\text{supp}\,\rho_t)} \right)$$

$$= -\log m(\text{supp}\,\rho_t) \geq -\log m\big(Z_t(A, B)\big).$$

We complete the proof. □

As a corollary, we find that m has no atom unless X consists of a single point.

Corollary 7.11. *If (X, d, m) contains more than two points and if it satisfies CD(K, N) with some $K \in \mathbb{R}$ and $N \in (1, \infty]$, then any one point set $\{x\} \subset X$ has null measure.*

Proof. It is sufficient to show the case of $N = \infty$ (due to Remark 3(c)). Put $A = \{x\}$ and assume that $m(\{x\}) > 1$ (due to Remark 3(a)). Take $r > 0$ with $m(B(x, 2r) \setminus B(x, r)) > 0$ (it is the case for small $r > 0$), and note that

$$Z_t\big(\{x\}, B(x, 2r) \setminus B(x, r)\big) \subset B(x, 2tr) \setminus B(x, tr)$$

for all $t \in (0, 1)$. Thus we apply Theorem 7.10(ii) with $t = 2^{-k}$, $k \in \mathbb{N}$, and find

$$\log m\big(B(x, 2^{1-k}r) \setminus B(x, 2^{-k}r)\big) \geq (1 - 2^{-k})\log\big(m(\{x\})\big)$$

$$+ 2^{-k}\log\big(m(B(x, 2r) \setminus B(x, r))\big)$$

$$- \frac{|K|}{2}(1 - 2^{-k})2^{-k}(2r)^2.$$

Summing this up in $k \in \mathbb{N}$, we observe

$$\sum_{k=1}^{\infty} \log m\big(B(x, 2^{1-k}r) \setminus B(x, 2^{-k}r)\big) = \infty.$$

This is a contradiction since we have

$$\sum_{k=1}^{\infty} \log m\big(B(x, 2^{1-k}r) \setminus B(x, 2^{-k}r)\big) \le \sum_{k=1}^{\infty} m\big(B(x, 2^{1-k}r) \setminus B(x, 2^{-k}r)\big)$$

$$= m\big(B(x, 2r) \setminus \{x\}\big) < \infty.$$

$\square$

Under $\mathrm{CD}(K, N)$ of the finite dimension $N < \infty$, applying Theorem 7.10(i) to thin annuli shows a generalization of the Bishop–Gromov volume comparison theorem (Theorem 7.1; see also (7.16)).

Theorem 7.12 (generalized Bishop–Gromov volume comparison). *Suppose that a metric measure space (X, d, m) satisfies $\mathrm{CD}(K, N)$ with $K \in \mathbb{R}$ and $N \in (1, \infty)$. Then we have*

$$\frac{m(B(x, R))}{m(B(x, r))} \le \frac{\int_0^R s_{K,N}(t)^{N-1}\, dt}{\int_0^r s_{K,N}(t)^{N-1}\, dt}$$

for all $x \in X$ and $0 < r < R\, (\le \pi\sqrt{(N-1)/K}$ if $K > 0)$.

Proof. The proof is essentially the same as the Riemannian case. We apply Theorem 7.10(i) to concentric thin annuli and obtain an estimate corresponding to the Bishop area comparison of concentric spheres (7.7). Then we take the sum and the limit (instead of integration), and obtain the theorem.

We give more detailed calculation for thoroughness. For any annulus $B(x, r_2) \setminus B(x, r_1)$ and $t \in (0, 1)$, Theorem 7.10(i) and Corollary 7.11 yield that

$$m\big(B(x, tr_2) \setminus B(x, tr_1)\big) \ge t^N \inf_{d \in [r_1, r_2]} \left(\frac{s_{K,N}(td)}{t s_{K,N}(d)}\right)^{N-1} m\big(B(x, r_2) \setminus B(x, r_1)\big)$$

$$\ge t \cdot \frac{\inf_{d \in [r_1, r_2]} s_{K,N}(td)^{N-1}}{\sup_{d \in [r_1, r_2]} s_{K,N}(d)^{N-1}} m\big(B(x, r_2) \setminus B(x, r_1)\big).$$

$$(7.29)$$

This corresponds to (7.7) in the Riemannian case. Set $h(t) := s_{K,N}(t)^{N-1}$ for brevity, and put $t_L := (r/R)^{1/L} < 1$ for $L \in \mathbb{N}$. Applying (7.29) to $r_1 = t_L r$,

$r_2 = r$, and $t = t_L^{l-1}$ for $l \in \mathbb{N}$, we have

$$m\big(B(x,r)\big) = \sum_{l=1}^{\infty} m\big(B(x, t_L^{l-1}r) \setminus B(x, t_L^{l}r)\big)$$

$$\geq \left\{ \sum_{l=1}^{\infty} t_L^{l-1} \frac{\inf_{d\in[t_Lr,r]} h(t_L^{l-1}d)}{\sup_{d\in[t_Lr,r]} h(d)} \right\} m\big(B(x,r) \setminus B(x, t_L r)\big).$$

We similarly deduce from (7.29) with $r_1 = t_L^{-L}r$, $r_2 = t_L^{l-1-L}r$, and $t = t_L^{L-l+1}$ for $l = 1, \ldots, L$ that

$$m\big(B(x,r) \setminus B(x, t_L r)\big) \sum_{l=1}^{L} t_L^{l-1} \sup_{d\in[t_L^{-L}r, t_L^{l-1-L}r]} h(d)$$

$$\geq t_L^{L} \inf_{d\in[t_Lr,r]} h(d) \sum_{l=1}^{L} m\big(B(x, t_L^{l-1-L}r) \setminus B(x, t_L^{l-L}r)\big)$$

$$= \frac{r}{R} \inf_{d\in[t_Lr,r]} h(d) \cdot m\big(B(x, R) \setminus B(x, r)\big).$$

Combining these, we obtain

$$m\big(B(x,r)\big) \cdot \sum_{l=1}^{L} (t_L^{l-1} - t_L^{l})R \sup_{d\in[t_L R, R]} h(t_L^{l-1}d)$$

$$= (1 - t_L)R \cdot m\big(B(x,r)\big) \cdot \sum_{l=1}^{L} t_L^{l-1} \sup_{d\in[t_L^{-L}r, t_L^{l-1-L}r]} h(d)$$

$$\geq (1 - t_L)R \cdot \left\{ \sum_{l=1}^{\infty} t_L^{l-1} \frac{\inf_{d\in[t_Lr,r]} h(t_L^{l-1}d)}{\sup_{d\in[t_Lr,r]} h(d)} \right\}$$

$$\cdot \frac{r}{R} \inf_{d\in[t_Lr,r]} h(d) \cdot m\big(B(x, R) \setminus B(x, r)\big)$$

$$\geq m\big(B(x, R) \setminus B(x, r)\big) \cdot \frac{\inf_{d\in[t_Lr,r]} h(d)}{\sup_{d\in[t_Lr,r]} h(d)} \sum_{l=1}^{\infty} (t_L^{l-1} - t_L^{l})r \inf_{d\in[t_Lr,r]} h(t_L^{l-1}d).$$

Letting L diverge to infinity shows

$$m\big(B(x,r)\big) \int_r^R s_{K,N}(t)^{N-1} \, dt \geq m\big(B(x, R) \setminus B(x, r)\big) \int_0^r s_{K,N}(t)^{N-1} \, dt.$$

$$(7.30)$$

This corresponds to (7.8) in the Riemannian case, and the same calculation as the last step of the proof of Theorem 7.1 completes the proof. $\qquad\square$

Theorem 7.12 shows that the doubling constant $\sup_{x \in X, r \le R} m(B(x, 2r))/m(B(x, r))$ is bounded for each $R \in (0, \infty)$; therefore, X is proper (see also the paragraph following Theorem 7.8).

Corollary 7.13. *Assume that* (X, d, m) *satisfies* $\mathsf{CD}(K, N)$ *for some* $K \in \mathbb{R}$ *and* $N \in (1, \infty)$. *Then* (X, d) *is proper.*

Next we generalize the Bonnet–Myers diameter bound (Corollary 7.3). We remark that the proof below uses Theorem 7.10(i) only for pairs of a point $A = \{x\}$ and a set $B \subset B(x, \pi \sqrt{(N-1)/K})$, so that it is consistent with Remark 3(b). The following proof is due to [Oh1].

Theorem 7.14 (generalized Bonnet–Myers diameter bound). *Suppose that a metric measure space* (X, d, m) *satisfies* $\mathsf{CD}(K, N)$ *with* $K > 0$ *and* $N \in (1, \infty)$. *Then we have the following*:

(i) *It holds that* $\operatorname{diam} X \le \pi \sqrt{(N-1)/K}$.
(ii) *Each* $x \in X$ *has at most one point of distance* $\pi \sqrt{(N-1)/K}$ *from* x.

Proof. It is enough to consider the case $K = N - 1$ thanks to the scaling property Remark 3(a).

(i) Suppose that there is a pair of points $x, y \in X$ with $d(x, y) > \pi$, set $\delta := d(x, y) - \pi > 0$ and take a minimal geodesic $\gamma : [0, \pi + \delta] \longrightarrow X$ from x to y. Choosing a different point on γ if necessary, we can assume $\delta < \pi/2$. For $\varepsilon \in (0, \delta)$, we apply Theorem 7.10(i) between $\{\gamma(\delta + \varepsilon)\}$ and $B(y, \varepsilon)$ with $t = (\pi - \delta - \varepsilon)/\pi$ and obtain

$$
\frac{m(Z_t(\{\gamma(\delta + \varepsilon)\}, B(y, \varepsilon)))}{m(B(y, \varepsilon))} \ge t^N \inf_{r \in (\pi - 2\varepsilon, \pi)} \left(\frac{\sin(tr)}{t \sin r} \right)^{N-1}
$$
$$
= t \left(\frac{\sin(t(\pi - 2\varepsilon))}{\sin(\pi - 2\varepsilon)} \right)^{N-1}.
$$

Then it follows from $t(\pi - 2\varepsilon) \le t\pi = \pi - \delta - \varepsilon$ that

$$
\frac{m(Z_t(\{\gamma(\delta + \varepsilon)\}, B(y, \varepsilon)))}{m(B(y, \varepsilon))} \ge \frac{\pi - \delta - \varepsilon}{\pi} \left(\frac{\sin(\delta + \varepsilon)}{\sin 2\varepsilon} \right)^{N-1} \to \infty \quad (7.31)
$$

as ε tends to zero. Given $z \in B(y, \varepsilon)$, we take a minimal geodesic $\eta : [0, 1] \longrightarrow X$ from $\gamma(\delta + \varepsilon)$ to z (see Figure 7.9), and derive from the triangle inequality that

$$
d(\gamma(\delta + \varepsilon), \eta(t)) = td(\gamma(\delta + \varepsilon), z) < t\{d(\gamma(\delta + \varepsilon), y) + \varepsilon\} = \pi - \delta - \varepsilon.
$$

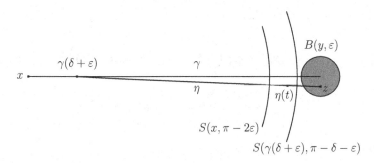

Figure 7.9

Moreover, we deduce from $d(\gamma(\delta + \varepsilon), z) < \pi$ that

$$d\big(x, \eta(t)\big) \geq d(x, z) - d\big(z, \eta(t)\big) > \pi + \delta - \varepsilon - (1-t)d\big(\gamma(\delta + \varepsilon), z\big)$$
$$> \pi + \delta - \varepsilon - (1-t)\pi = \pi - 2\varepsilon.$$

Thus, we have

$$Z_t\big(\{\gamma(\delta + \varepsilon)\}, B(y, \varepsilon)\big) \subset B\big(\gamma(\delta + \varepsilon), \pi - \delta - \varepsilon\big) \setminus B(x, \pi - 2\varepsilon)$$
$$\subset B(x, \pi) \setminus B(x, \pi - 2\varepsilon).$$

Combining this with (7.31), we conclude

$$\lim_{\varepsilon \downarrow 0} \frac{m(B(x, \pi) \setminus B(x, \pi - 2\varepsilon))}{m(B(y, \varepsilon))} = \infty.$$

Furthermore, (7.30) and Theorem 7.12 show that

$$m\big(B(x, \pi) \setminus B(x, \pi - 2\varepsilon)\big) \leq \frac{\int_{\pi-2\varepsilon}^{\pi} \sin^{N-1} r \, dr}{\int_0^{\pi-2\varepsilon} \sin^{N-1} r \, dr} m\big(B(x, \pi - 2\varepsilon)\big)$$

$$= \frac{\int_0^{2\varepsilon} \sin^{N-1} r \, dr}{\int_0^{\pi-2\varepsilon} \sin^{N-1} r \, dr} m\big(B(x, \pi - 2\varepsilon)\big) \leq m\big(B(x, 2\varepsilon)\big).$$

Hence we have, again due to Theorem 7.12 (with $K = 0$),

$$m\big(B(x, \pi) \setminus B(x, \pi - 2\varepsilon)\big) \leq m\big(B(x, 2\varepsilon)\big) \leq 2^N m\big(B(x, \varepsilon)\big). \qquad (7.32)$$

Therefore, we obtain $\lim_{\varepsilon \downarrow 0} m(B(x, \varepsilon))/m(B(y, \varepsilon)) = \infty$. This is a contradiction because we can exchange the roles of x and y.

(ii) We first see that $m(S(x, \pi)) = 0$, where $S(x, \pi) := \{y \in X \mid d(x, y) = \pi\}$. Given small $\varepsilon > 0$, take $\{x_i\}_{i=1}^k \subset S(x, 2\varepsilon)$ such that $S(x, 2\varepsilon) \subset \bigcup_{i=1}^k B(x_i, 2\varepsilon)$ and $d(x_i, x_j) \geq 2\varepsilon$ holds if $i \neq j$. Then, for any $y \in S(x, \pi)$, there is some x_i so that $d(y, x_i) < (\pi - 2\varepsilon) + 2\varepsilon = \pi$, while $d(y, x_i) \geq$

$\pi - 2\varepsilon$ holds in general. Thus, we have

$$m\big(S(x,\pi)\big) \leq m\left(\bigcup_{i=1}^{k} B(x_i,\pi) \setminus B(x_i,\pi-2\varepsilon)\right)$$

$$\leq \sum_{i=1}^{k} m\big(B(x_i,\pi) \setminus B(x_i,\pi-2\varepsilon)\big).$$

Then it follows from (7.32) that

$$m\big(S(x,\pi)\big) \leq \sum_{i=1}^{k} m\big(B(x_i,2\varepsilon)\big) \leq 2^N \sum_{i=1}^{k} m\big(B(x_i,\varepsilon)\big)$$

$$= 2^N m\left(\bigcup_{i=1}^{k} B(x_i,\varepsilon)\right) \leq 2^N m\big(B(x,3\varepsilon)\big).$$

Letting ε go to zero shows $m(S(x,\pi)) = 0$.

Now we suppose that there are two mutually distinct points $y, z \in X$ such that $d(x,y) = d(x,z) = \pi$. On the one hand, we derive from (7.32) that

$$m\big(B(x,r)\big) \geq m(B(x,\pi) \setminus B\big(x,\pi-r\big))$$

for $r \in (0, \pi/2)$. On the other hand, as $B(y,r) \subset X \setminus B(x,\pi-r)$ and $m(S(x,\pi)) = 0$, we find

$$m\big(B(y,r)\big) \leq m\big(B(x,\pi) \setminus B(x,\pi-r)\big).$$

Hence, we obtain $m(B(x,r)) \geq m(B(y,r))$ and similarly $m(B(y,r)) \geq m(B(x,r))$. This implies

$$m\big(B(x,r)\big) = m\big(B(y,r)\big) = m\big(B(z,r)\big) = m\big(B(x,\pi) \setminus B(x,\pi-r)\big).$$

Then we have, for $\varepsilon < d(y,z)/2$,

$$2m\big(B(x,\varepsilon)\big) = m\big(B(y,\varepsilon)\big) + m\big(B(z,\varepsilon)\big) = m\big(B(y,\varepsilon) \cup B(z,\varepsilon)\big)$$
$$\leq m\big(B(x,\pi) \setminus B(x,\pi-\varepsilon)\big) = m\big(B(x,\varepsilon)\big).$$

This is obviously a contradiction.   □

For (X,d,m) satisfying $\mathsf{CD}(K,\infty)$ with $K > 0$, though X is not necessarily bounded (see Example 1), we can verify that $m(X)$ is finite ([St3, Theorem 4.26]).

7.6.2 Maximal diameter

In Riemannian geometry, it is well known that the maximal diameter π among (unweighted) Riemannian manifolds of Ricci curvature $\geq n - 1$ is achieved only by the unit sphere $\mathbb{S}^n$. In our case, however, orbifolds $\mathbb{S}^n / \Gamma$ can also have the maximal diameter. Hence, what we can expect is a decomposition into a spherical suspension (in some sense). Owing to the scaling property (Remark 3(a)), we consider only the case of $K = N - 1 > 0$. See [Oh2] for a more precise discussion of the following theorem, and Section 7.8.1 for the definition of the nonbranching property.

Theorem 7.15. *Assume that* (X, d, m) *is nonbranching and satisfies* CD($N -$ $1, N$) *for some* $N \in (1, \infty)$ *as well as* diam $X = \pi$. *Then* (X, m) *is the spherical suspension of some topological measure space.*

Outline of proof. Fix $x_N, x_S \in X$ with $d(x_N, x_S) = \pi$. Then it follows from (7.32) that

$$m\big(B(x_N, r)\big) + m\big(B(x_S, \pi - r)\big) = m(X)$$

for all $r \in (0, \pi)$. This, together with the nonbranching property, shows that, for any $z \in X \setminus \{x_N, x_S\}$, there exists a unique minimal geodesic from x_N to x_S passing through z.

Now we introduce the set Y consisting of unit speed minimal geodesics from x_N to x_S, and equip it with the distance

$$d_Y(\gamma_1, \gamma_2) := \sup_{0 \leq t \leq \pi} d_X\big(\gamma_1(t), \gamma_2(t)\big).$$

We consider $SY := (Y \times [0, \pi]) / \sim$, where $(\gamma_1, t_1) \sim (\gamma_2, t_2)$ holds if $t_1 = t_2 = 0$ or $t_1 = t_2 = \pi$. We equip SY with the topology naturally induced from d_Y. Then the map $\Psi : SY \ni (\gamma, t) \longmapsto \gamma(t) \in X$ is well defined and continuous. Define the measures ν on Y and ω on SY by

$$\nu(W) := \left\{ \int_0^\pi \sin^{N-1} t \, dt \right\}^{-1} m\big(\Psi(W \times [0, \pi])\big),$$

$$d\omega := d\nu \times (\sin^{N-1} t \, dt).$$

Then one can prove that (SY, ω) is regarded as the spherical suspension of (Y, ν) as topological measure spaces, and that $\Psi : (SY, \omega) \longrightarrow (X, m)$ is homeomorphic and measure-preserving. We use the nonbranching property for the continuity of Ψ^{-1}. $\qquad \square$

184 S.-I. Ohta

7.6.3 Open questions

We close the section with a list of open questions.

7.6.3.1 Beyond Theorem 7.15

There is room for improvement of Theorem 7.15: Is the non-branching property necessary? Can one say anything about the relation between the distances of SY and X? Does (Y, d_Y, ν) satisfy $CD(N - 2, N - 1)$? We do not know any counterexample.

If X is an n-dimensional Alexandrov space of curvature ≥ 1 with diam $X = \pi$, then it is isometric to the spherical suspension of some $(n - 1)$-dimensional Alexandrov space of curvature ≥ 1. It is generally difficult to derive something about distance from the curvature-dimension condition.

7.6.3.2 Extremal case of Lichnerowicz inequality

Related to Theorem 7.14, we know the following ([LV1, Theorem 5.34]).

Theorem 7.16 (generalized Lichnerowicz inequality). *Assume that* (X, d, m) *satisfies* $CD(K, N)$ *for some* $K > 0$ *and* $N \in (1, \infty)$. *Then we have*

$$\int_X f^2 \, dm \leq \frac{N - 1}{KN} \int_X |\nabla^- f|^2 \, dm \qquad (7.33)$$

for any Lipschitz function $f : X \longrightarrow \mathbb{R}$ *with* $\int_X f \, dm = 0$.

Here $|\nabla^- f|$ is the *generalized gradient* of f defined by

$$|\nabla^- f|(x) := \limsup_{y \to x} \frac{\max\{f(x) - f(y), 0\}}{d(x, y)}.$$

The proof is done via careful calculation using (7.28) for S_N between $m(X)^{-1} \cdot m$ and its perturbation $(1 + \varepsilon f)m(X)^{-1} \cdot m$. The inequality (7.33) means that the lowest positive eigenvalue of the Laplacian is larger than or equal to $KN/(N - 1)$. The constant $(N - 1)/KN$ in (7.33) is sharp. Moreover, in Riemannian geometry, it is known that the best constant with $N = \dim M$ is achieved only by spheres.

In our general setting, it is not known whether the best constant is achieved only by spaces of maximal diameter $\pi \sqrt{(N - 1)/K}$. If so, then Theorem 7.15 provides us a decomposition into a spherical suspension (for nonbranching spaces).

7.6.3.3 Splitting

In Riemannian geometry, Cheeger and Gromoll's [CG] celebrated theorem asserts that, if a complete Riemannian manifold of nonnegative Ricci curvature admits an isometric embedding of the real line $\mathbb{R} \hookrightarrow M$, then M isometrically splits off $\mathbb{R}$; namely, M is isometric to a product space $M' \times \mathbb{R}$, where M' again has the nonnegative Ricci curvature. We can repeat this procedure if M' contains a line. This is an extremely deep theorem, and its generalization is a key tool of Cheeger and Colding's seminal work [CC] (see Further reading of Section 7.5.3).

Kuwae and Shioya [KS3] consider (weighted) Alexandrov spaces of curvature ≥ -1 with nonnegative Ricci curvature in terms of the measure contraction property (see Section 7.8.3). They show that if such an Alexandrov space contains an isometric copy of the real line, then it splits off $\mathbb{R}$ as topological measure spaces (compare this with Theorem 7.15). This is recently strengthened into an isometric splitting by [ZZ] under a slightly stronger notion of Ricci curvature bound in terms of Petrunin's second variation formula ([Pe1]).

For general metric measure spaces satisfying CD(0, N), the isometric splitting is false because n-dimensional Banach spaces satisfy CD(0, n) (Theorem 7.17) and do not split in general. The homeomorphic, measure-preserving splitting could be true, but it is open even for nonbranching spaces.

7.6.3.4 Lévy–Gromov isoperimetric inequality

Another challenging problem is to show (some appropriate variant of) the Lévy–Gromov isoperimetric inequality using optimal transport. Most known proofs in the Riemannian case appeal to the deep existence and regularity theory of minimal surfaces which cannot be expected in singular spaces.

For instance, let us consider the *isoperimetric profile* $I_M : (0, m(M)) \longrightarrow (0, \infty)$ of a weighted Riemannian manifold (M, g, m) with $m = e^{-\psi} \mathrm{vol}_g$; i.e., $I_M(V)$ is the least perimeter of sets with volume V. Bayle [Bay] shows that the differential inequality

$$(I_M^{N/(N-1)})'' \leq -\frac{KN}{N-1} I_M^{1/(N-1)-1} \tag{7.34}$$

holds if $\mathrm{Ric}_N \geq K$, which immediately implies the corresponding *Lévy–Gromov isoperimetric inequality*

$$\frac{I_M(t \cdot m(M))}{m(M)} \geq I_{K,N}(t)$$

for $t \in [0, 1]$, where $I_{K,N}$ is the isoperimetric profile of the N-dimensional space form of constant sectional curvature $K/(N-1)$ equipped with the normalized measure (extended to noninteger N numerically). The concavity estimate (7.34) seems to be related to the Brunn–Minkowski inequality; however, Bayle's proof of (7.34) is based on the variational formulas of minimal surfaces (see also [Mo, Chapter 18]). A more analytic approach could work in metric measure spaces, but we need a new idea for it.

7.6.4 Further reading

The generalized Brunn–Minkowski inequality (Theorem 7.10) is essentially contained in the proof of [vRS, Theorem 1.1] for $N = \infty$, and due to Sturm [St4] for $N < \infty$ (also the Brascamp–Lieb inequality in [CMS1] implies it in the unweighted Riemannian situation with $N = n$). It is used as a key tool in the proof of the derivation of the Ricci curvature bound from the curvature-dimension condition (see Theorem 7.6). Some more related interpolation inequalities can be found in [CMS1]; these all were new even for Riemannian manifolds.

The relation between $\mathrm{CD}(K, \infty)$ and functional inequalities such as the Talagrand, logarithmic Sobolev, and the global Poincaré inequalities are studied by Otto and Villani [OV] and Lott and Villani [LV2, Section 6]. We refer to [Ol], [BoS] for related work on discrete spaces (see also Section 7.7.3.2), and to [St2], [Vi2, Chapter 25], [OT] for the relation between variants of these functional inequalities and the displacement convexity of generalized entropies. Theorems 7.14 and 7.15 are due to [Oh1] and [Oh2], where the proof is given in terms of the measure contraction property (see Section 7.8.3).

7.7 The curvature-dimension condition in Finsler geometry

In this section, we demonstrate that almost everything so far works well also in the Finsler setting. In fact, the equivalence between $\mathrm{Ric}_N \geq K$ and $\mathrm{CD}(K, N)$ is extended by introducing an appropriate notion of the weighted Ricci curvature. Then we explain why this is significant and discuss two potential applications. We refer to [BCS] and [Sh2] for the fundamentals of Finsler geometry, and the main reference of the section is [Oh5].

7.7.1 A brief introduction to Finsler geometry

Let M be an n-dimensional connected C^∞-manifold. Given a local coordinate $(x^i)_{i=1}^n$ on an open set $U \subset M$, we always consider the coordinate $(x^i, v^i)_{i=1}^n$ on TU given by

$$v = \sum_{i=1}^n v^i \frac{\partial}{\partial x^i}\bigg|_x \in T_x M.$$

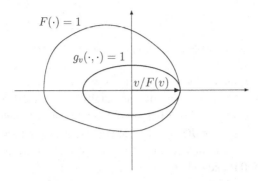

Figure 7.10

Definition 4 (Finsler structures). A C^∞-*Finsler structure* is a nonnegative function $F : TM \longrightarrow [0, \infty)$ satisfying the following three conditions:

(1) (regularity) F is C^∞ on $TM \setminus 0$, where 0 stands for the zero section;
(2) (positive homogeneity) $F(\lambda v) = \lambda F(v)$ holds for all $v \in TM$ and $\lambda \geq 0$;
(3) (strong convexity) given a local coordinate $(x^i)_{i=1}^n$ on $U \subset M$, the $n \times n$ matrix

$$\left(g_{ij}(v)\right)_{i,j=1}^n := \left(\frac{1}{2}\frac{\partial^2(F^2)}{\partial v^i \partial v^j}(v)\right)_{i,j=1}^n \tag{7.35}$$

is positive-definite for all $v \in T_x M \setminus 0, x \in U$.

In other words, each $F|_{T_x M}$ is a C^∞-*Minkowski norm* (see Example 2(a) below for the precise definition) and it varies C^∞-smoothly also in the horizontal direction. We remark that the homogeneity (2) is imposed only in the positive direction, so that $F(-v) \neq F(v)$ is allowed. The positive-definite symmetric matrix $(g_{ij}(v))_{i,j=1}^n$ in (7.35) defines the Riemannian structure g_v on $T_x M$ through

$$g_v\left(\sum_{i=1}^n v_1^i \frac{\partial}{\partial x^i}\bigg|_x, \sum_{j=1}^n v_2^j \frac{\partial}{\partial x^j}\bigg|_x\right) := \sum_{i,j=1}^n g_{ij}(v)v_1^i v_2^j. \tag{7.36}$$

Note that $F(v)^2 = g_v(v, v)$. If F is coming from a Riemannian structure, then g_v always coincides with the original Riemannian metric. In general, the inner product g_v is regarded as the best approximation of F in the direction v. More precisely, the unit spheres of F and g_v are tangent to each other at $v/F(v)$ up to the second order (that is possible thanks to the strong convexity; see Figure 7.10).

The *distance* between $x, y \in M$ is naturally defined by

$$d(x, y) := \inf \left\{ \int_0^1 F(\dot{\gamma})\, dt \,\bigg|\, \gamma : [0, 1] \longrightarrow M, \ C^1, \ \gamma(0) = x, \ \gamma(1) = y \right\}.$$

One remark is that the nonsymmetry $d(x, y) \neq d(y, x)$ may come up as F is only positively homogeneous. Thus, it is not totally correct to call d a distance; it might be called cost or action as F is a sort of Lagrangian cost function. Another remark is that the function $d(x, \cdot)^2$ is C^2 at the origin x if and only if $F|_{T_x M}$ is Riemannian. Indeed, the squared norm $|\cdot|^2$ of a Banach (or Minkowski) space $(\mathbb{R}^n, |\cdot|)$ is C^2 at 0 if and only if it is an inner product.

A C^∞-curve $\gamma : [0, l] \longrightarrow M$ is called a *geodesic* if it has constant speed $(F(\dot{\gamma}) \equiv c \in [0, \infty))$ and is locally minimizing (with respect to d). The reverse curve $\bar{\gamma}(t) := \gamma(l - t)$ is not necessarily a geodesic. We say that (M, F) is *forward complete* if any geodesic $\gamma : [0, \varepsilon] \longrightarrow M$ is extended to a geodesic $\bar{\gamma} : [0, \infty) \longrightarrow M$. Then any two points $x, y \in M$ are connected by a minimal geodesic from x to y.

7.7.2 Weighted Ricci curvature and the curvature-dimension condition

We introduced distance and geodesics in a natural (metric geometric) way, but the definition of curvature is more subtle. The flag and Ricci curvatures on Finsler manifolds, corresponding to the sectional and Ricci curvatures in Riemannian geometry, are defined via some connection as in the Riemannian case. The choice of connection is not unique in the Finsler setting; nevertheless, all connections are known to give rise to the same curvature. In these notes, however, we shall follow Shen's idea [Sh2, Chapter 6] of introducing the flag curvature using vector fields and corresponding Riemannian structures (via (7.36)). This intuitive description is not only geometrically understandable, but also useful and inspiring.

Fix a unit vector $v \in T_x M \cap F^{-1}(1)$, and extend it to a C^∞-vector field V on an open neighborhood U of x in such a way that every integral curve of V is geodesic. In particular, $V(\gamma(t)) = \dot{\gamma}(t)$ along the geodesic $\gamma : (-\varepsilon, \varepsilon) \longrightarrow M$ with $\dot{\gamma}(0) = v$. Using (7.36), we equip U with the Riemannian structure g_V. Then the *flag curvature* $\mathcal{K}(v, w)$ of v and a linearly independent vector $w \in T_x M$ coincides with the sectional curvature with respect to g_V of the 2-plane $v \wedge w$ spanned by v and w. Similarly, the *Ricci curvature* $\mathrm{Ric}(v)$ of v (with respect to F) coincides with the Ricci curvature of v with respect to g_V. This contains the fact that $\mathcal{K}(v, w)$ is independent of the choice of the extension V of v. We remark that $\mathcal{K}(v, w)$ depends not only on the *flag* $v \wedge w$, but also on the choice of the *flagpole* v in the flag $v \wedge w$. In particular, $\mathcal{K}(v, w) \neq \mathcal{K}(w, v)$ may happen.

As for measure, on Finsler manifolds, there is no constructive measure as good as the Riemannian volume measure. Therefore, as the theory of weighted Riemannian manifolds, we equip (M, F) with an arbitrary positive C^∞-measure m on M. Now, the weighted Ricci curvature is defined as follows [Oh5]. We extend given a unit vector $v \in T_x M$ to a C^∞-vector field V on a neighborhood $U \ni x$ such that every integral curve is geodesic (or it is sufficient to consider only the tangent vector field $\dot\gamma$ of the geodesic $\gamma : (-\varepsilon, \varepsilon) \longrightarrow M$ with $\dot\gamma(0) = v$), and decompose m as $m = e^{-\Psi(V)} \text{vol}_{g_V}$ on U. We remark that the weight Ψ is not a function on M, but a function on the unit tangent sphere bundle $SM \subset TM$. For simplicity, we set

$$\partial_v \Psi := \frac{d(\Psi \circ \dot\gamma)}{dt}(0), \qquad \partial_v^2 \Psi := \frac{d^2(\Psi \circ \dot\gamma)}{dt^2}(0). \tag{7.37}$$

Definition 5 (weighted Ricci curvature of Finsler manifolds). For $N \in [n, \infty]$ and a unit vector $v \in T_x M$, we define

(1) $\text{Ric}_n(v) := \begin{cases} \text{Ric}(v) + \partial_v^2 \Psi & \text{if } \partial_v \Psi = 0, \\ -\infty & \text{otherwise;} \end{cases}$

(2) $\text{Ric}_N(v) := \text{Ric}(v) + \partial_v^2 \Psi - \dfrac{(\partial_v \Psi)^2}{N - n}$ for $N \in (n, \infty)$;

(3) $\text{Ric}_\infty(v) := \text{Ric}(v) + \partial_v^2 \Psi$.

In other words, $\text{Ric}_N(v)$ of F is $\text{Ric}_N(v)$ of g_V (recall Definition 1), so that this curvature coincides with Ric_N in weighted Riemannian manifolds. We remark that the quantity $\partial_v \Psi$ coincides with Shen's S-curvature (also called the *mean covariance* or *mean tangent curvature*, see [Sh1], [Sh2], [Sh3]). Therefore, bounding Ric_n from below makes sense only when the S-curvature vanishes everywhere. This curvature enables us to extend Theorem 7.6 to the Finsler setting [Oh5]. Therefore, all results in the theory of the curvature-dimension condition are applicable to general Finsler manifolds.

Theorem 7.17. *A forward complete Finsler manifold (M, F, m) equipped with a positive C^∞-measure m satisfies $\text{CD}(K, N)$ for some $K \in \mathbb{R}$ and $N \in [n, \infty]$ if and only if $\text{Ric}_N(v) \geq K$ holds for all unit vectors $v \in TM$.*

We remark that, in the above theorem, the curvature-dimension condition is appropriately extended to nonsymmetric distances. The proof of Theorem 7.17 follows the same line as the Riemannian case; however, we should be careful about nonsymmetric distance and need some more extra discussion due to the fact that the squared distance function $d(x, \cdot)^2$ is only C^1 at x.

We present several examples of Finsler manifolds. The flag and Ricci curvatures are calculated in a number of situations, while the weighted Ricci curvature is still relatively much less investigated.

Example 2. (a) (Banach/Minkowski spaces with Lebesgue measures) A *Minkowski norm* $|\cdot|$ on $\mathbb{R}^n$ is a nonsymmetric generalization of usual norms. That is to say, $|\cdot|$ is a nonnegative function on $\mathbb{R}^n$ satisfying the positive homogeneity $|\lambda v| = \lambda |v|$ for $v \in \mathbb{R}^n$ and $\lambda > 0$; the convexity $|v + w| \le |v| + |w|$ for $v, w \in \mathbb{R}^n$; and the positivity $|v| > 0$ for $v \ne 0$. Note that the unit ball of $|\cdot|$ is a convex (but not necessarily symmetric to the origin) domain containing the origin in its interior (see Figure 7.10, where F is a Minkowski norm).

A Banach or Minkowski norm $|\cdot|$ which is uniformly convex and C^∞ on $\mathbb{R}^n \setminus \{0\}$ induces a Finsler structure in a natural way through the identification between $T_x \mathbb{R}^n$ and $\mathbb{R}^n$. Then $(\mathbb{R}^n, |\cdot|, \mathrm{vol}_n)$ has the flat flag curvature. Hence, a Banach or Minkowski space $(\mathbb{R}^n, |\cdot|, \mathrm{vol}_n)$ satisfies $\mathrm{CD}(0, n)$ by Theorem 7.17 for C^∞-norms, and by Theorem 7.9 via approximations for general norms.

(b) (Banach/Minkowski spaces with log-concave measures) A Banach or Minkowski space $(\mathbb{R}^n, |\cdot|, m)$ equipped with a measure $m = e^{-\psi} \mathrm{vol}_n$ such that ψ is K-convex with respect to $|\cdot|$ satisfies $\mathrm{CD}(K, \infty)$. Here the K-*convexity* means that

$$\psi\big((1-t)x + ty\big) \le (1-t)\psi(x) + t\psi(y) - \frac{K}{2}(1-t)t|y - x|^2$$

holds for all $x, y \in \mathbb{R}^n$ and $t \in [0, 1]$. This is equivalent to $\partial_v^2 \psi \ge K$ (in the sense of (7.37)) if $|\cdot|$ and ψ are C^∞ (on $\mathbb{R}^n \setminus \{0\}$ and $\mathbb{R}^n$, respectively). Hence, $\mathrm{CD}(K, \infty)$ again follows from Theorem 7.17 together with Theorem 7.9.

In particular, a Gaussian-type space $(\mathbb{R}^n, |\cdot|, e^{-|\cdot|^2/2} \mathrm{vol}_n)$ satisfies $\mathrm{CD}(0, \infty)$ independently of n. It also satisfies $\mathrm{CD}(K, \infty)$ for some $K > 0$ if (and only if) it is 2-*uniformly convex* in the sense that $|\cdot|^2/2$ is C^{-2}-convex for some $C \ge 1$ (see [BCL] and [Oh4]), and then $K = C^{-2}$. For instance, ℓ_p-spaces with $p \in (1, 2]$ are 2-uniformly convex with $C = 1/\sqrt{p-1}$, and hence satisfy $\mathrm{CD}(p - 1, \infty)$. Compare this with Example 1.

(c) (Randers spaces) A *Randers space* (M, F) is a special kind of Finsler manifold such that

$$F(v) = \sqrt{g(v, v)} + \beta(v)$$

for some Riemannian metric g and a one-form β. We suppose that $|\beta(v)|^2 < g(v, v)$ unless $v = 0$, then F is indeed a Finsler structure. Randers spaces are important in applications and reasonable for concrete calculations. In fact, we can see by calculation that $\mathbf{S}(v) = \partial_v \Psi \equiv 0$ holds if and only if β is a Killing form of constant length as well as m is the Busemann–Hausdorff measure (see [Oh7], [Sh2, Section 7.3] for more details). This means that there are many Finsler manifolds which do not admit any measures of $\mathrm{Ric}_n \geq K > -\infty$, and then we must consider Ric_N for $N > n$.

(d) (Hilbert geometry) Let $D \subset \mathbb{R}^n$ be a bounded open set with smooth boundary such that its closure $\overline{D}$ is strictly convex. Then the associated *Hilbert distance* is defined by

$$d(x_1, x_2) := \log \left(\frac{\|x_1 - x_2'\| \cdot \|x_2 - x_1'\|}{\|x_1 - x_1'\| \cdot \|x_2 - x_2'\|} \right)$$

for distinct $x_1, x_2 \in D$, where $\| \cdot \|$ is the standard Euclidean norm and x_1', x_2' are intersections of ∂D and the line passing through x_1, x_2 such that x_i' is on the side of x_i. Hilbert geometry is known to be realized by a Finsler structure with constant negative flag curvature. However, it is still unclear if it carries a (natural) measure for which the curvature-dimension condition holds.

(e) (Teichmüller space) A Teichmüller metric on a Teichmüller space is one of the most famous Finsler structures in differential geometry. It is known to be complete, while the Weil–Petersson metric is incomplete and Riemannian. The author does not know any investigation concerned with the curvature-dimension condition of Teichmüller space.

7.7.3 Remarks and potential applications

Owing to the celebrated work of Cheeger and Colding [CC], we know that a (non-Hilbert) Banach space cannot be the limit space of a sequence of Riemannian manifolds (with respect to the measured Gromov-Hausdorff convergence) with a uniform lower Ricci curvature bound. Therefore the fact that Finsler manifolds satisfy the curvature-dimension condition means that it is too weak to characterize limit spaces of Riemannian manifolds. This should be compared with the following facts:

(I) a Banach space can be an Alexandrov space only if it happens to be a Hilbert space (and then it has the nonnegative curvature);

(II) it is not known if all Alexandrov spaces X of curvature $\geq k$ can be approximated by a sequence of Riemannian manifolds $\{M_i\}_{i \in \mathbb{N}}$ of curvature $\geq k'$.

We know that there are counterexamples to (II) if we impose the non-collapsing condition $\dim M_i \equiv \dim X$ (see [Ka]), but the general situation admitting collapsing ($\dim M_i > \dim X$) is still open and is one of the most important and challenging questions in Alexandrov geometry. Thus, the curvature-dimension condition is not as good as the Alexandrov–Toponogov triangle comparison condition from the purely Riemannian geometric viewpoint.

From a different viewpoint, Cheeger and Colding's observation means that the family of Finsler spaces is properly much wider than the family of Riemannian spaces. Therefore, the validity of the curvature-dimension condition for Finsler manifolds opens the door to broader applications. Here we mention two of them.

7.7.3.1 The geometry of Banach spaces

Although their interested spaces are common to some extent, there is almost no connection between the geometry of Banach spaces and Finsler geometry (as far as the author knows). We believe that our differential geometric technique would be useful in the geometry of Banach spaces. For instance, Theorem 7.17 (together with Theorem 7.9) could recover and generalize Gromov and Milman's normal concentration of unit spheres in 2-uniformly convex Banach spaces (see [GM2] and [Le, Section 2.2]). To be precise, as an application of Theorem 7.17, we know the normal concentration of Finsler manifolds such that Ric_∞ goes to infinity (see [Oh5], and see [GM1], [Le, Section 2.2] for the Riemannian case). This seems to imply the concentration of unit spheres mentioned above.

7.7.3.2 Approximations of graphs

Generally speaking, Finsler spaces give much better approximations of graphs than Riemannian spaces, when we impose a lower Ricci curvature bound. For instance, Riemannian spaces into which the $\mathbb{Z}^n$-lattice is nearly isometrically embedded should have very negative curvature, while the $\mathbb{Z}^n$-lattice is isometrically embedded in flat ℓ_1^n. This kind of technique seems useful for investigating graphs with Ricci curvature bounded below (in some sense), and provides a different point of view on variants of the curvature-dimension condition for discrete spaces (e.g., see [Ol], [BoS]).

7.7.4 Further reading

We refer to [BCS] and [Sh2] for the fundamentals of Finsler geometry and important examples. The interpretation of the flag curvature using vector fields can be found in [Sh2, Chapter 6]. We also refer to [Sh1] and [Sh3] for the **S**-curvature and its applications including a volume comparison theorem different from Theorem 7.12 (which has some topological applications). The **S**-curvature of Randers spaces and the characterization of its vanishing (Example 2(c)) are studied in [Sh2, Section 7.3] and [Oh7].

Definition 5 and Theorem 7.17 are due to [Oh5], while the weight function Ψ on SM has already been considered in the definition of **S**-curvature. See also

[OhS] for related work concerning heat flow on Finsler manifolds, and [Oh6] for a survey on these subjects. The curvature-dimension condition $CD(0, n)$ of Banach spaces (Example 2(a)) is first demonstrated by Cordero-Erausquin (see [Vi2, page 908]).

7.8 Related topics

We briefly comment on further related topics.

7.8.1 Nonbranching spaces

We say that a geodesic space (X, d) is *nonbranching* if geodesics do not branch in the sense that each quadruple of points $z, x_0, x_1, x_2 \in X$ with $d(x_0, x_1) = d(x_0, x_2) = 2d(z, x_i)$ $(i = 0, 1, 2)$ must satisfy $x_1 = x_2$. This is a quite useful property. For instance, in such a space satisfying $CD(K, N)$ for some K and N, a.e. $x \in X$ has a unique minimal geodesic from x to a.e. $y \in X$ ([St4, Lemma 4.1]). Therefore, we can localize the inequality (7.27), and then (7.27) for single $U = S_N$ implies that for all $U \in \mathcal{DC}_N$ (see Remark 3(e), (f), [St4, Proposition 4.2]). There are some more results known only in nonbranching spaces (e.g., see [St4, Section 4], [FV] and also Section 7.8.3).

Riemannian (or Finsler) manifolds and Alexandrov spaces are clearly nonbranching. However, as n-dimensional Banach and Minkowski spaces satisfy $CD(0, n)$, the curvature-dimension condition does not prevent the branching phenomenon. One big open problem after Cheeger and Colding's work [CC] is whether any limit space of Riemannian manifolds with a uniform lower Ricci curvature bound is nonbranching or not.

7.8.2 Alexandrov spaces

As was mentioned in Remark 1, Alexandrov spaces are metric spaces whose sectional curvature is bounded from below in terms of the triangle comparison property (see [BGP], [OtS], [BBI, Chapters 4, 10] for more details). One interesting fact is that a compact geodesic space (X, d) is an Alexandrov space of nonnegative curvature if and only if the Wasserstein space $(\mathcal{P}(X), d_2^W)$ over it is an Alexandrov space of nonnegative curvature ([St3, Proposition 2.10], [LV2, Theorem A.8]). This is a metric geometric explanation of Otto's formal calculation of the sectional curvature of $(\mathcal{P}_2(\mathbb{R}^n), d_2^W)$ [Ot]. We remark that this relation cannot be extended to positive or negative curvature bounds. In fact, if (X, d) is not an Alexandrov space of nonnegative curvature, then $(\mathcal{P}(X), d_2^W)$ is not an Alexandrov space of curvature $\geq k$ even for negative k ([St3, Proposition 2.10]). Optimal transport in Alexandrov spaces is further studied in [Be], [Oh3], [Sav], [Gi] and [GO].

Since the Ricci curvature is the trace of the sectional curvature, it is natural to expect that Alexandrov spaces satisfy the curvature-dimension condition. Petrunin [Pe3] recently claimed that it is indeed the case for $K = 0$, and is extended to the general case $K \neq 0$ by [ZZ]. They use the second variation formula in [Pe1] and the gradient flow technique developed in [PP] and [Pe2], instead of calculations as in Sections 7.3 and 7.4 involving Jacobi fields.

7.8.3 The measure contraction property

For $K \in \mathbb{R}$ and $N \in (1, \infty)$, a metric measure space is said to satisfy the *measure contraction property* MCP(K, N) if the Bishop inequality (7.6) holds in an appropriate sense. More precisely, MCP(K, N) for (X, d, m) means that any $x \in X$ admits a measurable map $\Phi : X \longrightarrow \Gamma(X)$ satisfying $e_0 \circ \Phi \equiv x$, $e_1 \circ \Phi = \mathrm{Id}_X$ and

$$ dm \geq (e_t \circ \Phi)_\sharp \left(t^N \beta_{K,N}^t \big(d(x, y) \big) \, dm(y) \right) $$

for all $t \in (0, 1)$ as measures (compare this with Theorem 7.10(i)). As we mentioned in Further reading in Section 7.3.3, this kind of property was suggested in [CC, I, Appendix 2] and [Gr, Section 5.I], and systematically studied in [Oh1], [Oh2], and [St4, Sections 5 and 6]. Some variants have also been studied in [KS1] and [St1] before them.

MCP(K, N) can be regarded as the curvature-dimension condition CD(K, N) applied only for each pair of a Dirac measure and a uniform distribution on a set, and CD actually implies MCP in nonbranching spaces. It is known that Alexandrov spaces satisfy MCP (see [Oh1], [KS2]). For n-dimensional (unweighted) Riemannian manifolds, MCP(K, n) is equivalent to Ric $\geq K$; however, MCP(K, N) with $N > n$ does not imply Ric $\geq K$. In fact, a sufficiently small ball in $\mathbb{R}^n$ equipped with the Lebesgue measure satisfies MCP($1, n + 1$). This is one drawback of MCP. On the other hand, an advantage of MCP is its simplicity; there are several facts known for MCP and unknown for CD. We shall compare these properties in more detail.

7.8.3.1 Product spaces (L^2-tensorization property)

If (X_i, d_i, m_i) satisfies MCP(K_i, N_i) for $i = 1, 2$, then the product metric measure space $(X_1 \times X_2, d_1 \times d_2, m_1 \times m_2)$ satisfies MCP($\min\{K_1, K_2\}, N_1 + N_2$)

[Oh2]. The analogous property for CD is known only for $\min\{K_1, K_2\} = 0$ or $N_1 + N_2 = \infty$ in nonbranching spaces [St3].

Recently, Bacher and Sturm [BaS1] introduced a slightly weaker variant of CD, called the *reduced curvature-dimension condition* CD^* (recall (7.24) in Further reading of Section 7.4.3). They show that CD^* enjoys the tensorization property if the spaces in consideration are nonbranching.

7.8.3.2 Euclidean cones

If (X, d, m) satisfies $MCP(N - 1, N)$, then its Euclidean cone (CX, d_{CX}, m_{CX}) defined by

$$CX := (X \times [0, \infty))/\sim, \quad (x, 0) \sim (y, 0),$$
$$d_{CX}((x, s), (y, t)) := \sqrt{s^2 + t^2 - 2st \cos d(x, y)},$$
$$dm_{CX} := dm \times (t^N \, dt)$$

satisfies $MCP(0, N + 1)$ [Oh2]. This is recently established for the curvature-dimension condition by [BaS2] in the case where (X, d, m) is Riemannian.

7.8.3.3 Local-to-global property

Sturm [St3] shows that, if (X, d, m) is nonbranching and if every point in X admits an open neighborhood on which $CD(K, \infty)$ holds, then the whole space (X, d, m) globally satisfies $CD(K, \infty)$. In other words, $CD(K, \infty)$ is a local condition, as is the Ricci curvature bound on a Riemannian manifold. The same holds true also for $CD(0, N)$ with $N < \infty$. It is shown in [BaS1] that $CD^*(K, N)$ satisfies the local-to-global property for general $K \in \mathbb{R}$ and $N \in (1, \infty)$; however, it is still open and unclear if $CD(K, N)$ for $K \neq 0$ and $N < \infty$ is a local condition.

In contrast, the local-to-global property is known to be false for MCP. As we mentioned above, sufficiently small balls in $\mathbb{R}^n$ satisfy $MCP(1, n + 1)$, while the entire space $\mathbb{R}^n$ does not satisfy it by virtue of the Bonnet–Myers diameter bound (Theorem 7.14).

Acknowledgments

I would like to express my gratitude to the organizers for the kind invitation to the fascinating summer school, and to all the audience for their attendance and interest. I also thank the referee for careful reading and valuable suggestions. Partly supported by the Grant-in-Aid for Young Scientists (B) 20740036.

References

[AGS] L. Ambrosio, N. Gigli, and G. Savaré, *Gradient Flows in Metric Spaces and in the Space of Probability Measures*, Birkhäuser Verlag, Basel, 2005.

[BaS1] K. Bacher and K.-T. Sturm, Localization and tensorization properties of the curvature-dimension condition for metric measure spaces, *J. Funct. Anal.* **259** (2010), 28–56.

[BaS2] K. Bacher and K.-T. Sturm, Ricci bounds for Euclidean and spherical cones, Preprint (2010). Available at arXiv:1103.0197.

[BE] D. Bakry and M. Émery, Diffusions hypercontractives (French), *Séminaire de probabilités, XIX, 1983/84*, 177–206, Lecture Notes in Math. 1123, Springer, Berlin, 1985.

[BCL] K. Ball, E.A. Carlen, and E.H. Lieb, Sharp uniform convexity and smoothness inequalities for trace norms, *Invent. Math.* **115** (1994), 463–482.

[Bal] W. Ballmann, *Lectures on Spaces of Nonpositive Curvature*. With an appendix by Misha Brin, Birkhäuser Verlag, Basel, 1995.

[BCS] D. Bao, S.-S. Chern, and Z. Shen, *An Introduction to Riemann–Finsler Geometry*, Springer-Verlag, New York, 2000.

[Bay] V. Bayle, Propriétés de concavité du profil isopérimétrique et applications (French), Thèse de Doctorat, Institut Fourier, Université Joseph-Fourier, Grenoble, 2003.

[Be] J. Bertrand, Existence and uniqueness of optimal maps on Alexandrov spaces, *Adv. Math.* **219** (2008), 838–851.

[BoS] A.-I. Bonciocat and K.-T. Sturm, Mass transportation and rough curvature bounds for discrete spaces, *J. Funct. Anal.* **256** (2009), 2944–2966.

[Br] Y. Brenier, Polar factorization and monotone rearrangement of vector-valued functions, *Comm. Pure Appl. Math.* **44** (1991), 375–417.

[BBI] D. Burago, Yu. Burago, and S. Ivanov, *A Course in Metric Geometry*, American Mathematical Society, Providence, RI, 2001.

[BGP] Yu. Burago, M. Gromov, and G. Perel'man, A.D. Alexandrov spaces with curvatures bounded below, *Russian Math. Surveys* **47** (1992), 1–58.

[Ch] I. Chavel, *Riemannian Geometry. A Modern Introduction*, second edition, Cambridge University Press, Cambridge, 2006.

[CC] J. Cheeger and T.H. Colding, On the structure of spaces with Ricci curvature bounded below. I, II, III, *J. Differential Geom.* **46** (1997), 406–480; *ibid.* **54** (2000), 13–35; *ibid.* **54** (2000), 37–74.

[CE] J. Cheeger and D.G. Ebin, *Comparison Theorems in Riemannian Geometry*, revised reprint of the 1975 original, AMS Chelsea Publishing, Providence, RI, 2008.

[CG] J. Cheeger and D. Gromoll, The splitting theorem for manifolds of nonnegative Ricci curvature, *J. Differential Geom.* **6** (1971/72), 119–128.

[CMS1] D. Cordero-Erausquin, R.J. McCann, and M. Schmuckenschläger, A Riemannian interpolation inequality à la Borell, Brascamp and Lieb, *Invent. Math.* **146** (2001), 219–257.

[CMS2] D. Cordero-Erausquin, R.J. McCann, and M. Schmuckenschläger, Prékopa–Leindler type inequalities on Riemannian manifolds, Jacobi fields, and optimal transport, *Ann. Fac. Sci. Toulouse Math.* (6) **15** (2006), 613–635.

[FF] A. Fathi and A. Figalli, Optimal transportation on non-compact manifolds, *Israel J. Math.* **175** (2010), 1–59.

[FG] A. Figalli and N. Gigli, Local semiconvexity of Kantorovich potentials on non-compact manifolds, *ESAIM Control Optim. Calc. Var.* **17** (2011), 648–653.

[FV] A. Figalli and C. Villani, Strong displacement convexity on Riemannian manifolds, *Math. Z.* **257** (2007), 251–259.

[Fu] K. Fukaya, Collapsing of Riemannian manifolds and eigenvalues of Laplace operator, *Invent. Math.* **87** (1987), 517–547.

[Ga] R.J. Gardner, The Brunn–Minkowski inequality, *Bull. Am. Math. Soc. (N.S.)* **39** (2002), 355–405.

[Gi] N. Gigli, On the inverse implication of Brenier–McCann theorems and the structure of $(\mathcal{P}_2(M), W_2)$, *Methods Appl. Anal.* **18** (2011), 127–158.

[GO] N. Gigli and S. Ohta, First variation formula in Wasserstein spaces over compact Alexandrov spaces, *Can. Math. Bull.* **55** (2012), 723–735.

[Gr] M. Gromov, *Metric Structures for Riemannian and Non-Riemannian Spaces*, Birkhäuser, Boston, MA, 1999.

[GM1] M. Gromov and V.D. Milman, A topological application of the isoperimetric inequality, *Am. J. Math.* **105** (1983), 843–854.

[GM2] M. Gromov and V.D. Milman, Generalization of the spherical isoperimetric inequality to uniformly convex Banach spaces, *Compositio Math.* **62** (1987), 263–282.

[Ka] V. Kapovitch, Regularity of limits of noncollapsing sequences of manifolds, *Geom. Funct. Anal.* **12** (2002), 121–137.

[KS1] K. Kuwae and T. Shioya, On generalized measure contraction property and energy functionals over Lipschitz maps, ICPA98 (Hammamet). *Potential Anal.* **15** (2001), 105–121.

[KS2] K. Kuwae and T. Shioya, Infinitesimal Bishop–Gromov condition for Alexandrov spaces, *Adv. Stud. Pure Math.* **57** (2010), 293–302.

[KS3] K. Kuwae and T. Shioya, A topological splitting theorem for weighted Alexandrov spaces, *Tohoku Math. J. (2)* **63** (2011), 59–76.

[Le] M. Ledoux, *The Concentration of Measure Phenomenon*, American Mathematical Society, Providence, RI, 2001.

[Lo1] J. Lott, Some geometric properties of the Bakry–Émery–Ricci tensor, *Comment. Math. Helv.* **78** (2003), 865–883.

[Lo2] J. Lott, Optimal transport and Ricci curvature for metric-measure spaces, in *Surveys in Differential Geometry*, vol. XI, J. Cheeger and K. Grove, eds, 229–257, International Press, Somerville, MA, 2007.

[LV1] J. Lott and C. Villani, Weak curvature conditions and functional inequalities, *J. Funct. Anal.* **245** (2007), 311–333.

[LV2] J. Lott and C. Villani, Ricci curvature for metric-measure spaces via optimal transport, *Ann. Math.* **169** (2009), 903–991.

[Mc1] R.J. McCann, A convexity principle for interacting gases, *Adv. Math.* **128** (1997), 153–179.

[Mc2] R.J. McCann, Polar factorization of maps on Riemannian manifolds, *Geom. Funct. Anal.* **11** (2001), 589–608.

[Mo] F. Morgan, *Geometric Measure Theory. A Beginner's Guide*, fourth edition, Elsevier/Academic Press, Amsterdam, 2009.

[Oh1] S. Ohta, On the measure contraction property of metric measure spaces, *Comment. Math. Helv.* **82** (2007), 805–828.

[Oh2] S. Ohta, Products, cones, and suspensions of spaces with the measure contraction property, *J. Lond. Math. Soc. (2)* **76** (2007), 225–236.

[Oh3] S. Ohta, Gradient flows on Wasserstein spaces over compact Alexandrov spaces, *Am. J. Math.* **131** (2009), 475–516.

[Oh4] S. Ohta, Uniform convexity and smoothness, and their applications in Finsler geometry, *Math. Ann.* **343** (2009), 669–699.

[Oh5] S. Ohta, Finsler interpolation inequalities, *Calc. Var. Partial Dif. Equations* **36** (2009), 211–249.

[Oh6] S. Ohta, Optimal transport and Ricci curvature in Finsler geometry, *Adv. Stud. Pure Math.* **57** (2010), 323–342.

[Oh7] S. Ohta, Vanishing S-curvature of Randers spaces, *Dif. Geom. Appl.* **29** (2011), 174–178.

[OhS] S. Ohta and K.-T. Sturm, Heat flow on Finsler manifolds, *Comm. Pure Appl. Math.* **62** (2009), 1386–1433.

[OT] S. Ohta and A. Takatsu, Displacement convexity of generalized relative entropies, *Adv. Math.* **228** (2011), 1742–1787.

[Ol] Y. Ollivier, Ricci curvature of Markov chains on metric spaces, *J. Funct. Anal.* **256** (2009), 810–864.

[OtS] Y. Otsu, T. Shioya, The Riemannian structure of Alexandrov spaces, *J. Dif. Geom.* **39** (1994), 629–658.

[Ot] F. Otto, The geometry of dissipative evolution equations: the porous medium equation, *Comm. Partial Dif. Equations* **26** (2001), 101–174.

[OV] F. Otto and C. Villani, Generalization of an inequality by Talagrand and links with the logarithmic Sobolev inequality, *J. Funct. Anal.* **173** (2000), 361–400.

[PP] G. Perel'man and A. Petrunin, Quasigeodesics and gradient curves in Alexandrov spaces, Unpublished preprint (1994). Available at http://www.math.psu.edu/petrunin/.

[Pe1] A. Petrunin, Parallel transportation for Alexandrov space with curvature bounded below, *Geom. Funct. Anal.* **8** (1998), 123–148.

[Pe2] A. Petrunin, Semiconcave functions in Alexandrov's geometry, *Surveys in differential geometry*, vol. XI, J. Cheeger and K. Grove, eds, 137–201, International Press, Somerville, MA, 2007.

[Pe3] A. Petrunin, Alexandrov meets Lott–Villani–Sturm, *Münster J. Math.* **4** (2011), 53–64.

[Qi] Z. Qian, Estimates for weighted volumes and applications, *Quart. J. Math. Oxford Ser. (2)* **48** (1997), 235–242.

[vRS] M.-K. von Renesse and K.-T. Sturm, Transport inequalities, gradient estimates, entropy and Ricci curvature, *Comm. Pure Appl. Math.* **58** (2005), 923–940.

[Sak] T. Sakai, *Riemannian Geometry*, Translated from the 1992 Japanese original by the author. Translations of Mathematical Monographs, 149. American Mathematical Society, Providence, RI, 1996.

[Sav] G. Savaré, Gradient flows and diffusion semigroups in metric spaces under lower curvature bounds, *C. R. Math. Acad. Sci. Paris* **345** (2007), 151–154.

[Sh1] Z. Shen, Volume comparison and its applications in Riemann–Finsler geometry, *Adv. Math.* **128** (1997), 306–328.

[Sh2] Z. Shen, *Lectures on Finsler Geometry*, World Scientific Publishing Co., Singapore, 2001.

[Sh3] Z. Shen, Landsberg curvature, S-curvature and Riemann curvature, in *A sampler of Riemann–Finsler Geometry*, D. Bao, R.L. Bryant, S.-S. Chen, and Z. Shen, eds, 303–355, Mathematical Sciences Research Institute Publications, 50, Cambridge University Press, Cambridge, 2004.

[St1] K.-T. Sturm, Diffusion processes and heat kernels on metric spaces, *Ann. Probab.* **26** (1998), 1–55.

[St2] K.-T. Sturm, Convex functionals of probability measures and nonlinear diffusions on manifolds, *J. Math. Pures Appl.* **84** (2005), 149–168.

[St3] K.-T. Sturm, On the geometry of metric measure spaces. I, *Acta Math.* **196** (2006), 65–131.

[St4] K.-T. Sturm, On the geometry of metric measure spaces. II, *Acta Math.* **196** (2006), 133–177.

[Vi1] C. Villani, *Topics in Optimal Transportation*, American Mathematical Society, Providence, RI, 2003.

[Vi2] C. Villani, *Optimal Transport, Old and New*, Springer-Verlag, Berlin, 2009.

[We] G. Wei, Manifolds with a lower Ricci curvature bound, *Surveys in Differential Geometry*, vol. XI, J. Cheeger and K. Grove, eds, 203–227, International Press, Somerville, MA, 2007.

[ZZ] H.-C. Zhang and X.-P. Zhu, Ricci curvature on Alexandrov spaces and rigidity theorems, *Comm. Anal. Geom.* **18** (2010), 503–553.

PART TWO

Surveys and Research Papers

8

Computing a mass transport problem with a least-squares method

OLIVIER BESSON,[a] MARTINE PICQ,[b] AND
JÉRÔME POUSSIN[b,*]

[a] *Université de Neuchâtel, Institut de Mathématiques, 11, rue E. Argand,*
2009 Neuchâtel, Switzerland,
[b] *Université de Lyon CNRS INSA-Lyon ICJ UMR 5208, bat. L. de Vinci,*
20 Av. A. Einstein, F-69100 Villeurbanne Cedex France

Abstract

This work originates from a heart's image tracking, generating an apparent continuous motion, observable through intensity variation from one starting image to an ending one, both supposed segmented. Given two images ρ_0 and ρ_1, we calculate an evolution process $\rho(t, \cdot)$ which transports ρ_0 to ρ_1 by using the extended optical flow. In this chapter we propose an algorithm based on a fixed-point formulation and a space-time least-squares formulation of the mass conservation equation for computing a mass transport problem. The strategy is implemented in a 2D case and numerical results are presented with a first-order Lagrange finite element.

8.1 Introduction

Modern medical imaging modalities can provide a great amount of information to study the human anatomy and physiological functions in both space and time. In cardiac magnetic resonance imaging (MRI), for example, several slices can be acquired to cover the heart in 3D and at a collection of discrete time samples over the cardiac cycle. From these partial observations, the challenge is to extract the heart's dynamics from these input spatio-temporal data throughout the cardiac cycle [10, 12].

Image registration consists in estimating a transformation which insures the warping of one reference image onto another target image (supposed to present

* Corresponding Author
AMS Classification 35F40; 35L85; 35R05; 62–99;

some similarity). Continuous transformations are privileged; the sequence of transformations during the estimation process is usually not much considered. Most important is the final resulting transformation, not the way one image will be transformed to the other. Here, we consider a reasonable registration process to continuously map the image intensity functions between two images in the context of cardiac motion estimation and modeling.

The aim of this chapter is to present, in the context of extended optical flow (EOF), an algorithm to compute a time-dependent transportation plan without using Lagrangian techniques.

The chapter is organized as follows. The introduction continues by recalling the EOF model and its space-time least-squares formulation. In Section 8.2, the algorithm we propose for solving the space-time least-squares formulation of the EOF is presented. Its convergence is discussed. In Section 8.3 the time-dependent optimal mass transportation problem is presented in a formalism similar to the space-time least-squares formulation of the EOF. Section 8.4 deals with numerical results, where a 2D example of cardiac medical images is considered.

8.1.1 The extended optical flow method

Let us denote by ρ the intensity function and by v the velocity of the apparent motion of the brightness pattern. An image sequence is considered via the gray-value map $\rho : Q = (0, 1) \times \Omega \to \mathbb{R}$, where $\Omega \subset \mathbb{R}^d$ is a bounded regular domain, the support of images, for $d = 1, 2, 3$. If image points move according to the velocity field $v : Q \to \mathbb{R}^d$, then gray values $\rho(t, X(t, x))$ are constant along motion trajectories $X(t, x)$. One obtains the optical flow equation:

$$\frac{d}{dt}\rho(t, X(t, x)) = \partial_t \rho(t, X(t, x)) + \big(v \mid \nabla_X \rho(t, X(t, x)) \big)_{\mathbb{R}^d} = 0. \quad (8.1)$$

The assumption that the pixel intensity does not change during the movement is in some cases too restrictive. A weakened assumption, sometimes called EOF, can replace the intensity preservation by a mass preservation condition which reads

$$\partial_t \rho + \big(v \mid \nabla_x \rho \big)_{\mathbb{R}^d} + \operatorname{div}(v)\rho = 0. \quad (8.2)$$

The previous equations lead to an ill-posed problem for the unknown (ρ, v). Variational formulations or relaxed minimizing problems for computing jointly (ρ, v) were first proposed in [4] and afterwards by many other authors. Here, our concern is somewhat different. Finding (ρ, v) simultaneously is possible by solving (8.3) in a least-squares sense.

Let ρ_0 and ρ_1 be the cardiac images between two times arbitrarily fixed to zero and one; the mathematical problem reads: find ρ the gray-level function defined from Q with values in $[0, 1]$ and the velocity function v verifying

$$\begin{cases} \partial_t \rho(t, x) + \text{div}(v(t, x)\rho(t, x)) = 0, & \text{in } (0, 1) \times \Omega \\ \rho(0, x) = \rho_0(x); \quad \rho(1, x) = \rho_1(x) \end{cases} \tag{8.3}$$

Problem (8.3) is solved by minimizing the functional

$$\inf_{\substack{\rho, v \\ \rho(0)=\rho_0;\, \rho(1)=\rho_1 \text{ in } \Omega}} \int_0^1 \int_\Omega [\partial_t \rho(t, x) + \text{div}(v(t, x)\rho(t, x))]^2 \, dt dx. \tag{8.4}$$

Thus, we get an image sequence through the gray-value map ρ. Assume the velocity verifies $v = \nabla\varphi$, then the optimization problem (8.4) reads

$$\inf_{\substack{\{\varphi - C \in L^2((0,1);H_0^1(\Omega)),\, \rho \in L^2((0,1);L^2(\Omega)) \\ \partial_t \rho + \text{div}(\rho\nabla\varphi)) \in L^2((0,1);L^2(\Omega)) \\ \rho(0)=\rho_0;\, \rho(1)=\rho_1 \text{ in } \Omega\}}} \int_0^1 \int_\Omega [\partial_t \rho(t, x) + \text{div}(\rho(t, x)\nabla\varphi(t, x))]^2 \, dt dx.$$

$$\tag{8.5}$$

Let us mention [3], for example, where the optimal mass transportation approach is used in image processing. The mass transportation proposed here is based on a least-squares formulation for the EOF equation ([8] for example), which differs from the methods proposed in [3], or in [5], based on time-dependent optimal mass transportation. For general properties of optimal transportation, the reader is referred to the book by C. Villani [13].

8.2 Algorithm for solving the extended optical flow

In what follows, let us specify our hypotheses.

H1 Ω is a bounded $C^{2,\alpha}$ domain satisfying the exterior sphere condition.
H2 $\rho_i \in C^{1,\alpha}(\overline{\Omega})$ for $i = 0, 1$, and $\rho_0 = \rho_1$ on $\partial\Omega$. Moreover, there exist two constants such that $0 < \beta \le \rho_i \le \overline{\beta}$ in Ω.

Let $\rho^0 \in C^{1,\alpha}([0, 1] \times \overline{\Omega})$ be given by $\rho^0(t, x) = (1 - t)\rho_0(x) + t\rho_1(x)$. We have $\|\partial_t \rho^0\|_{C^{0,\alpha}([0,1]\times\overline{\Omega})} \le C(\rho_0, \rho_1)$ and $\partial_t \rho^0|_{\partial\Omega} = 0$.

Formally, the optimality conditions for the optimization problem (8.5) are given by

$$\int_0^1 \int_\Omega [\partial_t \rho + \text{div}(\rho\nabla\varphi)] [\partial_t \eta + \text{div}(\eta\nabla\varphi)] \, dt dx = 0, \quad \forall \eta$$
$$\int_0^1 \int_\Omega [\partial_t \rho + \text{div}(\rho\nabla\varphi)] \text{div}(\rho\nabla\psi) \, dt dx = 0, \quad \forall \psi. \tag{8.6}$$

If we assume that ρ is a positive function, since the domain Ω is bounded and regular, the eigenfunctions of the operator $-\operatorname{div}(\rho\nabla\cdot)$ with Dirichlet boundary condition constitute a basis of $H_0^1(\Omega)$. Thus, the second equation in (8.6) is reduced to

$$\int_\Omega \partial_t\rho\psi - \rho\nabla\varphi\nabla\psi \, dtdx = 0, \; \forall\psi \in H_0^1(\Omega). \tag{8.7}$$

For each $t \in [0, 1]$, our need for problem (8.3)–(8.4) is a velocity field vanishing on $\partial\Omega$. To do so, the following method is used.

• Compute

$$\begin{cases} -\operatorname{div}(\rho^n(t, \cdot)\nabla\eta) = 0 \text{ in } \Omega \\ \rho^n(t, \cdot)\partial_n\eta = 1 \quad \text{on } \partial\Omega, \end{cases} \tag{8.8}$$

and set $C^n(t) = \frac{1}{|\partial\Omega|} \int_\Omega \partial_t\rho^n\eta \, dx$.
• For each $t \in [0, 1]$ compute φ^{n+1} solution to

$$\begin{cases} -\operatorname{div}(\rho^n(t, \cdot)\nabla\varphi^{n+1}) = \partial_t\rho^n(t, \cdot), \text{ in } \Omega \\ \varphi^{n+1} = C^n(t) \quad \text{on } \partial\Omega. \end{cases} \tag{8.9}$$

• Set $v^{n+1} = \nabla\varphi^{n+1}$.
• Compute ρ^{n+1}, L^2-least-squares solution to

$$\begin{cases} \partial_t\rho^{n+1}(t, x) + \operatorname{div}(v^{n+1}(t, x)\rho^{n+1}(t, x)) = 0, \text{ in } (0, 1) \times \Omega, \\ \rho^{n+1}(0, x) = \rho_0(x); \quad \rho^{n+1}(1, x) = \rho_1(x), \end{cases} \tag{8.10}$$

subject to the constraint $\underline{\beta} \leq \rho^{n+1} \leq \overline{\beta}$.

Remark 1. If the requirement $\rho_0|_{\partial\Omega} = \rho_1|_{\partial\Omega}$ is canceled in Hypothesis H2, then the boundary condition of Problem (8.9) is replaced by $\partial_n\varphi^{n+1} = 0$.

For each $t \in [0, 1]$, since $\rho^n(t, \cdot)$, and $\partial_t\rho^n(t, \cdot) \in C^{0,\alpha}(\overline{\Omega})$, Theorem 6.14 (p. 107) of [9] applies, and there exists a unique $\varphi^{n+1}(t, \cdot) \in C^{2,\alpha}(\overline{\Omega})$ solution of problem (8.9). In problem (8.9) the time is a parameter. As the $\rho^n \in C^{1,\alpha}$; $\partial_t\rho^n \in C^{0,\alpha}$; $C^n \in C^{0,\alpha}$ regularities with respect to time are verified, the classical $C^{2,\alpha}(\overline{\Omega})$ a priori estimates for solutions to elliptic problems allow us to prove that φ^{n+1} is a $C^{0,\alpha}$ function with respect to time. So, we have

$$\|\varphi^{n+1}\|_{C^{0,\alpha}([0,1];C^{2,\alpha}(\overline{\Omega}))} \leq M(\|C^n\|_{C^{0,\alpha}([0,1])} + \|\partial_t\rho^n\|_{C^{0,\alpha}([0,1]\times\overline{\Omega})}).$$

Consider the extension of φ^{n+1} by C^n outside of the domain Ω, still denoted by φ^{n+1}. Since the right-hand side of Equation (8.9) vanishes on $\partial\Omega$, this extension is regular, and the function v^{n+1} vanishes outside Ω and belongs to $C^{0,\alpha}([0, 1]; C^{1,\alpha}(\mathbb{R}^2))$. Hence, the following flows will be constant for $x \in \partial\Omega$.

Define the two flows $X_{\pm}^{n+1}(s, t, x) \in C^{1,\alpha}([0, 1] \times [0, 1] \times \mathbb{R}^2; \mathbb{R}^2)$ by

$$\begin{cases} \frac{d}{ds}X_{\pm}^{n+1}(s, t, x) = \pm v^{n+1}(s, X_{\pm}^{n+1}(s, t, x)) \text{ in } (0, 1) \\ X_{\pm}^{n+1}(t, t, x) = x. \end{cases} \tag{8.11}$$

Theorem 2.5.2 (p. 61) in [11] claims solving the transport equation (8.10) in the least-squares sense is equivalent to solving Equation (8.12):

$$\begin{cases} \partial_t \rho^{n+1}(t, x) + \left(v^{n+1} \mid \nabla_x \rho^{n+1} \right) = -\text{div}(v^{n+1})\rho^{n+1}, \text{ in } (0, 1) \times \Omega, \\ \rho^{n+1}(0, x) = \rho_0(x); \quad \rho^{n+1}(1, x) = \rho_1(x), \end{cases} \tag{8.12}$$

subject to the constraint $\underline{\beta} \le \rho^{n+1} \le \overline{\beta}$, on each integral curve defined by (8.11).

Set $r(t) = \rho^{n+1}(s, X_+^{n+1}(s, t, x))$, and express Equation (8.12) along the integral curves of Equation (8.11). The equation is reduced to an ordinary differential equation with initial and final conditions

$$\begin{cases} \frac{d}{ds}r(s) = -\text{div}\left(v^{n+1}(s, X_+^{n+1}(s, t, x)) \right) r(s) \\ r(0) = \rho_0(X_+^{n+1}(0, t, x)); \ r(1) = \rho_1(X_+^{n+1}(1, t, x)) \end{cases} \tag{8.13}$$

subject to the constraint $\underline{\beta} \le \rho^{n+1} \le \overline{\beta}$. Introducing the change of the unknown, $R(s) = e^{\int_0^s \text{div}(v^{n+1}(\tau, X_+^{n+1}(\tau, t, x)))\, d\tau} r(s)$, we have

$$\begin{cases} \frac{d}{ds}R(s) = 0; \quad 0 < s < 1, \\ R(0) = \rho_0(X_+^{n+1}(0, t, x)); \\ R(1) = e^{\int_0^1 \text{div}(v^{n+1}(\tau, X_+^{n+1}(\tau, t, x)))\, d\tau} \rho_1(X_+^{n+1}(1, t, x)), \end{cases} \tag{8.14}$$

subject to the constraint $h(s, R) = e^{-\int_0^s \text{div}(v^{n+1}(\tau, X_+^{n+1}(\tau, t, x)))\, d\tau} R \in [\underline{\beta}, \overline{\beta}]$. Set $D = [\underline{\beta}, \overline{\beta}]$; $m = R(1) - R(0)$, if the initial and final conditions are extended linearly with respect to the s-variable, since the right-hand side of the ordinary differential equation is a zero mean function, the L^2-least-squares solution to the ordinary differential equation (8.14) with initial and final conditions is the solution to

$$\begin{cases} \frac{d}{ds}R(s) = m; \quad 0 < s < 1, \\ R(0) = \rho_0(X_+^{n+1}(0, t, x)), \end{cases} \tag{8.15}$$

subject to the constraint $h(s, R) \in D$. For the constraint to be taken into account, Nagumo's theorem [2, 11] requires the right-hand side of the differential equation to belong to the contingent cone of the constraint. The contingent cone to

D at point $h(s, R)$ is given by

$$T_D(h(s, R)) = \begin{cases} \mathbb{R} \text{ if } \underline{\beta} \leq h(s, R) \leq \overline{\beta}; \\ \mathbb{R}_+ \text{ if } \underline{\beta} = h(s, R); \\ \mathbb{R}_- \text{ if } h(s, R) = \overline{\beta}. \end{cases} \quad (8.16)$$

According to Lemma 3.3.2.1 (p. 87) in [11], since the constraint depends on the s variable, the contingent cone to the inverse image of a convex subset is

$$T_C(s, R) = \{w \in \mathbb{R}^2; \; Dh(s, R)w \in T_D(h(s, R))s\}.$$

Equation (8.15) is augmented by the unknown s, and the right-hand side of the system becomes $w = \begin{pmatrix} 1 \\ m \end{pmatrix}$. The previous condition is expressed with this vector w, which reads

$$Dh(s, R)\begin{pmatrix} 1 \\ m \end{pmatrix} = \begin{cases} -\operatorname{div}(v^{n+1}) + me^{-\int_0^s \operatorname{div}(v^{n+1}(\tau, X_+^{n+1}(\tau,t,x)))\,d\tau} \text{ if } \underline{\beta} \leq h(s, R) \leq \overline{\beta}; \\ \left[-\operatorname{div}(v^{n+1}) + me^{-\int_0^s \operatorname{div}(v^{n+1}(\tau, X_+^{n+1}(\tau,t,x)))\,d\tau} \right]^+ \text{ if } \underline{\beta} = h(s, R); \\ \left[-\operatorname{div}(v^{n+1}) + me^{-\int_0^s \operatorname{div}(v^{n+1}(\tau, X_+^{n+1}(\tau,t,x)))\,d\tau} \right]^- \text{ if } h(s, R) = \overline{\beta}; \end{cases}$$

$$(8.17)$$

where $[\cdot]^\pm$ denote the positive or negative parts. The strategy is to modify the right-hand side of Equation (8.15) in such a way that condition (8.17) is satisfied. So define

$$m_T(s, R)$$
$$= \text{if } \underline{\beta} \leq h(s, R) \leq \overline{\beta} \quad R_1 - R_0;$$

$$\text{if } \underline{\beta} = h(s, R) \begin{cases} R_1 - R_0 \text{ if } R_1 - R_0 \geq \dfrac{\operatorname{div}(v^{n+1})\underline{\beta}}{e^{-\int_0^s \operatorname{div}(v^{n+1}(\tau, X_+^{n+1}(\tau,t,x)))\,d\tau}} \\ \dfrac{\operatorname{div}(v^{n+1})\underline{\beta}}{e^{-\int_0^s \operatorname{div}(v^{n+1}(\tau, X_+^{n+1}(\tau,t,x)))\,d\tau}} \text{ if } R_1 - R_0 < \dfrac{\operatorname{div}(v^{n+1})\underline{\beta}}{e^{-\int_0^s \operatorname{div}(v^{n+1}(\tau, X_+^{n+1}(\tau,t,x)))\,d\tau}} \end{cases};$$

$$\text{if } h(s, R) = \overline{\beta} \begin{cases} R_1 - R_0 \text{ if } R_1 - R_0 \leq \dfrac{\operatorname{div}(v^{n+1})\overline{\beta}}{e^{-\int_0^s \operatorname{div}(v^{n+1}(\tau, X_+^{n+1}(\tau,t,x)))\,d\tau}} \\ \dfrac{\operatorname{div}(v^{n+1})\overline{\beta}}{e^{-\int_0^s \operatorname{div}(v^{n+1}(\tau, X_+^{n+1}(\tau,t,x)))\,d\tau}} \text{ if } R_1 - R_0 > \dfrac{\operatorname{div}(v^{n+1})\overline{\beta}}{e^{-\int_0^s \operatorname{div}(v^{n+1}(\tau, X_+^{n+1}(\tau,t,x)))\,d\tau}} \end{cases}.$$

$$(8.18)$$

The function m_T is piecewise C^1. The function is regularized according to the parameter $\frac{1}{n}$, so $m_T^n \in C^2$ is such that $Dh(s, R)\begin{pmatrix} 1 \\ m_T^n \end{pmatrix} \in T_D(h(s, R))$.

We have

$$
\rho^{n+1}(s, X_+^{n+1}(s, t, x))) =
\begin{cases}
(1-s)e^{-\int_0^s \operatorname{div}(v^{n+1}(\tau, X_+^{n+1}(\tau, t, x)))\,d\tau}\rho_0(X_+^{n+1}(0, t, x)) \\
\quad + se^{\int_s^1 \operatorname{div}(v^{n+1}(\tau, X_+^{n+1}(\tau, t, x)))\,d\tau}\rho_1(X_+^{n+1}(1, t, x)) \\
\quad \text{if the constraint is not active;} \\
\left(\rho_0(X_+^{n+1}(0, t, x)) + \int_0^s m_T^n(\tau, R(\tau))\,d\tau\right) \\
\quad \left(e^{-\int_0^s \operatorname{div}(v^{n+1}(\tau, X_+^{n+1}(\tau, t, x)))\,d\tau}\right) \text{ if the constraint is active.}
\end{cases}
$$

$$(8.19)$$

We have the following technical lemma, which provides a representation formula for function ρ^{n+1} when the constraint is not active.

Lemma 8.1. *The L^2-least-squares solution to problem* (8.10) *when the constraint is not active is given by*

$$
\rho^{n+1}(t, x) = (1-t)\frac{\rho_0^2(X_+^{n+1}(0, t, x))}{\rho^n(t, x)}
$$
$$
\quad + t\frac{\rho_1^2(X_+^{n+1}(1, t, x))}{\rho^n(t, x)}.
$$

$$(8.20)$$

Proof. We have $X_-^{n+1}(1-s, 1-t, x) = X_+^{n+1}(s, t, x)$ for every $(s, t, x) \in [0, 1] \times [0, 1] \times \mathbb{R}^2$ (for example, see [1]).

Starting from the representation formula (8.19), when the constraint is not active, we deduce

$$
\rho^{n+1}(s, X_+^{n+1}(s, t, x))) = (1-s)e^{-\int_0^s \operatorname{div}(v^{n+1}(\tau, X_+^{n+1}(\tau, t, x)))\,d\tau}\rho_0(X_+^{n+1}(0, t, x))
$$
$$
\quad + se^{\int_s^1 \operatorname{div}(v^{n+1}(\tau, X_+^{n+1}(\tau, t, x)))\,d\tau}\rho_1(X_+^{n+1}(1, t, x)).
$$

$$(8.21)$$

Equation (8.9) gives the following expression for the divergence:

$$
\operatorname{div}(v^{n+1}(s, X_+^{n+1}(s, t, x))) = \operatorname{div}(v^{n+1}(s, X_-^{n+1}(1-s, 1-t, x)))
$$
$$
= \frac{d}{ds}\ln(\rho^n(s, X_-^{n+1}(1-s, 1-t, x))). \quad (8.22)
$$

The representation formula (8.20) is straightforwardly deduced from (8.21). The regularity of the function ρ^{n+1} is a consequence of the regularity of the flow X_+^{n+1}. □

In the case where the constraint is active, the same regularity is true for function ρ^{n+1}, since in representation formula (8.19) the functions are C^1.

Remark 2. Representation formula (8.20) tells us if the algorithm (8.8)–(8.10) converges, then the constraint is not active, but we are solving problem (8.10) with $\rho \in D$.

Let us now consider the convergence of the algorithm (8.8)–(8.10).

Theorem 8.2. *There exist* $(\rho, \varphi) \in C^1([0, 1] \times \overline{\Omega}; \mathbb{R}_+^*) \times C^0([0, 1]; C^2(\overline{\Omega}))$, L^2*-least-squares solutions, respectively solutions to*

$$\begin{cases} \partial_t \rho(t, x) + \operatorname{div}(\nabla \varphi(t, x)\rho(t, x)) = 0, & \text{in } (0, 1) \times \Omega \\ \rho(0, x) = \rho_0(x); \quad \rho(1, x) = \rho_1(x) \text{ in } \Omega \end{cases} \tag{8.23}$$

$$\begin{cases} -\operatorname{div}(\rho(t, \cdot)\nabla \varphi) = \partial_t \rho(t, \cdot), & \text{in } \Omega \\ \varphi = C(t); \; \nabla \varphi = 0 \text{ on } \partial\Omega \end{cases} \tag{8.24}$$

with $C(t)$ *defined by*

$$\begin{cases} -\operatorname{div}(\rho(t, \cdot)\nabla \eta) = 0 \text{ in } \Omega \\ \rho(t, \cdot)\partial_n \eta = 1 \text{ on } \partial\Omega \\ C = \frac{1}{|\partial\Omega|} \int_\Omega \partial_t \rho \, \eta \, dx. \end{cases} \tag{8.25}$$

Proof. Since $\quad \|v^0\|_{C^{0,\alpha}([0,1])} + \|\partial_t \rho^0\|_{C^{0,\alpha}([0,1]\times\overline{\Omega})} \quad$ is bounded, $\|\varphi^{n+1}\|_{C^{0,\alpha}([0,1];C^{2,\alpha}(\overline{\Omega}))}$ and $\|v^{n+1}\|_{C^{0,\alpha}([0,1];C^{1,\alpha}(\mathbb{R}^2))}$ are uniformly bounded in n.

From Lemma 8.1 there exists a unique ρ^{n+1}, the L^2-least squares solution of (8.10). Let us give an estimate for $D_3 X_+^{n+1}$. Starting from

$$D_1 X_+^{n+1}(s, t, x)) = v^{n+1}(s, X_+^{n+1}(s, t, x)),$$

we deduce (see [1])

$$\begin{cases} D_3 D_1 X_+^{n+1}(s, t, x) = D_2 v^{n+1}(s, X_+^{n+1}(s, t, x)) D_3 X_+^{n+1}(s, t, x) \\ D_3 X_+^{n+1}(t, t, x) = Id. \end{cases} \tag{8.26}$$

Since $D_3 D_1 X_+^{n+1}(s, t, x) = D_1 D_3 X_+^{n+1}(s, t, x)$ we get

$$D_3 X_+^{n+1}(s, t, x) = e^{-\int_t^s D_2(v^{n+1}(\tau, X_+^{n+1}(\tau, t, x))) d\tau} Id. \tag{8.27}$$

Thus, $\|D_3 v_+^{n+1}\|_{C^{0,\alpha}([0,1]^2 \times \mathbb{R}^2)}$ is uniformly bounded in n.

Since we have [1]

$$D_2 X_+^{n+1}(s, t, x) = \left(v^{n+1}(s, t, x) \mid D_3 X_+^{n+1}(s, t, x) \right),$$

we obtain a bound for $\|D_2 v^{n+1}\|_{C^{0,\alpha}([0,1]^2 \times \mathbb{R}2)}$ independent of n.

From Lemma 8.1 we deduce that $\|\rho^{n+1}\|_{C^{1,\alpha}([0,1]\times\overline{\Omega})}$ is uniformly bounded. Since the embeddings

$$C^{0,\alpha}([0, 1]; C^{2,\alpha}(\overline{\Omega})) \hookrightarrow C^0([0, 1]; C^2(\overline{\Omega})) \text{ and}$$

$$C^{1,\alpha}([0, 1] \times \overline{\Omega}) \hookrightarrow C^1([0, 1] \times \overline{\Omega})$$

are relatively compact, there is a subsequence of (ρ^n, φ^n) solution to (8.8)–(8.10), still denoted by (ρ^n, φ^n) converging to (ρ, φ) in $C^1([0, 1] \times \overline{\Omega}) \times C^0([0, 1]; C^2(\overline{\Omega}))$, and (ρ, φ) is the solution of (8.23)–(8.25) provided the boundary conditions are justified. The condition $\nabla \varphi^n|_{\partial\Omega} = 0$ is valid for the approximations φ^n (since the functions can be extended by C^n outside of Ω). So the convergence in $C^0([0, 1]; C^2(\overline{\Omega}))$ yields the condition for the gradient of the limit function. For the approximations of function ρ, the formula given in Lemma 8.1 combined with the regularity result shows that the boundary conditions are exactly satisfied. These conditions are thus valid for the limit function due to the convergence in C^1. $\qquad\square$

8.3 Comparison with the time-dependent optimal transport solution

Assume (q, w) with $\underline{\beta} \le q \le \overline{\beta}$ is the solution to the time-dependent optimal transport problem

$$(q, w) = \underset{\substack{\{v\in L^2((0,1);(L^2(\Omega))^2), \ u\in L^2((0,1);L^2(\Omega)) \\ \underline{\beta} \le u \le \overline{\beta} \\ \partial_t u + \operatorname{div}(uv)=0; \ u(0)=\rho_0; \ u(1)=\rho_1 \text{ in } \Omega\}}}{\operatorname{Argmin}} \int_0^1 \int_\Omega u\|v\|^2 \, dx dt. \qquad (8.28)$$

Let $H = H_0^1(\Omega)$ be equipped with the following inner product:

$$(\theta, \psi) = \int_\Omega q \left(\nabla\theta \,|\, \nabla\psi\right) dx,$$

which induces a semi-norm that is equivalent to the H^1-norm since $0 < \underline{\beta} \le q \le \overline{\beta}$. Riez's theorem claims that for the linear continuous form

$$\mathcal{L}_w(\psi) = \langle -\operatorname{div}(qw), \psi \rangle_{H;H'} = \langle \partial_t q, \psi \rangle_{H;H'},$$

there is a unique $\theta \in H$ such that

$$\mathcal{L}_w(\psi) = \int_\Omega q \left(\nabla\theta \,|\, \nabla\psi\right) dx, \ \forall \psi \in H.$$

Therefore, $w = \nabla\theta$ for a $\theta \in H_0^1(\Omega)$ and problem (8.28) is reduced to

$$\underset{\substack{\{\psi\in L^2((0,1);H_0^1(\Omega)), \ u\in L^2((0,1);L^2(\Omega)) \\ \underline{\beta} \le u \le \overline{\beta} \\ \partial_t u + \operatorname{div}(u\nabla\psi)=0; \ u(0)=\rho_0; \ u(1)=\rho_1 \text{ in } \Omega\}}}{\operatorname{Argmin}} \int_0^1 \int_\Omega u\|\nabla\psi\|^2 \, dx dt. \qquad (8.29)$$

Accounting for the result

$$\int_\Omega u\|\nabla\psi\|^2\,dx = \|-\operatorname{div}(u\nabla\psi)\|_{H'}^2,$$

problem (8.29) reads

$$\operatorname*{Argmin}_{\substack{\{-\psi\in L^2((0,1);H_0^1(\Omega)),\ u\in L^2((0,1);L^2(\Omega)) \\ \underline{\beta}\le u\le\overline{\beta} \\ \partial_t u-\operatorname{div}(u\nabla\psi)=0;\,u(0)=\rho_0;\,u(1)=\rho_1\ \text{in}\ \Omega\}}} \int_0^1 \|\operatorname{div}(-u\nabla\psi)\|_{H^{-1}}^2\,dt. \qquad (8.30)$$

From the definition of $\mathcal{L}$, we observe that

$$\frac{1}{4}\|\operatorname{div}(-u\nabla\psi)+\partial_t u\|_{H'}^2 = \|\operatorname{div}(-u\nabla\psi)\|_{H'}^2$$

and we deduce that problem (8.28) reads

$$\operatorname*{Argmin}_{\substack{\{\psi\in L^2((0,1);H_0^1(\Omega)),\ u\in L^2((0,1);L^2(\Omega)) \\ \underline{\beta}\le u\le\overline{\beta} \\ \partial_t u+\operatorname{div}(u\nabla\psi)=0;\,u(0)=\rho_0;\,u(1)=\rho_1\ \text{in}\ \Omega\}}} \frac{1}{4}\int_0^1 \|\operatorname{div}(-u\nabla\psi)+\partial_t u\|_{H'}^2\,dt. \qquad (8.31)$$

8.4 Numerical approximation of the 2D extended optical flow

The numerical method is based on a finite-element space-time L^2 least-squares formulation (see [6]) of the linear conservation law (8.10) subject to the constraint $\rho \in D = [\underline{\beta}, \overline{\beta}]$.

Define $\tilde{v}^{n+1}$ as

$$\tilde{v}^{n+1} = (1, v_1^{n+1}, v_2^{n+1})^t$$

and, for a sufficiently regular function φ defined on Q, set

$$\tilde{\nabla}\varphi = \left(\frac{\partial\varphi}{\partial t}, \frac{\partial\varphi}{\partial x_1}, \frac{\partial\varphi}{\partial x_2}\right)^t,$$

and

$$\widetilde{\operatorname{div}}(\tilde{v}^{n+1}\,\varphi) = \frac{\partial\varphi}{\partial t} + \sum_{i=1}^{2}\frac{\partial}{\partial x_i}(v_i^{n+1}\,\varphi).$$

Let $\{\varphi_1 \cdots \varphi_N\}$ be a basis of a space-time finite-element subspace

$V_h = \{\varphi,$ piecewise regular polynomial functions, with $\varphi(0, \cdot) = \varphi(1, \cdot) = 0\}$;

for example, a brick Lagrange finite element of order one [7]. Let Π_h be the Lagrange interpolation operator. Let also W_h be the finite-element subspace of $H_0^1(\Omega)$, where the basis functions $\{\psi_1 \cdots \psi_M\}$ are the traces at $t = 0$ of basis functions $\{\varphi_i\}_{i=1}^N$. An approximation of problem (8.9) is: for a discrete sequence of time t compute

$$\int_\Omega \rho_h^n(t, \cdot) \left(\nabla(\varphi_h^{n+1} - C^n(t)) \mid \nabla \psi_h \right) dx = \int_\Omega \partial_t \rho_h^n(t, \cdot) \psi_h \, dx \quad \forall \psi_h \in W_h,$$

$$(8.32)$$

and define $\tilde{v}^{n+1} = \nabla \varphi_h^{n+1}$. The L^2 least-squares formulation of problem (8.10) is defined in the following way. Consider the functional

$$J(c_h) = \frac{1}{2} \int_Q \left(\widetilde{\operatorname{div}}(\tilde{v}^{n+1} c_h) + \widetilde{\operatorname{div}} \left[\tilde{v}^{n+1} \Pi_h \big((1 - t)\rho_0 + t\rho_1\big) \right] \right)^2 dx \, dt.$$

This functional is convex and coercive in an appropriate anisotropic Sobolev space $H = \{\varphi \in L^2(Q); \widetilde{\operatorname{div}}(\tilde{v}^{n+1}\varphi) \in L^2(Q); \varphi(0, \cdot) = \varphi(1, \cdot) = 0\}$ since the velocity field v^{n+1} is sufficiently regular. Moreover, $\left\| \widetilde{\operatorname{div}}(\tilde{v}^{n+1}\varphi) \right\|_{L^2(Q)}$ is a norm in H (see [6]). Set

$$\theta_h = \Pi_h\big((1 - t)\rho_0 + t\rho_1\big),$$

and introduce

$$K_h = \{\varphi_h \in V_h; \; \varphi_h + \theta_h \in D\},$$

which is a closed convex subspace. Thus, an approximation of (8.10) is defined with the following minimization problem:

$$\min_{c \in K_h} J(c). \qquad (8.33)$$

The minimizer of problem (8.33) is

$$c_h = \rho_h^{n+1} - \theta_h,$$

which is characterized by the Fermat rule [2]

$$\int_Q \widetilde{\operatorname{div}}(\tilde{v}^{n+1} c_h^{n+1}) \cdot \widetilde{\operatorname{div}}(\tilde{v}^{n+1} \psi_h) \, dx \, dt$$

$$\geq \int_Q \left(-\widetilde{\operatorname{div}} \left(\tilde{v}^{n+1} \theta_h \right) \right) \cdot \widetilde{\operatorname{div}}(\tilde{v}^{n+1} \psi_h) \, dx \, dt \qquad (8.34)$$

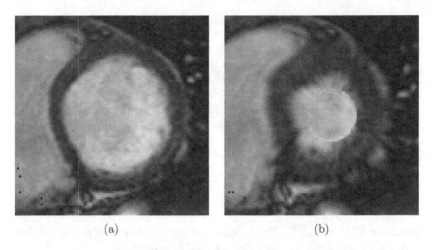

Figure 8.1 End of diastole of a left ventricle (a) and of systole (b).

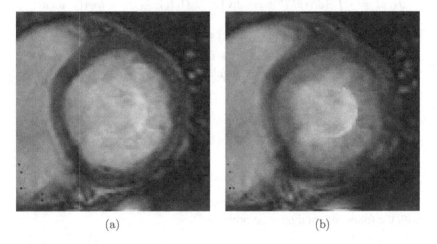

Figure 8.2 Time steps (a) 3 and (b) 6.

for all $\psi_h \in T_{K_h}$, the contingent cone to K_h, or by appropriate complementary conditions. Here, c_h^{n+1} is defined by

$$c_h^{n+1} = \sum_{i=1}^{N} c_i^{n+1} \varphi_i(t, x).$$

Thus, an approximation of the solution to problem (8.10) is

$$c_h^{n+1} + \Pi_h\big((1-t)\rho_0 + t\rho_1\big) \in V_h.$$

The iterative strategy described in Section 8.2 is used to compute an approximated solution, and to reconstruct the systole to diastole images of a slice of a left ventricle. Ten time steps have been used to compute the solution, and 10 000 degrees of freedom for the space-time least-squares finite element. The approximated fixed-point algorithm converges in about 10 iterations with an accuracy of about 10^{-7}. In Figure 8.1, the initial image and the final image are presented. In Figure 8.2, two intermediate times, $1/3$ and $2/3$, are shown.

To summarize, in this work, we present a fixed-point algorithm for the computation of the extended optical flow problem, allowing to handle the images tracking problem.

References

[1] L. Ambrosio, Transport equation and Cauchy problem for BV vector fields. *Invent. Math.*, 158, 227–260, (2004).

[2] J.P. Aubin, *Viability Theory*. Birkhauser, 1991.

[3] S. Angenent, S. Haker, and A. Tannenbaum, Minimizing flows for the Monge–Kantorovich problem, *SIAM J. Math. Anal.*, 35, 61–97, (2003).

[4] G. Aubert, R. Deriche, and P. Kornprobst, Computing optical flow problem via variational techniques. *SIAM J. Appl. Math.*, 80, 156–182, (1999).

[5] J. Benamou and Y. Brenier, A computational fluid mechanics solution to the Monge-Kantorovich mass transfer problem, *Numer. Math.*, 84, 375–393, (2000).

[6] O. Besson and J. Pousin, Solutions for linear conservation laws with velocity fields in L^∞. *Arch. Rational Mech. Anal.*, 186, 159–175, (2007).

[7] O. Besson and G. de Montmollin, Space-time integrated least squares: a time marching approach. *Int. J. Numer. Meth. Fluids*, 44, 525–543, (2004).

[8] P.B. Bochev and M.D. Gunzburger, *Least-Squares Finite Element Methods*, Applied Mathematical Sciences, volume 166, Springer, 2009.

[9] D. Gilbarg and N.S. Trundinger, *Elliptic Partial Differential Equations of Second Order*, Springer, 2001.

[10] M. Lynch, O. Ghita, and P.F. Whelan, Segmentation of the left ventricle of the heart in 3D+t MRI data using an optimized non-rigid temporal model, *IEEE Trans. Med. Imaging*, 27, 195–203, (2008).

[11] M. Picq, Résolution de l'équation du transport sous contraintes. Editions Universitaires, 2010.

[12] J. Schaerer, P. Clarysse, and J. Pousin, A new dynamic elastic model for cardiac image analysis, in *Proceedings of the 29th Annual International Conference of the IEEE EMBS*, Lyon, France, 4488–4491, 2007.

[13] C. Villani, Topics in Optimal Transportation, Graduate Studies in Mathematics 58, American Mathematical Society, Providence, RI, 2003.

9

On the duality theory for the
Monge–Kantorovich transport problem

MATHIAS BEIGLBÖCK, CHRISTIAN LÉONARD, AND
WALTER SCHACHERMAYER

9.1 Introduction

This chapter, which is an accompanying paper to [BLS09], consists of two
parts. In Section 9.2 we present a version of Fenchel's perturbation method for
the duality theory of the Monge–Kantorovich problem of optimal transport.
The treatment is elementary as we suppose that the spaces $(X, \mu), (Y, \nu)$, on
which the optimal transport problem [Vil03, Vil09] is defined, simply equal
the finite set $\{1, \ldots, N\}$ equipped with uniform measure. In this setting the
optimal transport problem reduces to a finite-dimensional linear programming
problem.

The purpose of this first part of the paper is rather didactic: it should stress
some features of the linear programming nature of the optimal transport problem, which carry over also to the case of general Polish spaces X, Y equipped
with Borel probability measures μ, ν, and general Borel measurable cost functions $c : X \times Y \to [0, \infty]$. This general setting is analysed in detail in [BLS09];
Section 9.2 may serve as a motivation for the arguments in the proof of Theorems 1.2 and 1.7 of [BLS09] which pertain to the general duality theory.

The second – and longer – part of the paper, consisting of Sections 9.3 and
9.4, is of a quite different nature.

Section 9.3 is devoted to illustrating a technical feature of [BLS09, Theorem
4.2] by an explicit example. The technical feature is the appearance of the
singular part $\widehat{h}^s$ of the dual optimizer $\widehat{h} \in L^1(X \times Y, \pi)^{**}$ obtained in [BLS09,
Theorem 4.2]. In Example 1 in Section 9.3 we show that, in general, the dual
optimizer $\widehat{h}$ does indeed contain a non-trivial singular part. In addition, this
example allows one to observe in a rather explicit way how this singular part
'builds up', for an optimizing sequence $(\varphi_n \oplus \psi_n)_{n=1}^{\infty} \in L^1(X \times Y, \pi)$ which
converges to $\widehat{h}$ with respect to the weak-star topology. The construction of this
example, which is a variant of an example due to Ambrosio and Pratelli [AP03],

216

is rather long and technical. Some motivation for this construction will be given at the end of Section 9.2.

Section 9.4 pertains to a modified version of the duality relation in the Monge–Kantorovich transport problem. Trivial counterexamples such as [BLS09, Example 1.1] show that in the case of a measurable cost function $c : X \times Y \to [0, \infty]$ there may be a duality gap. The main result (Theorem 1.2) of [BLS09] asserts that one may avoid this difficulty by considering a suitable relaxed form of the primal problem; if one does so, duality holds true in complete generality. In a different vein, one may leave the primal problem unchanged, and overcome the difficulties encountered in the above-mentioned simple example by considering a slightly modified dual problem (cf. [BLS09, Remark 3.4]). In the last part of the article we consider a certain twist of the construction given in Section 9.3, which allows us to prove that this dual relaxation does not lead to a general duality result.

9.2 The finite case

In this section we present the duality theory of optimal transport for the finite case. Let $X = Y = \{1, \ldots, N\}$ and let $\mu = \nu$ assign probability N^{-1} to each of the points $1, \ldots, N$. Let $c = (c(i, j))_{i,j=1}^{N}$ be an $\mathbb{R}_+$-valued $N \times N$ matrix.

The problem of optimal transport then becomes the subsequent linear optimization problem

$$\langle c, \pi \rangle := \sum_{i=1}^{N} \sum_{j=1}^{N} \pi(i, j)\, c(i, j) \to \min, \quad \pi \in \mathbb{R}^{N^2}, \tag{9.1}$$

under the constraints

$$\sum_{j=1}^{N} \pi(i, j) = N^{-1}, \quad i = 1, \ldots, N,$$

$$\sum_{i=1}^{N} \pi(i, j) = N^{-1}, \quad j = 1, \ldots, N,$$

$$\pi(i, j) \geq 0, \quad i, j = 1, \ldots, N.$$

Of course, this is an easy and standard problem of linear optimization; yet we want to treat it in some detail in order to develop intuition and concepts for the general case considered in [BLS09] as well as in Section 9.3.

For the two sets of *equality* constraints we introduce $2N$ Lagrange multipliers $(\varphi(i))_{i=1}^{N}$ and $(\psi(j))_{j=1}^{N}$ taking values in $\mathbb{R}$, and for the inequality constraints we introduce Lagrange multipliers $(\varrho_{ij})_{i,j=1}^N$ taking values in $\mathbb{R}_+$. The Lagrangian functional $L(\pi, \varphi, \psi, \varrho)$ is then given by

$$L(\pi, \varphi, \psi, \varrho) = \sum_{i=1}^N \sum_{j=1}^N c(i, j)\pi(i, j)$$

$$- \sum_{i=1}^N \varphi(i) \left(\sum_{j=1}^N \pi(i, j) - N^{-1} \right)$$

$$- \sum_{j=1}^N \psi(j) \left(\sum_{i=1}^N \pi(i, j) - N^{-1} \right)$$

$$- \sum_{i=1}^N \sum_{j=1}^N \varrho(i, j)\pi(i, j),$$

where the $\pi(i, j)$, $\varphi(i)$ and $\psi(j)$ range in $\mathbb{R}$, while the $\varrho(i, j)$ range in $\mathbb{R}_+$.

It is designed in such a way that

$$C(\pi) := \sup_{\varphi, \psi, \varrho} L(\pi, \varphi, \psi, \varrho) = \langle c, \pi \rangle + \chi_{\Pi(\mu, \nu)}(\pi),$$

where $\Pi(\mu, \nu)$ denotes the admissible set of π values; i.e., the probability measures on $X \times Y$ with marginals μ and ν, and $\chi_A(\ .\ )$ denote the indicator function of a set A in the sense of convex function theory, namely, taking the value 0 on A and the value $+\infty$ outside of A.

In particular, we have

$$P := \inf_{\pi \in \mathbb{R}^{N^2}} C(\pi) = \inf_{\pi \in \mathbb{R}^{N^2}} \sup_{\varphi, \psi, \varrho} L(\pi, \varphi, \psi, \varrho),$$

where P is the optimal value of the primal optimization problem (9.1).

To develop the duality theory of the primal problem (9.1) we pass from inf sup L to sup inf L. Denote by $D(\varphi, \psi, \varrho)$ the dual function

$$D(\varphi, \psi, \varrho) = \inf_{\pi \in \mathbb{R}^{N^2}} L(\pi, \varphi, \psi, \varrho)$$

$$= \inf_{\pi \in \mathbb{R}^{N^2}} \sum_{i=1}^N \sum_{j=1}^N \pi(i, j)[c(i, j) - \varphi(i) - \psi(j) - \varrho(i, j)]$$

$$+ N^{-1} \left[\sum_{i=1}^N \varphi(i) + \sum_{j=1}^N \psi(j) \right].$$

Hence we obtain as the optimal value of the dual problem

$$D := \sup_{\varphi, \psi, \varrho} D(\varphi, \psi, \varrho) = (\mathbb{E}_\mu[\varphi] + \mathbb{E}_\nu[\psi]) - \chi_\Psi(\varphi, \psi), \qquad (9.2)$$

where Ψ denotes the admissible set of φ, ψ, i.e. satisfying

$$\varphi(i) + \psi(j) + \varrho(i, j) = c(i, j), \quad 1 \le i, j \le N,$$

for some non-negative 'slack variables' $\varrho(i, j)$.

Let us show that there is no duality gap, i.e. the values of P and D coincide. Of course, in the present finite-dimensional case, this equality as well as the fact that the inf sup (resp. sup inf) above is a min max (resp. a max min) easily follows from general compactness arguments. Yet we want to verify things directly using the idea of 'complementary slackness' of the primal and the dual constraints (e.g. good references are [PSU88, ET99, AE06]).

We apply 'Fenchel's perturbation map' to explicitly show the equality $P = D$. Let $T : \mathbb{R}^{N^2} \to \mathbb{R}^N \times \mathbb{R}^N$ be the linear map defined as

$$T\left((\pi(i, j))_{1 \le i, j \le N} \right) = \left(\left(\sum_{j=1}^{N} \pi(i, j) \right)_{i=1}^{N}, \left(\sum_{i=1}^{N} \pi(i, j) \right)_{j=1}^{N} \right)$$

so that the problem (9.1) now can be phrased as

$$\langle c, \pi \rangle = \sum_{i=1}^{N} \sum_{j=1}^{N} c(i, j) \pi(i, j) \to \min, \quad \pi \in \mathbb{R}_+^{N^2},$$

under the constraint

$$T(\pi) = \left((N^{-1}, \ldots, N^{-1}), (N^{-1}, \ldots, N^{-1}) \right).$$

The range of the linear map T is the subspace $E \subseteq \mathbb{R}^N \times \mathbb{R}^N$, of codimension 1, formed by the pairs (f, g) such that $\sum_{i=1}^{N} f(i) = \sum_{j=1}^{N} g(j)$, in other words $\mathbb{E}_\mu[f] = \mathbb{E}_\nu[g]$. We consider T as a map from $\mathbb{R}^{N^2}$ to E and denote by E_+ the positive orthant of E.

Let $\Phi : E_+ \to [0, \infty]$ be the map

$$\Phi(f, g) = \inf \left\{ \langle c, \pi \rangle, \ \pi \in \mathbb{R}_+^{N^2}, \ T(\pi) = (f, g) \right\}.$$

We shall verify explicitly that Φ is an $\mathbb{R}_+$-valued, convex, lower semi-continuous, positively homogeneous map on E_+.

The finiteness and positivity of Φ follow from the fact that, for $(f, g) \in E_+$, the set of $\pi \in \mathbb{R}_+^{N^2}$ with $T(\pi) = (f, g)$ is non-empty and from the non-negativity of c. As regards the convexity of Φ, let $(f_1, g_1), (f_2, g_2) \in E_+$ and find $\pi_1, \pi_2 \in \mathbb{R}_+^{N^2}$ such that $T(\pi_1) = (f_1, g_1)$, $T(\pi_2) = (f_2, g_2)$ and $\langle c, \pi_1 \rangle <$ $\Phi(f_1, g_1) + \varepsilon$ as well as $\langle c, \pi_2 \rangle < \Phi(f_2, g_2) + \varepsilon$. Then

$$\Phi\left(\frac{(f_1, g_1) + (f_2, g_2)}{2}\right) \leq \left\langle c, \frac{\pi_1 + \pi_2}{2} \right\rangle < \frac{\Phi(f_1, g_1) + \Phi(f_2, g_2)}{2} + \varepsilon,$$

which proves the convexity of Φ.

If $((f_n, g_n))_{n=1}^{\infty} \in E_+$ converges to (f, g), find $(\pi_n)_{n=1}^{\infty}$ in $\mathbb{R}_+^{N^2}$ such that $T(\pi_n) = (f_n, g_n)$ and $\langle c, \pi_n \rangle < \Phi(f_n, g_n) + n^{-1}$. Note that $(\pi_n)_{n=1}^{\infty}$ is bounded in $\mathbb{R}_+^{N^2}$, so that there is a subsequence $(\pi_{n_k})_{k=1}^{\infty}$ converging to $\pi \in \mathbb{R}_+^{N^2}$. Hence, $\Phi(f, g,) \leq \langle c, \pi \rangle$ showing the lower semi-continuity of Φ. Finally note that Φ is positively homogeneous, i.e. $\Phi(\lambda f, \lambda g) = \lambda \Phi(f, g)$, for $\lambda \geq 0$.

The point (f_0, g_0) with $f_0 = g_0 = (N^{-1}, \ldots, N^{-1})$ is in E_+ and Φ is bounded in a neighbourhood V of (f_0, g_0). Indeed, fixing any $0 < a < N^{-1}$, the subsequent set V does the job

$$V = \{(f, g) \in E : |f(i) - N^{-1}| < a, \ |g(j) - N^{-1}| < a, \ \text{for } 1 \leq i, j \leq N\}.$$

The boundedness of the lower semi-continuous convex function Φ on V implies that the subdifferential of Φ at (f_0, g_0) is non-empty. Considering Φ as a function on $\mathbb{R}^{2N}$ (by defining it to equal $+\infty$ on $\mathbb{R}^{2N} \setminus E_+$) we may find an element $(\widehat{\varphi}, \widehat{\psi}) \in \mathbb{R}^N \times \mathbb{R}^N$ in this subdifferential. By the positive homogeneity of Φ we have

$$\Phi(f, g) \geq \langle (\widehat{\varphi}, \widehat{\psi}), (f, g) \rangle = \langle \widehat{\varphi}, f \rangle + \langle \widehat{\psi}, g \rangle, \quad \text{for } (f, g) \in \mathbb{R}^N \times \mathbb{R}^N,$$

and

$$P = \Phi(f_0, g_0) = \langle \widehat{\varphi}, f_0 \rangle + \langle \widehat{\psi}, g_0 \rangle.$$

By the definition of Φ we therefore have, for each $\pi \in \mathbb{R}_+^{N^2}$,

$$\langle c, \pi \rangle \geq \inf_{\tilde{\pi} \in \mathbb{R}_+^{N^2}} \{\langle c, \tilde{\pi} \rangle : T(\pi) = T(\tilde{\pi})\}$$

$$= \Phi(T(\pi))$$

$$\geq \langle T(\pi), (\hat{\varphi}, \hat{\psi}) \rangle$$

$$= \sum_{i=1}^{N} \sum_{j=1}^{N} \pi(i, j) [\widehat{\varphi}(i) + \widehat{\psi}(j)]$$

so that

$$c(i, j) \geq \widehat{\varphi}(i) + \widehat{\psi}(j), \quad \text{for } 1 \leq i, j \leq n. \tag{9.3}$$

By compactness, there is $\widehat{\pi} \in \Pi(\mu, \nu)$, i.e. there is an element $\widehat{\pi} \in \mathbb{R}_+^{N^2}$ verifying $T(\widehat{\pi}) = (f_0, g_0)$ such that

$$\langle c, \widehat{\pi} \rangle = \langle \widehat{\varphi} + \widehat{\psi}, \widehat{\pi} \rangle. \tag{9.4}$$

Summing up, we have shown that $\widehat{\pi}$ and $(\widehat{\varphi}, \widehat{\psi})$ are primal and dual optimizers and that the value of the primal problem equals the value of the dual problem, namely $\langle \widehat{\varphi} + \widehat{\psi}, \widehat{\pi} \rangle$.

To finish this elementary treatment of the finite case, let us consider the case when we allow the cost function c to take values in $[0, \infty]$ rather than in $[0, \infty[$. In this case the primal problem simply loses some dimensions: for the (i, j) values where $c(i, j) = \infty$ we must have $\pi(i, j) = 0$ so that we consider

$$\langle c, \pi \rangle := \sum_{i=1}^{N} \sum_{j=1}^{N} \pi(i, j) c(i, j) \to \min, \ \pi \in \mathbb{R}_+^{N^2},$$

where we now optimize over $\pi \in \mathbb{R}_+^{N^2}$ with $\pi(i, j) = 0$ if $c(i, j) = \infty$. For the problem to make sense we clearly must have that there is at least one $\pi \in \Pi(\mu, \nu)$ with $\langle c, \pi \rangle < \infty$. If this non-triviality condition is satisfied, the above arguments carry over without any non-trivial modification.

We now analyse explicitly the well-known 'complementary slackness conditions' and interpret them in the present context. For a pair $\widehat{\pi}$ and $(\widehat{\varphi}, \widehat{\psi})$ of primal and dual optimizers we have

$$c(i, j) > \widehat{\varphi}(i) + \widehat{\psi}(j) \ \Rightarrow \ \widehat{\pi}(i, j) = 0$$

and

$$\widehat{\pi}(i, j) > 0 \ \Rightarrow \ c(i, j) = \widehat{\varphi}(i) + \widehat{\psi}(j).$$

Indeed, these relations follow from the admissibility condition $c \geq \widehat{\varphi} + \widehat{\psi}$ and the duality relation $\langle \widehat{\pi}, c - (\widehat{\varphi} + \widehat{\psi}) \rangle = 0$.

This motivates the following definitions in the theory of optimal transport (e.g. see [RR96] for (a) and [ST08] for (b).)

Definition 1. Let $X = Y = \{1, \ldots, N\}$ and $\mu = \nu$ the uniform distribution on X and Y respectively, and let $c : X \times Y \to \mathbb{R}_+$ be given.

(a) A subset $\Gamma \subseteq X \times Y$ is called 'cyclically c-monotone' if, for $(i_1, j_1), \ldots, (i_n, j_n) \in \Gamma$ we have

$$\sum_{k=1}^{n} c(i_k, j_k) \leq \sum_{k=1}^{n} c(i_k, j_{k+1}), \tag{9.5}$$

where $j_{n+1} = j_1$.

(b) A subset $\Gamma \subseteq X \times Y$ is called 'strongly cyclically c-monotone' if there are
functions φ, ψ such that $\varphi(i) + \psi(j) \le c(i, j)$, for all $(i, j) \in X \times Y$, with
equality holding true for $(i, j) \in \Gamma$.

In the present finite setting, the following facts are rather obvious (assertion
(iii) following from the above discussion):

(i) The support of each primal optimizer $\widehat{\pi}$ is cyclically c-monotone.
(ii) Every $\pi \in \Pi(\mu, \nu)$ which is supported by a cyclically c-monotone set Γ
is a primal optimizer.
(iii) A set $\Gamma \subseteq X \times Y$ is cyclically c-monotone iff it is strongly cyclically
c-monotone.

In general, one may ask, for a given Monge–Kantorivich transport optimization problem, defined on Polish spaces X, Y, equipped with Borel probability
measures μ, ν, and a Borel measurable cost function $c : X \times Y \to [0, \infty]$, the
following natural questions:

(P) Does there exist a primal optimizer to (9.1), i.e. a Borel measure $\widehat{\pi} \in$
$\Pi(\mu, \nu)$ with marginals μ, ν, such that

$$\int_{X \times Y} c \, d\widehat{\pi} = \inf_{\pi \in \Pi(\mu, \nu)} \int_{X \times Y} c \, d\pi =: P$$

holds true?

(D) Do there exist dual optimizers to (9.2), i.e. Borel functions $(\widehat{\varphi}, \widehat{\psi})$ in
$\Psi(\mu, \nu)$, such that

$$\int_X \widehat{\varphi} \, d\mu + \int_Y \widehat{\psi} \, d\nu = \sup_{(\varphi, \psi) \in \Psi(\mu, \nu)} \left(\int_X \varphi \, d\mu + \int_Y \psi \, d\nu \right) =: D, \quad (9.6)$$

where $\Psi(\mu, \nu)$ denotes the set of all pairs of $[-\infty, +\infty[$-valued integrable
Borel functions (φ, ψ) on X, Y such that $\varphi(x) + \psi(y) \le c(x, y)$, for all $(x, y) \in$
$X \times Y$?

(DG) Is there a duality gap, or do we have $P = D$, as it should – morally
speaking – hold true?

These are three natural questions which arise in every convex optimization
problem. In addition, one may ask the following two questions pertaining to
the special features of the Monge–Kantorovich transport problem.

(CC) Is every cyclically c-monotone transport plan $\pi \in \Pi(\mu, \nu)$ optimal,
where we call $\pi \in \Pi(\mu, \nu)$ cyclically c-monotone if there is a Borel subset

$\Gamma \subseteq X \times Y$ of full support $\pi(\Gamma) = 1$, verifying condition (9.5), for any $(x_1, y_1), \ldots, (x_n, y_n) \in \Gamma$?

(SCC) Is every strongly cyclically c-monotone transport plan $\pi \in \Pi(\mu, \nu)$ optimal, where we call $\pi \in \Pi(\mu, \nu)$ strongly cyclically c-monotone if there are Borel functions $\varphi : X \to [-\infty, +\infty[$ and $\psi : Y \to [-\infty, +\infty[$, satisfying $\varphi(x) + \psi(y) \leq c(x, y)$, for all $(x, y) \in X \times Y$, and $\pi\{\varphi + \psi = c\} = 1$?

Much effort has been made over the past decades to provide increasingly general answers to the questions above. We mention the work of Rüschendorf [Rüs96] who adapted the notion of cyclical monotonicity from Rockafellar [Roc66]. Rockafellar's work pertains to the case $c(x, y) = -\langle x, y \rangle$, for $x, y \in \mathbb{R}^n$, while Rüschendorf's work pertains to the present setting of general cost functions c, thus arriving at the notion of cyclical c-monotonicity. Intimately related is the notion of the c-conjugate φ^c of a function φ.

We also mention Kellerer's fundamental work on the duality theory; in [Kel84] he established that $P = D$ provided that $c : X \times Y \to [0, \infty]$ is lower semi-continuous, or merely Borel-measurable and uniformly bounded.

The seminal paper [GM96] proves (among many other results) that we have a positive answer to question (CC) above in the following situation: every cyclically c-monotone transport plan is optimal provided that the cost function c is continuous and X, Y are compact subsets of $\mathbb{R}^n$. In [Vil03, Problem 2.25] it is asked whether this extends to the case $X = Y = \mathbb{R}^n$ with the squared Euclidian distance as cost function. This was answered independently in [Pra08] and [ST08]: the answer to (CC) is positive for general Polish spaces X and Y, provided that the cost function $c : X \times Y \to [0, \infty]$ is continuous [Pra08] or lower semi-continuous and finitely valued [ST08]. Indeed, in the latter case, a transport plan is optimal if and only if it is strongly c-monotone.

Let us briefly summarize the state of the art pertaining to the five questions above.

As regards the most basic issue, namely (DG) pertaining to the question whether duality makes sense at all, this is analysed in detail – building on a lot of previous literature – in Section 2 of the accompanying paper [BLS09]: it is shown there that, for a properly relaxed version of the primal problem, question (DG) has an affirmative answer in a perfectly general setting, i.e. for arbitrary Borel-measurable cost functions $c : X \times Y \to [0, \infty]$ defined on the product of two Polish spaces X, Y, equipped with Borel probability measures μ, ν.

As regards question (P) we find the following situation: if the cost function $c : X \times Y \to [0, \infty]$ is *lower semi-continuous*, the answer to question (P) is always positive. Indeed, for an optimizing sequence $(\pi_n)_{n=1}^{\infty}$ in $\Pi(\mu, \nu)$, one may apply Prokhorov's theorem to find a weak limit $\widehat{\pi} = \lim_{k\to\infty} \pi_{n_k}$. If c is lower semi-continuous, we get

$$\int_{X\times Y} c\,d\widehat{\pi} \le \lim_{k\to\infty} \int_{X\times Y} c\,d\pi_{n_k},$$

which yields the optimality of $\widehat{\pi}$.

On the other hand, if c fails to be lower semi-continuous, there is little reason why a primal optimizer should exist (e.g. see [Kel84, Example 2.20]).

As regards (D), the question of the existence of a dual optimizer is more delicate than for the primal case (P): it was shown in [AP03, Theorem 3.2] that, for $c : X \times Y \to \mathbb{R}_+$, satisfying a certain moment condition, one may assert the existence of *integrable* optimizers $(\widehat{\varphi}, \widehat{\psi})$. However, if one drops this moment condition, there is little reason why, for an optimizing sequence $(\varphi_n, \psi_n)_{n=1}^{\infty}$ in (D) above, the L^1-norms should remain bounded. Hence there is little reason why one should be able to find *integrable* optimizers $(\widehat{\varphi}, \widehat{\psi})$ as shown by easy examples (e.g. [BS09, Examples 4.4, 4.5]), arising in rather regular situations.

Yet one would like to be able to pass to *some kind of limit* $(\widehat{\varphi}, \widehat{\psi})$, whether these functions are integrable or not. In the case when $\widehat{\varphi}$ and/or $\widehat{\psi}$ fail to be integrable, special care then has to be taken to give a proper sense to (9.6).

This situation was the motivation for the introduction of the notion of *strong cyclical c-monotonicity* in [ST08]: this notion (see (SCC) above) characterizes the optimality of a given $\pi \in \Pi(\mu, \nu)$ in terms of a 'complementary slackness condition', involving some $(\varphi, \psi) \in \Psi(\mu, \nu)$, playing the role of a dual optimizer $(\widehat{\varphi}, \widehat{\psi})$. The crucial feature is that *we do not need any integrability of the functions φ and ψ* for this notion to make sense. It was shown in [BS09] that, also in situations where there are no integrable optimizers $(\widehat{\varphi}, \widehat{\psi})$, one may find Borel measurables functions (φ, ψ), taking their roles in the setting of (SCC) above.

This theme was further developed in [BS09], where it was shown that, for $\mu \otimes \nu$-a.s. finite, Borel measurable $c : X \times Y \to [0, \infty]$, one may find Borel functions $\widehat{\varphi} : X \to [-\infty, +\infty)$ and $\widehat{\psi} : Y \to [-\infty, \infty)$, which are dual optimizers if we interpret (9.6) properly; instead of considering

$$\int_X \widehat{\varphi}\,d\mu + \int_Y \widehat{\psi}\,d\nu, \tag{9.7}$$

which needs integrability of $\widehat{\psi}$ and $\widehat{\psi}$ in order to make sense, we consider

$$\int_{X\times Y} (\widehat{\varphi}(x) + \widehat{\psi}(y))\,d\pi(x, y), \tag{9.8}$$

where the transport plan $\pi \in \Pi(\mu, \nu)$ is assumed to have finite transport cost $\int_{X \times Y} c(x, y) d\pi(x, y) < \infty$. If (9.7) makes sense, then its value coincides with the value of (9.8); the crucial feature is that (9.8) also makes sense in cases when (9.7) does not make sense any more, as shown in [BS09, Lemma 1.1]. In particular, the value of (9.8) does not depend on the choice of the transport plan $\pi \in \Pi(\mu, \nu)$, provided π has finite transport cost $\int_{X \times Y} c(x, y) d\pi(x, y) < \infty$.

Summing up the preceding discussion on the existence (D) of a dual optimizer $(\widehat{\varphi}, \widehat{\psi})$: this question has a – properly interpreted – positive answer provided that the cost function $c : X \times Y \to [0, \infty]$ is $\mu \otimes \nu$-a.s. finite [BS09, Theorem 2].

But things become much more complicated if we pass to cost functions $c : X \times Y \to [0, \infty]$ assuming the value $+\infty$ on possibly 'large' subsets of $X \times Y$.

In [BLS09, Example 4.1] we exhibit an example, which is a variant of an example due to Ambrosio and Pratelli [AP03, Example 3.5], of a lower semi-continuous cost function $c : [0, 1) \times [0, 1) \to [0, \infty]$, where $(X, \mu) = (Y, \nu)$ equals $[0, 1)$ equipped with Lebesgue measure, for which there are no *Borel measurable* functions $\widehat{\varphi}, \widehat{\psi}$ verifying $\widehat{\varphi}(x) + \widehat{\psi}(y) \leq c(x, y)$, maximizing (9.8) above.

In this example, the cost function c equals the value $+\infty$ on 'many' points of $X \times Y = [0, 1) \times [0, 1)$. In fact, for each $x \in [0, 1[$, there are precisely two points $y_1, y_2 \in [0, 1[$ such that $c(x, y_1) < \infty$ and $c(x, y_2) < \infty$, while for all other $y \in [0, 1[$, we have $c(x, y) = \infty$. In addition, there is an optimal transport plan $\widehat{\pi} \in \Pi(\mu, \nu)$ whose support equals the set $\{(x, y) \in [0, 1) \times [0, 1) : c(x, y) < \infty\}$.

In this example one may observe the following phenomenon: while there *do not* exist Borel measurable functions $\widehat{\varphi} : [0, 1) \to [-\infty, +\infty)$ and $\widehat{\psi} : [0, 1) \to [-\infty, \infty)$ such that $\widehat{\varphi}(x) + \widehat{\psi}(y) = c(x, y)$ on $\{c(x, y) < \infty\}$, there *does* exist a Borel function $\widehat{h} : [0, 1) \times [0, 1) \to [-\infty, \infty)$ such that $\widehat{h}(x, y) = c(x, y)$ on $\{c(x, y) < \infty\}$ and such that $\widehat{h}(x, y) = \lim_{n \to \infty}(\varphi_n(x) + \psi_n(y))$ where $(\varphi_n, \psi_n)_{n=1}^{\infty}$ are properly chosen, bounded Borel functions. The point is that the limit holds true (only) in the norm of $L^1([0, 1[\times [0, 1[, \widehat{\pi})$ as well as $\widehat{\pi}$-a.s.

In other words, in this example we are able to identify some kind of dual optimizer $\widehat{h} \in L^1([0, 1) \times [0, 1), \widehat{\pi})$ which, however, is not of the form $\widehat{h}(x, y) = \widehat{\varphi}(x) + \widehat{\psi}(y)$ for some Borel functions $(\widehat{\varphi}, \widehat{\psi})$, but only a $\widehat{\pi}$-a.s. limit of such functions $(\varphi_n(x) + \psi_n(y))_{n=1}^{\infty}$.

In [BLS09, Theorem 4.2] we established a result which shows that much of the positive aspect of this phenomenon, i.e. the existence of an optimal $\widehat{h} \in L^1(\widehat{\pi})$, encountered in the context of the above example, can be carried over to a general setting. For the convenience of the reader we restate this theorem and the notations required to formulate it.

Fix a finite transport plan $\pi_0 \in \Pi(\mu, \nu, c) := \{\pi \in \Pi(\mu, \nu) : \int_{X \times Y} c \, d\pi < \infty\}$. We denote by $\Pi^{(\pi_0)}(\mu, \nu)$ the set of elements $\pi \in \Pi(\mu, \nu)$ such that $\pi \ll \pi_0$ and $\|\frac{d\pi}{d\pi_0}\|_{L^\infty(\pi_0)} < \infty$. Note that $\Pi^{(\pi_0)}(\mu, \nu) = \Pi(\mu, \nu) \cap L^\infty(\pi_0) \subseteq \Pi(\mu, \nu, c)$. We shall replace the usual Kantorovich optimization problem over the set $\Pi(\mu, \nu, c)$ by the optimization over the smaller set $\Pi^{(\pi_0)}(\mu, \nu)$. Its value is

$$P^{(\pi_0)} = \inf\{\langle c, \pi \rangle = \int c \, d\pi : \pi \in \Pi^{(\pi_0)}(\mu, \nu)\}. \tag{9.9}$$

As regards the dual problem, we define, for $\varepsilon > 0$,

$$D^{(\pi_0, \varepsilon)} = \sup \left\{ \int \varphi \, d\mu + \int \psi \, d\nu : \varphi \in L^1(\mu), \psi \in L^1(\nu), \right.$$
$$\left. \int_{X \times Y} (\varphi(x) + \psi(y) - c(x, y))_+ \, d\pi_0 \leq \varepsilon \right\} \quad \text{and}$$
$$D^{(\pi_0)} = \lim_{\varepsilon \to 0} D^{(\pi_0, \varepsilon)}. \tag{9.10}$$

Define the 'summing' map S by

$$S : L^1(X, \mu) \times L^1(Y, \nu) \to L^1(X \times Y, \pi_0)$$
$$(\varphi, \psi) \mapsto \varphi \oplus \psi,$$

where $\varphi \oplus \psi$ denotes the function $\varphi(x) + \psi(y)$ on $X \times Y$. Denote by $L^1_S(X \times Y, \pi_0)$ the $\|.\|_1$-closed linear subspace of $L^1(X \times Y, \pi_0)$ spanned by $S(L^1(X, \mu) \times L^1(Y, \nu))$. Clearly $L^1_S(X \times Y, \pi_0)$ is a Banach space under the norm $\|.\|_1$ induced by $L^1(X \times Y, \pi_0)$.

We shall also need the bi-dual $L^1_S(X \times Y, \pi_0)^{**}$ which may be identified with a subspace of $L^1(X \times Y, \pi_0)^{**}$. In particular, an element $h \in L^1_S(X \times Y, \pi_0)^{**}$ can be decomposed into $h = h^r + h^s$, where $h^r \in L^1(X \times Y, \pi_0)$ is the regular part of the finitely additive measure h and h^s its purely singular part.

Theorem 9.1. *Let $c : X \times Y \to [0, \infty]$ be Borel measurable, and let $\pi_0 \in \Pi(\mu, \nu, c)$ be a finite transport plan. We have*

$$P^{(\pi_0)} = D^{(\pi_0)}. \tag{9.11}$$

There is an element $\hat{h} \in L^1_S(X \times Y, \pi_0)^{**}$ such that $\hat{h} \leq c$ and

$$D^{(\pi_0)} = \langle \hat{h}, \pi_0 \rangle.$$

If $\pi \in \Pi^{(\pi_0)}(\mu, \nu)$ (identifying π with $\frac{d\pi}{d\pi_0}$) satisfies $\int c \, d\pi \leq P^{(\pi_0)} + \alpha$ for some $\alpha \geq 0$, then

$$|\langle \hat{h}^s, \pi \rangle| \leq \alpha. \tag{9.12}$$

In particular, if π is an optimizer of (9.9), then $\hat{h}^s$ vanishes on the set $\{\frac{d\pi}{d\pi_0} > 0\}$.

In addition, we may find a sequence of elements $(\varphi_n, \psi_n) \in L^1(\mu) \times L^1(\nu)$ such that

$$\varphi_n \oplus \psi_n \to \hat{h}^r, \; \pi_0\text{-a.s.}, \qquad \|(\varphi_n \oplus \psi_n - \hat{h}^r)_+\|_{L_1(\pi_0)} \to 0$$

and

$$\lim_{\delta \to 0} \sup_{A \subseteq X \times Y, \pi_0(A) < \delta} \lim_{n \to \infty} -\langle (\varphi_n \oplus \psi_n)1_A, \pi_0 \rangle = \|\hat{h}^s\|_{L_1(\pi_0)^{**}}. \tag{9.13}$$

The assertion of the theorem extends the phenomenon of [BLS09, Example 4.1] to a general setting. There is, however, one additional complication, as compared to the situation of this specific example: in the above theorem we only can assert that we find the optimizer $\hat{h}$ in $L^1(\hat{\pi})^{**}$ rather than in $L^1(\hat{\pi})$. The question arises whether this complication is indeed unavoidable. The purpose of the subsequent section is to construct an example showing that the phenomenon of a non-vanishing singular part $\hat{h}^s$ of $\hat{h} = \hat{h}^r + \hat{h}^s$ may indeed arise in the above setting. In addition, the example gives a good illustration of the subtleties of the situation described by the theorem above.

9.3 The singular part of the dual optimizer

In this section we refine the construction of Examples 4.1 and 4.3 in [BLS09] (which in turn are variants of an example due to Ambrosio and Pratelli [AP03, Example 3.2]). We assume that the reader is familiar with these examples and freely use the notation from this paper. In particular, for an irrational $\alpha \in [0, 1)$ we write, for $k \in \mathbb{Z}$,[1]

$$\begin{aligned} \varrho_k(x) = 1 &+ \#\{0 \leq i < k : x \oplus i\alpha \in [0, \tfrac{1}{2})\} \\ &- \#\{0 \leq i < k : x \oplus i\alpha \in [\tfrac{1}{2}, 1)\}, \end{aligned} \tag{9.14}$$

where, for $k < 0$, we mean by $0 \leq i < k$ the set $\{k + 1, k + 2, \ldots, 0\}$ and $\oplus$ denotes addition modulo 1. We also recall that the function $h : [0, 1) \times [0, 1) \to$

---

[1] In [BLS09] the constructions are carried out for $\mathbb{N}$ instead of $\mathbb{Z}$, but for our purposes the latter choice turns out to be better suited.

$\mathbb{Z}$ is defined in [BLS09, Example 4.3] as

$$h(x, y) = \begin{cases} \varrho_k(x), & k \in \mathbb{Z} \text{ and } y = x \oplus k\alpha \\ \infty, & \text{otherwise.} \end{cases} \tag{9.15}$$

In [BLS09, Example 4.3] we considered the $[0, \infty]$-valued cost function $c(x, y) := h_+(x, y)$, the positive part of the function h. We now construct an example restricting $h_+(x, y)$ to a certain subset of $[0, 1) \times [0, 1)$.

Example 1. Consider $X = Y = [0, 1)$ and denote by μ resp. ν the Lebesgue measure on X, resp. Y. There is an irrational $\alpha \in [0, 1)$ and a map $\tau : [0, 1) \to \mathbb{Z}$ such that, for

$$\Gamma_0 = \{(x, x), x \in [0, 1)\},$$
$$\Gamma_1 = \{(x, x \oplus \alpha) : x \in [0, 1)\},$$
$$\Gamma_\tau = \{(x, x \oplus \tau(x)\alpha) : x \in [0, 1)\}$$

and letting

$$c(x, y) = \begin{cases} h_+(x, y), & \text{for } x \in \Gamma_0 \cup \Gamma_1 \cup \Gamma_\tau \\ \infty, & \text{otherwise,} \end{cases}$$

the following properties are satisfied.

(i) The maps

$$T_\alpha^0(x) = x, \qquad T_\alpha^1(x) = x \oplus \alpha, \qquad T_\alpha^{(\tau)}(x) = x \oplus (\tau(x)\alpha)$$

are measure preserving bijections from $[0, 1)$ to $[0, 1)$ with respect to the Lebesgue measure (μ in the present setting). Denote by π_0, π_1, π_τ the corresponding transport plans in $\Pi(\mu, \mu)$, i.e.

$$\pi_0 = (id, id)_{\#}\mu, \qquad \pi_1 = (id, T_\alpha)_{\#}\mu, \qquad \pi_\tau = (id, T_\alpha^{(\tau)})_{\#}\mu,$$

and let $\pi = (\pi_0 + \pi_1 + \pi_\tau)/3$.

(ii) The transport plans π_0 and π_1 are optimal while π_τ is not. In fact, we have

$$\langle c, \pi_0 \rangle = \langle c, \pi_1 \rangle = 1 \text{ while } \langle c, \pi_\tau \rangle \geq \langle h, \pi_\tau \rangle > 1. \tag{9.16}$$

(iii) There is a sequence $(\varphi_n, \psi_n)_{n=1}^{\infty}$ of bounded Borel functions such that

(a) $\varphi_n(x) + \psi_n(y) \leq c(x, y), \quad \text{for } x \in X, y \in Y,$ \hfill (9.17)

(b) $\displaystyle \lim_{n \to \infty} \left(\int_X \varphi_n(x) \, d\mu(x) + \int_Y \psi_n(y), dv(y) \right) = 1,$ \hfill (9.18)

(c) $\displaystyle \lim_{n \to \infty} (\varphi_n(x) + \psi_n(y)) = h(x, y), \quad \pi\text{-almost surely.}$ \hfill (9.19)

(iv) Using the notation of [BLS09, Theorem 4.2] we find that for each dual optimizer $\widehat{h} \in L^1(\pi)^{**}$, which decomposes as $\widehat{h} = \widehat{h}^r + \widehat{h}^s$ into its regular part $\widehat{h}^r \in L^1(\pi)$ and its purely singular part $\widehat{h}^s \in L^1(\pi)^{**}$, we have

$$\widehat{h}^r = h, \ \pi\text{-a.s.}, \tag{9.20}$$

and the singular part $\widehat{h}^s$ satisfies $\|\widehat{h}^s\|_{L^1(\pi)^{**}} = \langle h, \pi_\tau \rangle - 1 > 0$. In particular, the singular part $\widehat{h}^s$ of $\widehat{h}$ does not vanish. The finitely additive measure $\widehat{h}^s$ is supported by Γ_τ, i.e. $\langle \widehat{h}^s, \mathbf{1}_{\Gamma_0} + \mathbf{1}_{\Gamma_1} \rangle = 0$.

We shall use a special irrational $\alpha \in [0, 1)$, namely

$$\alpha = \sum_{j=1}^{\infty} \frac{1}{M_j},$$

where $M_j = m_1 m_2 \ldots m_j = M_{j-1} m_j$, and $(m_j)_{j=1}^{\infty}$ is a sequence of prime numbers $m_j \geq 5$ tending sufficiently fast to infinity, to be specified below. We let

$$\alpha_n := \sum_{j=1}^{n} \frac{1}{M_j},$$

which, of course, is a rational number.

We will need the following lemma. We thank Leonhard Summerer for showing us the proof of Lemma 9.2.

Lemma 9.2. *It is possible to choose a sequence $m_1, m_2, \ldots$ of primes growing arbitrarily fast to infinity, such that with $M_1 = m_1$, $M_2 = m_1 \cdot m_2, \ldots, M_n = m_1 \cdots m_n, \ldots$ we have, for each $n \in \mathbb{N}$,*

$$\sum_{j=1}^{n} \frac{1}{M_j} = \frac{P_n}{M_n},$$

with P_n and M_n relatively prime.

Proof. We have

$$\sum_{j=1}^{n} \frac{1}{M_j} = \frac{m_2 \ldots m_n + \cdots + m_n + 1}{M_n} =: \frac{P_n}{M_n},$$

thus P_n and M_n are relatively prime, if and only if

$$m_1 \nmid \qquad m_2 \cdots m_n + m_3 \cdots m_n + \cdots + \qquad m_n + 1 \qquad (9.21)$$
$$m_2 \nmid \qquad \qquad m_3 \cdots m_n + \cdots + \qquad m_n + 1 \qquad (9.22)$$
$$\vdots \qquad\qquad\qquad\qquad\qquad \vdots \qquad\qquad\qquad (9.23)$$
$$m_{n-1} \nmid \qquad\qquad\qquad\qquad\qquad\qquad m_n + 1. \qquad (9.24)$$

We claim that these conditions are, for example, satisfied provided that we choose $m_1, m_2, \ldots$ such that $m_i \geq 3$ and

$$m_{i+1} \equiv +1 \, (m_i) \qquad (9.25)$$
$$m_{i+j} \equiv -1 \, (m_i) \text{ if } j \geq 2. \qquad (9.26)$$

for all $i \geq 1$. Indeed (9.25), (9.26) imply that for $k \in \{1, \ldots, n-1\}$ we have modulo (m_k)

$$m_{k+1} \cdots m_n + \quad m_{k+2} \cdots m_n + \quad m_{k+3} \cdots m_n + \cdots + \qquad m_n + \qquad 1 \equiv$$
$$(\pm 1) + \qquad\quad (\pm 1) + \qquad\qquad (\mp 1) + \cdots + \quad (-1) + \quad (+1),$$

where in the second line the $(n - k + 1)$ summands start to alternate after the second term. Thus, for even $n - k$, this amounts to

$$m_{k+1} \cdots m_n + \quad m_{k+2} \cdots m_n + \quad m_{k+3} \cdots m_n + \cdots + \qquad m_n + \qquad 1 \equiv$$
$$(-1) + \qquad\quad (-1) + \qquad\qquad (+1) + \cdots + \quad (-1) + \quad (+1) \equiv -1,$$

while we obtain, for odd $n - k$,

$$m_{k+1} \cdots m_n + \quad m_{k+2} \cdots m_n + \quad m_{k+3} \cdots m_n + \cdots + \qquad m_n + \qquad 1 \equiv$$
$$(+1) + \qquad\quad (+1) + \qquad\qquad (-1) + \cdots + \quad (-1) + \quad (+1) \equiv +2.$$

Hence, (9.21)–(9.24) are satisfied as the m_n where chosen such that $m_n > 2$.

We use induction to construct a sequence of primes satisfying (9.25) and (9.26). Assume that $m_1, \ldots, m_i$ have been defined. By the Chinese remainder theorem the system of congruences

$$x \equiv -1 \, (m_1), \ldots, \qquad x \equiv -1 \, (m_{i-1}), \qquad x \equiv +1 \, (m_i)$$

has a solution $x_0 \in \{1, \ldots, m_1 \ldots m_i\}$. By Dirichlet's theorem, the arithmetic progression $x_0 + k m_1 \ldots m_i, k \in \mathbb{N}$, contains infinitely many primes, so we may pick one which is as large as we please. The induction continues.   □

For $\beta \in [0, 1)$, denote by $T_\beta : [0, 1) \to [0, 1)$, $T_\beta(x) := x \oplus \beta$ the addition of β modulo 1. With this notation we have $T_{\alpha_n}^{M_n} = id$ and, by Lemma 9.2, it is possible to choose $m_1, \ldots, m_n$ in such a way that M_n is the smallest such number in $\mathbb{N}$. Our aim is to construct a function $\tau : [0, 1) \to \mathbb{Z}$ such that the map

$$T_\alpha^{(\tau)} : \begin{cases} [0, 1) \to [0, 1) \\ x \mapsto T_\alpha^{(\tau)}(x) = T_\alpha^{\tau(x)}(x) \end{cases}$$

defines, up to a μ-null set, a measure preserving bijection on $[0, 1)$, and such that the corresponding transport plan $\pi_\tau \in \Pi(\mu, \nu)$, given by $\pi_\tau = (id, T_\alpha^{(\tau)})_\# \mu$, has the properties listed above with respect to the cost function $c(x, y)$ which is the restriction of the function $h_+(x, y)$ to $\Gamma_0 \cup \Gamma_1 \cup \Gamma_\tau$. We shall do so by an inductive procedure, defining bounded $\mathbb{Z}$-valued functions τ_n on $[0, 1)$ such that the maps $T_{\alpha_n}^{(\tau_n)}$ are measure preserving bijections on $[0, 1)$. The map $T_\alpha^{(\tau)}$ then will be the limit of these $T_{\alpha_n}^{(\tau_n)}$.

Step $n = 1$: Fix a prime $M_1 = m_1 \geq 5$, so that $\alpha_1 = \frac{1}{M_1}$. Define

$$I_{k_1} := \left[\frac{k_1 - 1}{M_1}, \frac{k_1}{M_1} \right), \quad k_1 = 1, \ldots, M_1,$$

so that $(I_{k_1})_{k_1=1}^{M_1}$ forms a partition of $[0, 1)$ and T_{α_1} maps I_{k_1} to I_{k_1+1}, with the convention $I_{M_1+1} = I_1$. We also introduce the notations

$$L^1 := [0, \tfrac{1}{2} - \tfrac{1}{2M_1}) \text{ and } R^1 := [\tfrac{1}{2} + \tfrac{1}{2M_1}, 1)$$

for the segments left and right of the middle interval

$$I_{\text{middle}}^1 := I_{(M_1+1)/2} = [\tfrac{1}{2} - \tfrac{1}{2M_1}, \tfrac{1}{2} + \tfrac{1}{2M_1}).$$

We define the functions φ^1, ψ^1 on $[0, 1)$ such that $\varphi^1(x) + \psi^1(x) \equiv 1$ and

$$\varphi^1(x) + \psi^1(T_{\alpha_1}(x)) = \begin{cases} 0 & x \in L^1, \\ 1 & x \in I_{\text{middle}}^1, \\ 2 & x \in R^1, \end{cases}$$

which leads to the relation

$$\varphi^1(T_{\alpha_1}(x)) = \varphi^1(x) + \begin{cases} 1, & x \in L^1, \\ 0, & x \in I_{\text{middle}}^1, \\ -1, & x \in R^1. \end{cases}$$

Making the choice $\varphi^1 \equiv 0$ on I_1 this leads to

$$\varphi^1(x) = \begin{cases} k_1 - 1, & x \in I_{k_1}, k_1 \in \{1, \ldots, (M_1 + 1)/2\}, \\ M_1 + 1 - k_1, & x \in I_{k_1}, k_1 \in \{(M_1 + 3)/2, M_1\}, \end{cases} \quad (9.27)$$

$$\psi^1(x) = 1 - \varphi^1(x).$$

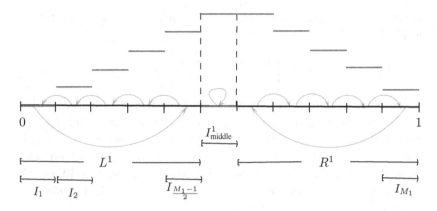

Figure 9.1 Representations of φ^1 and τ^1. The step function is φ^1 and the arrows indicate the action of $T_{\alpha_1}^{(\tau_1)}$. This figure corresponds to the value $M_1 = 11$.

The function φ^1 starts at 0, increases until the middle interval, stays constant when stepping to the interval right of the middle, and then decreases, reaching 1 on the final interval I_{M_1}.

The idea is to define the map $\tau_1 : [0, 1) \to \mathbb{Z}$ in such a way that the map

$$T_{\alpha_1}^{(\tau_1)} : \begin{cases} [0, 1) \to [0, 1) \\ x \mapsto T_{\alpha_1}^{\tau_1(x)}(x) \end{cases}$$

is a measure-preserving bijection enjoying the following property: the map

$$x \mapsto \varphi^1(x) + \psi^1(T_{\alpha_1}^{(\tau_1)}(x))$$

equals the value 2 on a large set while it has concentrated a negative mass which is close to -1 on a small set.

This can be done, for example, by shifting the first interval I_1 to the interval $I_{(M_1-1)/2}$, which is left of the middle one, while we shift the intervals $I_2, \ldots, I_{(M_1-1)/2}$ by one interval to the left. On the right-hand side of $[0, 1)$ we proceed symmetrically, while the middle interval simply is not moved.

More precisely, we set

$$\tau_1(x) = \begin{cases} \frac{M_1-3}{2}, & x \in I_1, \\ -1, & x \in I_{k_1}, k_1 \in \{2, \ldots, (M_1 - 1)/2\}, \\ 0, & x \in I_{(M_1+1)/2}, \\ 1, & x \in I_{k_1}, k_1 \in \{(M_1 + 3)/2, \ldots, M_1\}, \\ -\frac{M_1-3}{2}, & x \in I_{M_1}. \end{cases} \qquad (9.28)$$

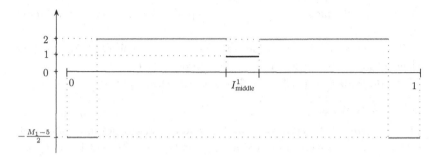

Figure 9.2 Representation of $\varphi^1 + \psi^1 \circ T_{\alpha_1}^{(\tau_1)}$.

Then $T_{\alpha_1}^{(\tau_1)}$ induces a permutation of the intervals $(I_{k_1})_{k_1=1}^{M_1}$ and a short calculation shows that

$$\varphi^1(x) + \psi^1(T_{\alpha_1}^{(\tau_1)}(x)) = \begin{cases} 2, & x \in I_{k_1}, k_1 \in \{2, \ldots, (M_1-1)/2, \\ & \qquad (M_1+3)/2, \ldots, M_1-1\}, \\ -\frac{M_1-5}{2}, & x \in I_{k_1}, k_1 = 1, M_1, \\ 1, & x \in I_{(M_1+1)/2}. \end{cases}$$

$$(9.29)$$

Figure 9.2 is a representation of this 'quasi-cost' at level $n = 1$, with the same value $M_1 = 11$ as in Figure 9.1.

Assessment of Step $n = 1$. Let us resume what we have achieved in the first induction step. For later use we formulate things only in terms of $\varphi^1(\cdot)$ rather than $\psi^1(\cdot) = 1 - \varphi^1(\cdot)$.

For the set $J_1^g = \{2, \ldots, \frac{M-1}{2}\} \cup \{\frac{M+3}{2}, \ldots, M_1-1\}$ of *good*[2] indices we have

$$\varphi^1(x) - \varphi^1(T_{\alpha_1}^{(\tau_1)}(x)) = 1, \quad x \in I_{k_1}, k_1 \in J_1^g, \tag{9.30}$$

while for the set $J_1^s = \{1, M_1\}$ of *singular* indices we have

$$\varphi^1(x) - \varphi^1(T_{\alpha_1}^{(\tau_1)}(x)) = -\frac{M_1-3}{2}, \quad x \in I_{k_1}, k_1 \in J^s, \tag{9.31}$$

so that

$$\sum_{k_1 \in J_1^s} \int_{I_{k_1}} [\varphi^1(x) - \varphi^1(T_{\alpha_1}^{(\tau_1)}(x))] \, dx = -\frac{M_1-3}{2} \frac{2}{M_1} = -1 + \frac{3}{M_1}.$$

---

[2] We use the term 'good' rather than 'regular' as the abbreviation r is already taken by the word 'right'.

For the middle interval $I^1_{\text{middle}} = I_{(M_1+1)/2}$ we have $\varphi^1(x) - \varphi^1(T^{\tau_1}_{\alpha_1}(x)) = 0$.

We also note for later use that, for $x \in [0, 1)$, the orbit $(T^i_{\alpha_1}(x))^{\tau_1(x)}_{i=1}$ never visits I^1_{middle}. Here we mean that i runs through $\{\tau_1(x), \tau_1(x) + 1, \ldots, -1\}$ when $\tau_1(x) < 0$ and runs through the empty set when $\tau_1(x) = 0$.

Step $n = 2$: We now pass from $\alpha_1 = \frac{1}{M_1}$ to $\alpha_2 = \frac{1}{M_1} + \frac{1}{M_2}$, where $M_2 = M_1 m_2 = m_1 m_2$ and where m_2, to be specified below, satisfies the relations of Lemma 9.2 and is large compared to M_1. For $1 \le k_1 \le M_1$ and $1 \le k_2 \le m_2$ we denote by I_{k_1,k_2} the interval

$$I_{k_1,k_2} = \left[\tfrac{k_1-1}{M_1} + \tfrac{k_2-1}{M_2}, \tfrac{k_1-1}{M_1} + \tfrac{k_2}{M_2} \right).$$

Similarly, as above, we will also use the notations $L^2 = [0, \frac{1}{2} - \frac{1}{2M_2})$, $R^2 = [\frac{1}{2} + \frac{1}{2M_2}, 1)$, and $I^2_{\text{middle}} = I_{(M_1+1)/2,(m_2+1)/2} = [\frac{1}{2} - \frac{1}{2M_2}, \frac{1}{2} + \frac{1}{2M_2})$.

We now define functions φ^2, ψ^2 such that $\varphi^2(x) + \psi^2(x) \equiv 1$ and

$$\varphi^2(x) + \psi^2(T_{\alpha_2}(x)) = \begin{cases} 0, & x \in L^2, \\ 1, & x \in I^2_{\text{middle}}, \\ 2, & x \in R^2. \end{cases}$$

This is achieved if, for example, we set $\varphi^2 \equiv 0$ on $I_{1,1}$, and

$$\varphi^2(T_{\alpha_2}(x)) = \varphi^2(x) + \begin{cases} 1 & x \in L^2, \\ 0 & x \in I^2_{\text{middle}}, \\ -1 & x \in R^2, \end{cases} \tag{9.32}$$

$$\psi^2(x) = 1 - \varphi^2(x).$$

Yet another way to express this is to say that for $j \in \{0, \ldots, M_2 - 1\}$ we have

$$\varphi^2(T^j_{\alpha_2}(x)) = \begin{aligned} &\#\{i \in \{0, \ldots, j-1\} : T^i_{\alpha_2}(x) \in L^2\} \\ &- \#\{i \in \{0, \ldots, j-1\} : T^i_{\alpha_2}(x) \in R^2\} \end{aligned}, \quad x \in I_{1,1}, \tag{9.33}$$

in analogy to (9.14).

While the function $\varphi^1(x)$ in the first induction step was increasing from I_1 to $I_{(M_1+1)/2}$ and then decreasing from $I_{(M_1+3)/2}$ to I_{M_1}, the function $\varphi^2(x)$ displays a similar feature on each of the intervals I_{k_1}: roughly speaking, i.e. up to terms controlled by M_1, it increases on the left half of each such interval and then decreases again on the right half. The next lemma makes this fact precise. We keep in mind, of course, that m_2 will be much bigger than M_1.

Lemma 9.3 (oscillations of φ^2). *The function φ^2 defined in (9.32) has the following properties:*

(i) $|\varphi^2(x) - \varphi^2(x \oplus \frac{1}{M_2})| \leq 4M_1^2$, $\quad x \in [0, 1)$.

(ii) For each $1 \leq k_1', k_1'' \leq M_1$ we have

$$\varphi^2|_{I_{k_1',(m_2+1)/2}} - \varphi^2|_{I_{k_1'',1}} \geq \frac{m_2}{2M_1} - 10M_1^3.$$

Proof. Let us begin with the proof of (i).

• *Proof of (i).* While $T_{\alpha_1}^{M_1} = id$ holds true, we have that $T_{\alpha_2}^{M_1}$ is only close to the identity map. In fact, as $T_{\alpha_2}(x) = x \oplus \frac{m_2+1}{M_2}$, we have

$$T_{\alpha_2}^{M_1}(x) = x \oplus \frac{M_1}{M_2}. \tag{9.34}$$

Somewhat less obvious is the fact that $T_{\alpha_2}^{m_2-2}$ also is close to the identity map. In fact,

$$T_{\alpha_2}^{m_2-2}(x) = x \ominus \frac{2}{M_2}. \tag{9.35}$$

Indeed, by (9.25) applied to $i = 1$, there is $c \in \mathbb{N}$ such that $m_2 = cM_1 + 1$. Hence

$$T_{\alpha_2}^{m_2-2}(x) = x \oplus (m_2 - 2)\frac{m_2 + 1}{M_2}$$

$$= x \oplus (cM_1 - 1)\frac{m_2 + 1}{M_2}$$

$$= x \oplus \frac{cM_2 - m_2 + (m_2 - 2)}{M_2} = x \ominus \frac{2}{M_2}.$$

Here is one more remarkable feature of the map $T_{\alpha_2}^{m_2-2}$.

Claim 9.4. *For $x \in [0, 1)$ the orbit $(T_\alpha^i(x))_{i=1}^{m_2-2}$ visits the intervals $L^2 = [0, \frac{1}{2} - \frac{1}{2M_2})$ and $R^2 = [\frac{1}{2} + \frac{1}{2M_2}, 1)$ approximately equally often. More precisely, the difference of the visits of these two intervals is bounded in absolute value by $4M_1$.*

Indeed, by Lemma 9.2, the orbit $(T_{\alpha_2}^i(x))_{i=1}^{M_2}$ visits each of the intervals I_{k_1,k_2} exactly one time so that it visits L^2 and R^2 equally often, namely $\frac{M_2-1}{2}$ times. The M_1 many disjoint subsets $(T_{\alpha_2}^{j(m_2-2)}(T_{\alpha_2}^i(x))_{i=1}^{m_2-2})_{j=1}^{M_1}$ of this orbit are obtained by shifting them successively by $2/M_2$ to the left (9.35). As the difference $(T_{\alpha_2}^i(x))_{i=1}^{M_2} \setminus (T_{\alpha_2}^{j(m_2-2)}(T_{\alpha_2}^i(x))_{i=1}^{m_2-2})_{j=1}^{M_1}$ consists only of $2M_1$ many points we have that the difference of the visits of $(T_{\alpha_2}^{j(m_2-2)}(T_{\alpha_2}^i(x))_{i=1}^{m_2-2})_{j=1}^{M_1}$ to L^2 and R^2 is bounded by $4M_1$. This implies that the difference of the visits of $(T_{\alpha_2}^i(x))_{i=1}^{m_2-2}$ to L^2 and R^2 can be estimated by $4M_1$ too; indeed, if this orbit visits $4M_1 + k$ many times L^2 more often then R^2 (or vice versa) for some

$k \geq 0$, then $(T_{\alpha_2}^{m_2-2}(T_{\alpha_2}^i(x)))_{i=1}^{m_2-2}$ visits L^2 at least $4M_1 + k - 4$ many times more often than R^2, etc. and finally $(T_{\alpha_2}^{M_1(m_2-2)}(T_{\alpha_2}^i(x)))_{i=1}^{m_2-2}$ visits L^2 at least k many times more often than R^2, which yields a contradiction. Hence we have proved the claim.

To prove assertion (i) note that by (9.34) and (9.35)

$$T_{\alpha_2}^{\frac{M_1-1}{2}(m_2-2)} \circ T_{\alpha_2}^{M_1}(x) = x \oplus \frac{1}{M_2} \tag{9.36}$$

We deduce from the claim that the difference of the visits of the orbit $(T_{\alpha_2}^i)_{i=0}^{\frac{M_1-1}{2}(m_2-2)+M_1}$ to L^2 and R^2 is bounded in absolute value by $\frac{M_1-1}{2}(4M_1) + M_1$ which proves (i).

• *Proof of (ii).* As regards (ii) suppose first $k_1' = k_1'' =: k_1$. Note that, for $x \in I_{k_1}^{\text{left}} := [\frac{k_1-1}{M_1}, \frac{k_1-1}{M_1} + \frac{1}{2M_1} - \frac{2M_1+1}{2M_2})$, we have that the orbit $(T_{\alpha_2}^i(x))_{x=0}^{M_1-1}$ visits L^2 one time more often than R^2, namely $\frac{M_1+1}{2}$ versus $\frac{M_1-1}{2}$ times. If we start with $x \in I_{k_1,1}$ then, for $1 \leq j < \frac{m_2}{2M_1} - 1$ we have that $T_{\alpha_2}^{jM_1}(x) \in I_{k_1}^{\text{left}}$. Hence, for the orbit $(T_{\alpha_2}^i)_{i=0}^{(\lfloor\frac{m_2}{2M_1}\rfloor-1)M_1}$, the difference of the visits to the interval L^2 and R^2 equals $\lfloor\frac{m_2}{2M_1}\rfloor - 1$, the integer part of $\frac{m_2}{2M_1} - 1$. Combining this estimate with the estimate (i) as well as the fact that the distance between $x \oplus (\lfloor\frac{m_2}{2M_1}\rfloor - 1)\frac{M_1}{M_2}$ and $x \oplus \frac{m_2-1}{2M_2}$ is bounded by $\frac{2M_1-1}{M_2}$, we obtain, for $x \in I_{k_1,1}$ and $y \in I_{k_1,\frac{m_2+1}{2}}$, that

$$\varphi^2(y) - \varphi^2(x) \geq \varphi^2(T_{\alpha_2}^{(\lfloor\frac{m_2}{2M_1}\rfloor-1)M_1}(x)) - \varphi^2(x) - \left|\varphi^2(y) - \varphi^2(T_{\alpha_2}^{(\lfloor\frac{m_2}{2M_1}\rfloor-1)M_1}(x))\right|$$

$$\geq (\lfloor\tfrac{m_2}{2M_1}\rfloor - 1) - (2M_1 - 1)(4M_1^2)$$

$$\geq \tfrac{m_2}{2M_1} - 8M_1^3.$$

Passing to the general case $1 \leq k_1', k_1'' \leq M_1$, observe that $T_{\alpha_2}^{k_1''-k_1'}$ maps $I_{k_1', \frac{m_2+1}{2}}$ to $I_{k_1'', \frac{m_2+1}{2}+k_1''-k_1'}$. Using again (i) we obtain estimate (ii). $\qquad\square$

We now are ready to do the inductive construction for $n = 2$. For m_2 satisfying the conditions of Lemma 9.3 and to be specified below, we shall define $\tau_2 : [0, 1) \rightarrow \{-\frac{M_2-1}{2}, \ldots, 0, \ldots, \frac{M_2-1}{2}\}$, where $M_2 = m_2 m_1$, such that the map

$$T_{\alpha_2}^{(\tau_2)} : \begin{cases} [0, 1) \rightarrow [0, 1) \\ x \mapsto T_{\alpha_2}^{(\tau_2)}(x) := T_{\alpha_2}^{\tau_2(x)}(x) \end{cases}$$

has the following properties.

(i) The measure-preserving bijection $T_{\alpha_2}^{(\tau_2)} : [0, 1) \to [0, 1)$ maps each interval I_{k_1} onto $T_{\alpha_1}^{(\tau_1)}(I_{k_1})$. It induces a permutation of the intervals I_{k_1,k_2}, where $1 \le k_1 \le M_1, 1 \le k_2 \le m_2$.

(ii) When $\tau_2(x) > 0$, we have

$$T_{\alpha_2}^i(x) \notin I_{\text{middle}}^2, \quad i = 0, \ldots, \tau_2(x), \tag{9.37}$$

and, when $\tau_2(x) < 0$, we have

$$T_{\alpha_2}^i(x) \notin I_{\text{middle}}^2, \quad i = \tau_2(x), \ldots, 0. \tag{9.38}$$

(iii) On the *good* intervals I_{k_1}, where $k_1 \in J_1^g = \{2, \ldots, \frac{M_1-1}{2}\} \cup \{\frac{M_1+3}{2}, \ldots, M_1 - 1\}$, for which we have, by (9.30),

$$\varphi^1(x) - \varphi^1(T_{\alpha_1}^{(\tau_1)}(x)) = 1,$$

the function τ_2 will satisfy the estimates

$$\mu[I_{k_1} \cap \{\tau_2 \ne \tau_1\}] \le \tfrac{M_1}{m_2}\mu[I_{k_1}], \tag{9.39}$$

and

$$\sum_{k_1 \in J_1^g} \int_{I_{k_1}} |1 - \varphi^2(x) + \varphi^2(T_{\alpha_2}^{(\tau_2)}(x))|\, dx < \frac{4M_1^2}{m_2}. \tag{9.40}$$

(iv) On the *singular* intervals I_{k_1}, where $k_1 \in J_1^s = \{1, M_1\}$, for which we have, by (9.31),

$$\varphi^1(x) - \varphi^1(T_{\alpha_1}^{(\tau_1)}(x)) = -\frac{M_1 - 3}{2},$$

we split $\{1, \ldots, m_2\}$ into a set $J^{k_1,g}$ of *good* indices, and a set $J^{k_1,s}$ of *singular* indices, such that

$$\varphi^2(x) - \varphi^2(T_{\alpha_2}^{(\tau_2)}(x)) = 0, \quad \text{for } x \in I_{k_1,k_2}, k_2 \in J^{k_1,g},$$

while

$$\varphi^2(x) - \varphi^2(T_{\alpha_2}^{(\tau_2)}(x)) < -\tfrac{m_2}{2M_1} + 20M_1^3 \quad \text{for } x \in I_{k_1,k_2}, k_2 \in J^{k_1,s},$$

where $J^{k_1,s}$ consists of $M_1(M_1 - 3)$ many elements of $\{1, \ldots, m_2\}$. Hence we have a total 'singular mass' of

$$\sum_{k_1 \in J_1^s} \sum_{k_2 \in J^{k_1,s}} \int_{I_{k_1,k_2}} [\varphi^2(x) - \varphi^2(T_{\alpha_2}^{(\tau_2)}(x))]\, dx < -1 + \tfrac{3}{M_1} + \tfrac{c(M_1)}{m_2},$$

$$\tag{9.41}$$

where $c(M_1)$ is a constant depending only on M_1.

(v) On the middle interval $I_{\text{middle}}^1 = I_{\frac{M_1+1}{2}}$ we simply let $\tau_2 = \tau_1 = 0$.

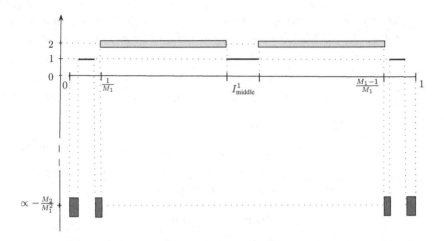

Figure 9.3 Shape of the quasi-cost $\varphi^2 + \psi^2 \circ T^{(\tau_2)}_{\alpha_2}$. The strips in this graphic representation symbolize the oscillations of the function $\varphi^2 + \psi^2 \circ T^{(\tau_2)}_{\alpha_2}$. On the *singular* set, it achieves values of order $-M_2/M_1^2$.

Let us illustrate graphically (Figure 9.3) an interesting property of this construction, namely the shape of the quasi-cost function $\varphi^2 + \psi^2 \circ T^{(\tau_2)}_{\alpha_2}$.

It will sometimes be more convenient to specify to which interval I_{l_1,l_2} the interval I_{k_1,k_2} is mapped under $T^{(\tau_2)}_{\alpha_2}$, instead of spelling out the value of τ_2 on the interval I_{k_1,k_2}. Note that by Lemma 9.2, for each map associating with (k_1, k_2) a pair (l_1, l_2), there corresponds precisely one value $\tau_2|_{I_{k_1,k_2}} : I_{k_1,k_2} \to \{-M_2 + 1, \ldots, 0, \ldots, M_2 - 1\}$ such that (9.37) (resp. (9.38)) is satisfied and $T^{(\tau_2)}_{\alpha_2}(I_{k_1,k_2}) = I_{l_1,l_2}$.

Let us start with a *good* interval I_{k_1}, with $k_1 \in J_1^g$ as in (iii) above (Figure 9.4), say $k_1 \in \{2, \ldots, \frac{M_1-1}{2}\}$, for which we have $\tau_1(x) = -1$. Then the intervals $I_{k_1,2}, \ldots, I_{k_1,m_2}$ are mapped under $T^{\tau_1(x)}_{\alpha_2}(x) = T^{-1}_{\alpha_2}(x)$ onto the intervals $I_{k_1-1,1}, \ldots, I_{k_1-1,m_2-1}$. Defining $\tau_2(x) = \tau_1(x)$ on these intervals we get for $x \in I_{k_1,k_2}$, where $2 \le k_1 \le \frac{M-1}{2}, 2 \le k_2 \le m_2$,

$$1 = \varphi^1(x) - \varphi^1(T^{(\tau_1)}_{\alpha_1}(x)) = \varphi^2(x) - \varphi^2(T^{(\tau_2)}_{\alpha_2}(x)). \tag{9.42}$$

We still have to define the value of $\tau_2(x)$, for $x \in I_{k_1,1}$. The map $T^{(\tau_2)}_{\alpha_2}$ has to map $I_{k_1,1}$ to the remaining gap I_{k_1-1,m_2}, which happens to be its left neighbour. We do not explicitly calculate the unique number $\tau_2|_{I_{k_1,1}} \in \{-M_2 + 1, \ldots, M_2 - 1\}$, satisfying (9.37) (resp. (9.38)), which does the job, but only use the conclusion of Lemma 9.3 to find that, for $x \in I_{k_1,1}$ such that

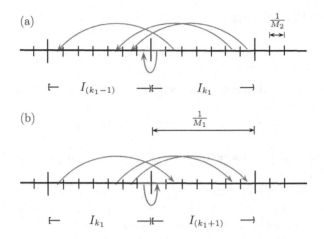

Figure 9.4 (a) $k_1 \in J_1^g$ on the left side.[3] (b) $k_1 \in J_1^g$ on the right side.

$$T_{\alpha_2}^{(\tau_2)}(x) \in I_{k_1-1,m_2},$$

$$|1 - [\varphi^2(x) - \varphi^2(T_{\alpha_2}^{(\tau_2)}(x))]| \leq 4M_1^2 + 1. \tag{9.43}$$

This takes care of the *good* intervals I_{k_1}, where $k_1 \in \{2, \ldots, \frac{M_1-1}{2}\}$.

For the *good* intervals I_{k_1}, where $k_1 \in \{\frac{M_1+3}{2}, \ldots, M_1 - 1\}$ we have $\tau_1(x) = 1$ so that $T_{\alpha_2}^{(\tau_1)}$ maps the intervals $I_{k_1,1}, \ldots, I_{k_1,m_2-1}$ to $I_{k_1+1,2}, \ldots, I_{k_1+1,m_2}$. Again we define $\tau_2(x) = \tau_1(x) = 1$, for x in these intervals so that we obtain the identity (9.42), for $\frac{M_1+3}{2} \leq k_1 \leq M_1 - 1$ and $1 \leq k_2 \leq m_2 - 1$. Finally, $T_{\alpha_2}^{(\tau_2)}$ has to map I_{k_1,m_2} to the interval $I_{k_1+1,1}$ so that again we derive an estimate as in (9.43).

This finishes item (iii), i.e. the definition of τ_2 on the *good* intervals I_{k_1}. Noting that on this set we have $\tau_1 \neq \tau_2$ only on $M_1 - 3$ many intervals of length $\frac{1}{M_2}$, we obtain the estimate (9.40).

To show (iv), let us first consider the *singular* interval I_1, on which we have $\tau_1(x) = \frac{M_1-3}{2}$ and $\varphi^1(T_{\alpha_1}^{(\tau_1)}(x)) = \varphi^1(T_{\alpha_1}^{(\tau_1)}(x)) - \varphi^1(x) = \frac{M_1-3}{2}$. For the subintervals I_{1,k_2} of I_1, define the set of good indices as $J^{1,g} = J^{1,g,l} \cup J^{1,g,r}$, where

$$J^{1,g,l} = \{\frac{(M_1-3)(M_1-1)}{2} + 1, \ldots, \frac{m_2-1}{2}\},$$

$$J^{1,g,r} = \{\frac{m_2+1}{2}, \ldots, m_2 - \frac{(M_1-3)(M_1+1)}{2}\}.$$

---

[3] Figure 9.4(a) is built with the small value $m_2 = 7$ for the sake of clarity of the drawing. But this value is not feasible since with the lowest $m_1 = 5$, (9.25) implies that m_2 is at least equal to 11; other requirements of the construction imply that it has to be even larger.

Let us start by considering $k_2 \in J^{1,g,r}$. We define

$$\tau_2(x) = \tau_1(x) + \frac{M_1 - 3}{2} M_1 = \frac{(M_1 - 3)(M_1 + 1)}{2}, \quad x \in I_{1,k_2}, k_2 \in J^{1,g,r}.$$

First note that $T_{\alpha_2}^{(\tau_2)}$ then maps the intervals I_{1,k_2}, for $k_2 \in J^{1,g,r}$, to the intervals

$$I_{\frac{M_1-1}{2}, \frac{m_2+1}{2} + \frac{(M_1-3)(M_1+1)}{2}}, \quad \ldots \quad, I_{\frac{M_1-1}{2}, m_2}.$$

Observe that, for x as above, the orbit $(T_{\alpha_2}^i(x))_{i=0}^{\tau_2(x)-1}$ always lies in the right halves of the respective intervals I_{k_1}.

Let us count how often the orbit $(T_{\alpha_2}^i(x))_{i=0}^{\tau_2(x)-1}$ visits L^2 and R^2 respectively, for $x \in I_{1,k_2}$ and $k_2 \in J^{1,g,r}$. The first $\tau_1(x) = \frac{M_1-3}{2}$ elements of this orbit are all in L^2 which yields, similarly as in the induction step $n = 1$,

$$\varphi^2(T_{\alpha_2}^{(\tau_1)}(x)) - \varphi^2(x) = \varphi^1(T_{\alpha_1}^{(\tau_1)}(x)) - \varphi^1(x) = \frac{M_1 - 3}{2}.$$

But the next M_1 many elements of this orbit, namely

$$(T_{\alpha_2}^i(x))_{i=\tau_1(x)}^{\tau_1(x)+M_1-1},$$

visit R^2 one time more often than L^2 as the unique element of this orbit which lies in I_{middle}^1 belongs to the right half of I_{middle}^1.

This phenomenon repeats on the orbit $(T_{\alpha_2}^i(x))_{i=0}^{\tau_1(x)+\frac{M_1-3}{2}M_1-1}$ for $\frac{M_1-3}{2}$ many times so that

$$\varphi^2(x) - \varphi^2(T_{\alpha_2}^{(\tau_2)}(x)) = \varphi^2(x) - \varphi^2(T_{\alpha_2}^{(\tau_1)}(x))) + \varphi^2(T_{\alpha_2}^{(\tau_1)}(x)) - \varphi^2(T_{\alpha_2}^{(\tau_2)}(x)$$

$$= -\frac{M_1 - 3}{2} + \frac{M_1 - 3}{2} \tag{9.44}$$

$$= 0, \quad \text{for } x \in I_{1,k_2} \text{ and } k_2 \in J^{1,g,r}.$$

This takes care of I_{1,k_2} with $k_2 \in J^{1,g,r}$.

For $x \in I_{1,k_2}$ with $k_2 \in J^{1,g,l}$, the left half of the *good* intervals, we define symmetrically

$$\tau_2(x) = \tau_1(x) - \frac{M_1-3}{2} M_1 = -\frac{(M_1-3)(M_1-1)}{2}.$$

A similar analysis as above shows that $T_{\alpha_2}^{(\tau_2)}$ maps the intervals I_{1,k_2}, where $k_2 \in J^{1,g,l}$, to the intervals $I_{\frac{M_1-1}{2},1}, \ldots, I_{\frac{M_1-1}{2}, \frac{m_2-1}{2} - \frac{(M_1-3)(M_1-1)}{2}}$. Hence, by a symmetric reasoning we again obtain equality (9.44) for x in the intervals I_{1,k_2}, and for $k_2 \in J^{1,g,r}$ too.

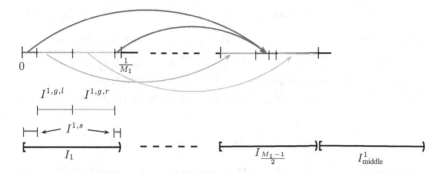

Figure 9.5 τ_2 for the *singular* indices on the left side. In this figure, the interval $I^{1,g,l}$ is the union of the intervals I_{1,k_2} with $k_2 \in J^{1,g,l}$. A similar convention holds for $I^{1,g,r}$ and $I^{1,s}$ (which is not an interval anymore).

Now we have to deal with the *singular* subintervals I_{1,k_2}, where $k_2 \in J^{1,s}$, and the singular indices are given by

$$J^{1,s} = \{1, \ldots, m_2\} \setminus J^{1,g}$$
$$= \{1, \ldots, \tfrac{(M_1-3)(M_1-1)}{2}\} \cup \{m_2 - \tfrac{(M_1-3)(M_1+1)}{2} + 1, \ldots, m_2\},$$

which consists of $M_1(M_1 - 3)$ many indices.

The map $T_{\alpha_2}^{(\tau_2)}$ has to map these intervals I_{1,k_2}, where $k_2 \in J^{1,s}$, to the 'remaining gaps' $I_{\frac{M_1-1}{2}, l_2}$ in the interval $I_{\frac{M_1-1}{2}}$, where $l_2 \in \{\frac{m_2+1}{2} - \frac{(M_1-3)(M_1-1)}{2}, \ldots, \frac{m_2+1}{2} + \frac{(M_1-3)(M_1+1)}{2} - 1\}$. Note that the corresponding intervals $I_{\frac{M_1-1}{2}, l_2}$ are – roughly speaking – in the middle of the interval $I_{\frac{M_1-1}{2}}$, while the intervals I_{1,k_2}, with $k_2 \in J^{1,s}$, are at the boundary of I_1.

To define τ_2 on I_{1,k_2}, for $k_2 \in J^{1,s}$, choose any function τ_2 taking values in $\{-M_2 + 1, \ldots, M_2 - 1\}$, satisfying (9.37) (resp. (9.38)) as above, which induces a bijection between the intervals $(I_{1,k_2})_{k_2 \in J^{1,s}}$ and the intervals $I_{\frac{M_1-1}{2}, l_2}$ considered above.

For each such τ_2 we obtain, for $x \in I_{1,k_2}$, $k_2 \in J^{1,s}$, from Lemma 9.3

$$\varphi^2(x) - \varphi^2(T_{\alpha_2}^{(\tau_2)}(x)) \leq -\frac{m_2}{2M_1} + 10M_1^3 + 2\frac{(M_1 - 3)(M_1 - 1)}{2}4M_1^2$$
$$\leq -\frac{m_2}{2M_1} + 20M_1^4. \tag{9.45}$$

Indeed, the leading term $\frac{-m_2}{2M_1}$ and the first error term $10M_1^3$ in the first line above come from Lemma 9.3(ii) when comparing the difference of the value of φ^2 on the interval $I_{1,1}$ with that of $I_{\frac{M_1-1}{2},\frac{m_2+1}{2}}$. For the difference of the value of φ^2 on I_{1,k_2} and $I_{\frac{M_1-1}{2},l_2}$, for arbitrary $k_2 \in J^{1,s}$ and $l_2 \in \{\frac{m_2+1}{2} - \frac{(M_1-3)(M_1-1)}{2}, \ldots, \frac{m_2+1}{2} + \frac{(M_1-3)(M_1-1)}{2}\}$ we apply for both cases at most $\frac{(M_1-3)(M_1+1)}{2}$ times estimate (i) of Lemma 9.3, which gives (9.45).

In particular, for $m_2 > 40M_1^5$, which of course we shall assume, we have that

$$\varphi^2(x) - \varphi^2(T^{(\tau_2)}_{\alpha_2}(x)) \leq 0, \qquad \text{for } x \in I_{1,k_2}, k_2 \in J^{1,s}.$$

There are $M_1(M_1 - 3) = M_1^2 - 3M_1$ many intervals I_{1,k_2} with $k_2 \in J^{1,s}$ each of length $1/M_2$. Hence we may estimate the 'singular mass' on the interval I_1 by

$$\sum_{k_2 \in J^{1,s}} \int_{I_{1,k_2}} [\varphi^2(x) - \varphi^2(T^{(\tau_2)}_{\alpha_2}(x))]\, dx \leq \left(-\frac{m_2}{2M_1} + 20M_1^4\right)(M_1^2 - 3M_1)\frac{1}{M_2}$$

$$\leq -\frac{1}{2} + \frac{3}{2M_1} + \frac{c(M_1)}{2m_2}, \qquad (9.46)$$

where $c(M_1)$ is a constant depending on M_1 only.[4]

We still have another *singular* interval at the present induction step $n = 2$, namely I_{M_1}. The analysis for this case is symmetric to the analysis of I_1 and – after properly defining τ_2 on this interval I_{M_1} – we arrive at the same estimate (9.46). In total, we thus obtain (9.41) by doubling the right-hand side of (9.46), showing that the 'singular mass' essentially equals -1.

Finally, define the sets J_2^g (resp. J_2^s) of *good* (resp. *singular*) indices at level 2 as

$$J_2^g = \{(k_1, k_2) : (k_1 \in J_1^g \text{ and } 1 \leq k_2 \leq m_2), \text{ or } (k_1 \in J_1^s \text{ and } k_2 \in J^{k_1,g})\},$$
$$J_2^s = \{(k_1, k_2) : k_1 \in J_1^s \text{ and } k_2 \in J^{k_1,2}\}.$$

This finishes the inductive step for $n = 2$.

General inductive step. Suppose that the prime numbers $m_1, \ldots, m_{n-1}$ have been defined. We use the notation $\alpha_{n-1} = \frac{1}{M_1} + \cdots + \frac{1}{M_{n-1}}$, where $M_{n-1} = m_1 \cdot m_2 \cdot \ldots \cdot m_{n-1}$.

---

[4] We shall find it convenient in the following to write $c(M_1, M_2, \ldots, M_i)$ for constants depending only on the choice of the numbers $M_1, M_2, \ldots, M_i$. The concrete numerical value of this expression may change, i.e. become bigger, from one line of reasoning to the next one, but at every stage it will be clear that an explicit bound for the respective meaning of the constant $c(M_1, M_2, \ldots, M_i)$ could be given, at least in principle. In fact, we shall always have that the constants $c(M_1, M_2, \ldots, M_i)$ used in the following are dominated by a polynomial in the variables $M_1, M_2, \ldots, M_i$.

For a prime m_n satisfying the condition of Lemma 9.2, and to be specified below, let $M_n = m_1 \cdot \ldots \cdot m_n$ and

$$L^n = \left[0, \frac{1}{2} - \frac{1}{2M_n}\right), \quad R^n = \left[\frac{1}{2} + \frac{1}{2M_n}, 1\right),$$

$$I^n_{\text{middle}} = \left[\frac{1}{2} - \frac{1}{2M_n}, \frac{1}{2} + \frac{1}{2M_n}\right).$$

For $1 \le k_1 \le m_1, \ldots, 1 \le k_n \le m_n$, let

$$I_{k_1,\ldots,k_n} = \left[\frac{k_1-1}{M_1} + \frac{k_2-1}{M_2} + \cdots + \frac{k_n-1}{M_n}, \frac{k_1-1}{M_1} + \frac{k_2-1}{M_2} + \cdots + \frac{k_n}{M_n}\right).$$

For $x \in I_{1,\ldots,1}$ and $j \in \{0, \ldots, M_n\}$ we define, similarly as in (9.33), $\varphi^n(x) = 0$ and

$$\varphi^n(T^j_{\alpha_n}(x)) = \#\{i \in \{0, \ldots, j-1\} : T^i_{\alpha_2}(x) \in L^n\}$$
$$- \#\{i \in \{0, \ldots, j-1\} : T^i_{\alpha_2}(x) \in R^n\}, \tag{9.47}$$

where $\alpha_n = \alpha_{n-1} + \frac{1}{M_n}$ and $M_n = M_{n-1} m_n$. We also let $\psi^n(x) = 1 - \varphi^n(x)$, for $x \in [0, 1)$.

Lemma 9.5 (oscillations of φ^n). *For given $M_1, \ldots, M_{n-1}$ there is a constant $c(M_1, \ldots, M_{n-1})$ depending only on $M_1, \ldots, M_{n-1}$, such that for all m_n as above we have*

(i) $|\varphi^n(x) - \varphi^n(x \oplus \frac{1}{M_n})| \le c(M_1, \ldots, M_{n-1})$,
(ii) for each $1 \le k'_1, k''_1 \le M_1, \ldots, 1 \le k'_{n-1}, k''_{n-1} \le m_{n-1}$,

$$\varphi^n|_{I_{k'_1,\ldots,k'_{n-1},(m_n+1)/2}} - \varphi^n|_{I_{k''_1,\ldots,k''_{n-1},1}} \ge \frac{m_n}{2M_{n-1}} - c(M_1, \ldots, M_{n-1}),$$

(iii) for each $1 \le k'_1, k''_1 \le M_1, \ldots, 1 \le k'_{n-1}, k''_{n-1} \le m_{n-1}$, and $1 \le k'_n, k''_n \le m_n$, with $\min\{k'_n, m_n - k'_n\} < M_{n-1}$ and $\min\{k''_n, m_n - k''_n\} < M_{n-1}$ we have

$$\left|\varphi^n|_{I_{k'_1,\ldots,k'_{n-1},k'_n}} - \varphi^n|_{I_{k''_1,\ldots,k''_{n-1},k''_n}}\right| \le c(M_1, \ldots, M_{n-1}). \tag{9.48}$$

Proof. We may and do assume that $m_n \ge 5M_{n-1}$.
• *Proof of (i).* We have $T_{\alpha_n}(x) = T_{\alpha_{n-1}}(T_{1/M_n}(x))$ so that

$$T^{M_{n-1}}_{\alpha_n}(x) = x \oplus \frac{M_{n-1}}{M_n} = x \oplus \frac{1}{m_n}, \tag{9.49}$$

in perfect analogy to (9.34). As regards the analogue to (9.35) things now are somewhat more complicated. First note that there is a unique number

$1 \leq q_{n-1} \leq M_{n-1} - 1$ such that

$$T_{\alpha_{n-1}}^{q_{n-1}}(x) = x \ominus \frac{1}{M_{n-1}}, \quad x \in [0, 1). \tag{9.50}$$

Indeed, by Lemma 9.2, when q_{n-1} runs through $\{1, \ldots, M_{n-1} - 1\}$, the left-hand side assumes the values $x \ominus \frac{l_{n-1}}{M_{n-1}}$, where l_{n-1} also runs through $\{1, \ldots, M_{n-1} - 1\}$.

Claim 9.6. *Letting* $r_n = \lfloor \frac{m_n}{M_{n-1}} \rfloor$, *the integer part of* $\frac{m_n}{M_{n-1}}$, *and taking* q_{n-1} *as in* (9.50), *we have*

$$T_{\alpha_n}^{r_n M_{n-1} + q_{n-1}}(x) = x \oplus \frac{d_{n-1}}{M_n},$$

where $|d_{n-1}| < M_{n-1}$.

Indeed, write m_n as $m_n = r_n M_{n-1} + e_{n-1}$, for some $1 \leq e_{n-1} \leq M_{n-1}$ to obtain

$$T_{\alpha_n}^{r_n M_{n-1} + q_{n-1}}(x) = (T_{\alpha_n}^{M_{n-1}})^{r_n} \circ T_{\alpha_{n-1}}^{q_{n-1}} \circ T_{\frac{1}{M_n}}^{q_{n-1}}(x)$$

$$= x \oplus r_n \frac{M_{n-1}}{M_n} \ominus \frac{1}{M_{n-1}} \oplus \frac{q_{n-1}}{M_n}$$

$$= x \oplus \frac{m_n}{M_n} \ominus \frac{e_{n-1}}{M_n} \ominus \frac{1}{M_{n-1}} \oplus \frac{q_{n-1}}{M_n}$$

$$= x \oplus \frac{q_{n-1} - e_{n-1}}{M_n} =: x \oplus \frac{d_{n-1}}{M_n},$$

which proves the claim.

Define $s_{n-1}^{(1)} = q_{n-1}$ if $d_{n-1} = q_{n-1} - e_{n-1} > 0$ and $s_{n-1}^{(1)} = q_{n-1} + M_{n-1}$ otherwise, to obtain by (9.49) and (9.50) that

$$T_{\alpha_n}^{r_n M_{n-1} + s_{n-1}^{(1)}}(x) = x \oplus \frac{l_{n-1}^{(1)}}{M_n},$$

for some $l_{n-1}^{(1)} \in \{1, \ldots, M_{n-1}\}$. We also deduce from (9.49) that $l_{n-1}^{(1)}$ must actually be in $\{1, \ldots, M_{n-1} - 1\}$.

Repeat the above argument to find $s_{n-1}^{(2)}$ with $-2M_{n-1} < s_{n-1}^{(2)} < 2M_{n-1}$ such that

$$T_{\alpha_n}^{2r_n M_{n-1} + s_{n-1}^{(2)}}(x) = x \oplus \frac{l_{n-1}^{(2)}}{M_n},$$

for some $l_{n-1}^{(2)} \in \{1, \ldots, M_{n-1} - 1\}$. Continuing in the same way, we find numbers $s_{n-1}^{(j)}$, for $j = 1, 2, \ldots, M_{n-1} - 1$ verifying $-jM_{n-1} < s_{n-1}^{(j)} < jM_{n-1}$ such that

$$T_{\alpha_n}^{jr_n M_{n-1} + s_{n-1}^{(j)}}(x) = x \oplus \frac{l_{n-1}^{(j)}}{M_n}, \tag{9.51}$$

for some $l_{n-1}^{(j)} \in \{1, \ldots, M_{n-1} - 1\}$. Note that, under the assumption $m_n \gg M_{n-1}$ so that $r_n \gg M_{n-1}$, the elements in (9.51) are all different. Therefore, $(l_{n-1}^{(j)})_{j=1}^{M_{n-1}-1}$ runs through all elements of $\{1, \ldots, M_{n-1} - 1\}$ when j runs through $\{1, \ldots, M_{n-1} - 1\}$; in particular, there must be some j_0 such that

$$T_{\alpha_n}^{j_0 r_n M_{n-1} + s_{n-1}^{(j_0)}}(x) = x \oplus \tfrac{1}{M_n},$$

in analogy to (9.36).

Now observe that there is a constant $c(M_1, \ldots, M_{n-1})$, depending only on $M_1, \ldots, M_{n-1}$, such that, for $x \in [0, 1)$, the difference of the number of visits of the orbit $(T_{\alpha_n}^{i}(x))_{i=0}^{r_n M_{n-1} + q_{n-1}}$ to L^n and R^n is bounded in absolute value by the constant $c(M_1, \ldots, M_{n-1})$. The argument is analogous to the corresponding one in the proof of the claim which is part of the proof of Lemma 9.3(i), and therefore skipped.

The numbers j_0 as well as $s_{n-1}^{(j_0)}$ are bounded in absolute value by M_{n-1}^2 so that the difference of the visits of the orbits $(T_{\alpha_n}^{i}(x))_{i=0}^{j_0 r_n M_{n-1} + s_{n-1}^{(j_0)}}$ to L^n and R^n are bounded in absolute value by some constant $c(M_1, \ldots, M_{n-1})$. This finishes the proof of assertion (i).

- *Proof of (ii).* Suppose first, as in the proof of Lemma 9.3(ii), that $(k_1', \ldots, k_{n-1}') = (k_1'', \ldots, k_{n-1}'') =: (k_1, \ldots, k_{n-1})$. For $x \in I_{k_1, \ldots, k_{n-1}, 1}$ we have that each of the orbits $(T_{\alpha_n}^{j M_{n-1} + i}(x))_{i=0}^{M_{n-1}-1}$, for $j = 0, \ldots, \lfloor \tfrac{m_n}{2M_{n-1}} \rfloor - 1$, visits L^n one time more often than R^n. Hence,

$$\varphi^n(T_{\alpha_n}^{\lfloor \frac{m_n}{2M_{n-1}} \rfloor M_{n-1}}(x)) - \varphi^n(x) = \lfloor \tfrac{m_n}{2M_{n-1}} \rfloor > \tfrac{m_n}{2M_{n-1}} - 1.$$

Noting that

$$T_{\alpha_n}^{\lfloor \frac{m_n}{2M_{n-1}} \rfloor M_{n-1}}(x) = x \oplus \lfloor \tfrac{m_n}{2M_{n-1}} \rfloor \tfrac{M_{n-1}}{M_n}$$

and

$$\tfrac{m_n+1}{2M_n} - \lfloor \tfrac{m_n}{2M_{n-1}} \rfloor \tfrac{M_{n-1}}{M_n} \leq \tfrac{M_{n-1}}{M_n},$$

we obtain (ii) by using assertion (i), and possibly passing to a bigger constant $c(M_1, \ldots, M_{n-1})$. Finally, the passage to general $(k_1', \ldots, k_{n-1}')$ and $(k_1'', \ldots, k_{n-1}'')$ is done again, similarly as in the proof of Lemma 9.3, by repeated application of (i) and by passing once more to a bigger constant $c(M_1, \ldots, M_{n-1})$.

- *Proof of (iii).* Fix $1 \leq k_1', k_1'' \leq M_1, \ldots, 1 \leq k_{n-1}', k_{n-1}'' \leq m_{n-1}$ and $1 \leq k_n', k_n'' \leq m_n$ as above. Suppose, for example, $k_n' \leq M_{n-1}$ and $m_n - k_n'' \leq M_{n-1}$, the other three cases being similar. Denote by $(k_1''', \ldots, k_{n-1}''')$ the index so that $I_{k_1''', \ldots, k_{n-1}'''} = I_{k_1'', \ldots, k_{n-1}'', k_n''} \oplus \frac{1}{M_{n-1}}$, i.e. $I_{k_1''', \ldots, k_{n-1}'''}$ is the right neighbour of $I_{k_1'', \ldots, k_{n-1}''}$. Now find $0 \le q_{n-1} < M_{n-1}$ such that $T_{\alpha_{n-1}}^{q_{n-1}}$ maps $I_{k_1', \ldots, k_{n-1}'}$ onto $I_{k_1''', \ldots, k_{n-1}'''}$. Hence, $T_{\alpha_n}^{q_{n-1}}$ maps $I_{k_1', \ldots, k_{n-1}', k_n'}$ onto $I_{k_1''', \ldots, k_{n-1}''', k_n'+q_{n-1}}$.

Finally, note that the distance from the latter interval to $I_{k_1'', \ldots, k_{n-1}'', k_n''}$ is bounded by $(2M_{n-1} + M_{n-1})\frac{1}{M_n}$. Hence, we obtain (9.48) by applying $2M_{n-1} + M_{n-1}$ times assertion (i) and using $0 \le q_{n-1} < M_{n-1}$. $\qquad\square$

After this preparation we are ready for the inductive step from $n-1$ to n. Suppose that the following inductive hypotheses are satisfied, for $1 \le l \le n-1$, functions $\tau_l : [0, 1) \to \{-M_l + 1, \ldots, M_l - 1\}$ and index sets J_l^g, J_l^s contained in $\{(k_1, \ldots, k_l) : 1 \le k_1 \le m_1, \ldots, 1 \le k_l \le m_l\}$.

(i) The measure-preserving bijection $T_{\alpha_{n-1}}^{(\tau_{n-1})} : [0, 1) \to [0, 1)$ maps the intervals $I_{k_1, \ldots, k_l}$, for $1 \le l < n-1$, and $1 \le k_1 \le m_1, \ldots, 1 \le k_l \le m_l$, onto the intervals $T_{\alpha_l}^{(\tau_l)}(I_{k_1, \ldots, k_l})$. It induces a permutation of the intervals $I_{k_1, \ldots, k_{n-1}}$, where $1 \le k_1 \le m_1, \ldots, 1 \le k_{n-1} \le m_{n-1}$.

(ii) When $\tau_{n-1}(x) > 0$, we have

$$T_{\alpha_{n-1}}^i(x) \notin I_{\text{middle}}^{n-1}, \quad i = 0, \ldots, \tau_{n-1}(x), \tag{9.52}$$

and, when $\tau_{n-1}(x) < 0$, we have

$$T_{\alpha_{n-1}}^i(x) \notin I_{\text{middle}}^{n-1}, \quad i = \tau_{n-1}(x), \ldots, 0. \tag{9.53}$$

(iii) There is a set of *good* indices $J_{n-1}^g \subseteq \{1 \le k_1 \le m_1, \ldots, 1 \le k_{n-1} \le m_{n-1}\}$. For $(k_1, \ldots, k_{n-2}) \in J_{n-2}^g$ we have that $(k_1, \ldots, k_{n-2}, k_{n-1}) \in J_{n-1}^g$ as well as

$$\mu[I_{k_1, \ldots, k_{n-2}} \cap \{\tau_{n-2} \ne \tau_{n-1}\}] \le \tfrac{M_{n-2}}{m_{n-1}} \mu[I_{k_1, \ldots, k_{n-2}}], \tag{9.54}$$

and

$$\sum_{(k_1, \ldots, k_{n-2}) \in J_{n-2}^g} \int_{I_{k_1, \ldots, k_{n-2}}} \Big| [\varphi^{n-2}(x) - \varphi^{n-2}(T_{\alpha_{n-2}}^{(\tau_{n-2})}(x))] \tag{9.55}$$

$$- [\varphi^{n-1}(x) - \varphi^{n-1}(T_{\alpha_{n-1}}^{(\tau_{n-1})}(x))] \Big| \, dx$$

$$\le \tfrac{c(M_1, \ldots, M_{n-2})}{m_{n-1}}.$$

(iv) There is a set of *singular* indices $J_{n-1}^s \subseteq \{(k_1, \ldots, k_{n-1}) : 1 \le k_1 \le m_1, \ldots, 1 \le k_{n-1} \le m_{n-1}\}$, disjoint from J_{n-1}^g, such that J_{n-1}^s consists

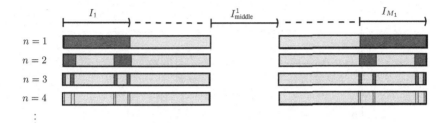

Figure 9.6 The fractal structure of the *singular* set. For the sake of simplicity of the drawing, the dark shaded area which represents the singular set is thicker than it should be. Note also that the effective singular set is not perfectly balanced.

of less than $2M_{n-1}^2$ many elements and such that

$$\varphi^{n-1}(x) - \varphi^{n-1}(T_{\alpha_{n-1}}^{(\tau_{n-1})}(x)) \leq 0, \quad \text{for } x \in I_{k_1,\dots,k_{n-1}} \tag{9.56}$$
$$\text{and } (k_1,\dots,k_{n-1}) \in J_{n-1}^s,$$

and

$$\sum_{(k_1,\dots,k_{n-1})\in J_{n-1}^s} \int_{I_{k_1,\dots,k_{n-1}}} [\varphi^{n-1}(x) - \varphi^{n-1}(T_{\alpha_{n-1}}^{(\tau_{n-1})}(x))]\,dx \tag{9.57}$$
$$\leq -1 + \frac{3}{m_1} + \frac{c(M_1)}{m_2} + \dots + \frac{c(M_1,\dots,M_{n-2})}{m_{n-1}},$$

where $c(\cdot)$ are constants depending only on $(\cdot)$.
(v) On the middle interval $I_{\text{middle}}^1 = I_{\frac{M_1+1}{2}}^1$ we have $\tau_1 = \tau_2 = \dots = \tau_{n-1} = 0$ and I_{middle}^1 together with the intervals $(I_{k_1,\dots,k_{n-1}})_{(k_1,\dots,k_{n-1})\in J_{n-1}^g \cup J_{n-1}^s}$ form a partition of $[0, 1)$.

We have to define τ_n as well as J_n^g and J_n^s so that the above list is satisfied with $n - 1$ replaced by n.

Let us illustrate graphically some features of this construction. Namely, the fractal structure of the singular set, Figure 9.6, and the resulting quasi-cost, Figure 9.7.

We start with a *good* interval $I_{k_1,\dots,k_{n-1}}$, i.e. $(k_1,\dots,k_{n-1}) \in J_{n-1}^g$ and simply write τ for $\tau_{n-1}|_{I_{k_1,\dots,k_{n-1}}}$. If $\tau > 0$, define $J^{k_1,\dots,k_{n-1},c}$, where c stands for 'change', as $\{m_n - \tau + 1,\dots,m_n\}$. This set consists of those indices k_n such that the interval $I_{k_1,\dots,k_n}$ is not mapped into $T_{\alpha_{n-1}}^{(\tau_{n-1})}(I_{k_1,\dots,k_{n-1}})$ under $T_{\alpha_n}^{(\tau_{n-1})}$. If $\tau < 0$, we define $J^{k_1,\dots,k_{n-1},c}$ as $\{1,\dots,|\tau|\}$. The complement $\{1,\dots,m_n\}\setminus J^{k_1,\dots,k_{n-1},c}$ is denoted by $J^{k_1,\dots,k_{n-1},u}$, where u stands for 'unchanged'.

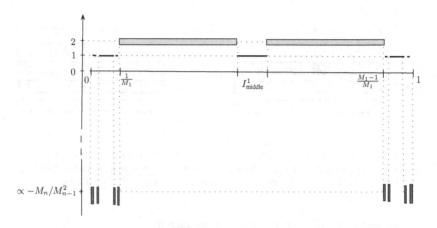

Figure 9.7 Shape of the quasi-cost $\varphi^n + \psi^n \circ T_{\alpha_n}^{(\tau_n)}$. The strips on this graphic representation symbolize the oscillations of the function $\varphi^n + \psi^n \circ T_{\alpha_n}^{(\tau_n)}$. On the *singular* set, this function achieves values of order $-M_n/M_{n-1}^2$. Of course, the effective singular set is much more fragmented than it appears on this figure.

Define $\tau_n := \tau_{n-1} = \tau$ on the intervals $I_{k_1,\dots,k_{n-1},k_n}$, for $k_n \in J^{k_1,\dots,k_{n-1},u}$. For x in one of those intervals we have by (9.52), (9.53) and (9.47) that

$$\varphi^n(x) - \varphi^n(T_{\alpha_n}^{(\tau_n)}(x)) = \varphi^{n-1}(x) - \varphi^{n-1}(T_{\alpha_{n-1}}^{(\tau_{n-1})}(x)),$$

which yields (9.54) with $n-1$ replaced by n.

On the remaining intervals $I_{k_1,\dots,k_n}$ with $k_n \in J^{k_1,\dots,k_{n-1},c}$ we define τ_n such that it takes constant values in $\{-M_n + 1, \dots, M_n - 1\}$ on each of these intervals, such that (9.52) (resp. (9.53)) is satisfied, and such that these intervals $I_{k_1,\dots,k_n}$ are mapped onto the 'remaining gaps' in $T_{\alpha_{n-1}}^{(\tau_{n-1})}(I_{k_1,\dots,k_{n-1}})$.

The crucial observation is that the intervals $I_{k_1,\dots,k_{n-1},k_n}$ where we have $\tau_n \neq \tau_{n-1}$, i.e. where $k_n \in J^{k_1,\dots,k_{n-1},c}$, are all on the 'boundary' of $I_{k_1,\dots,k_{n-1}}$: they are the $|\tau|$ many intervals on the left or right end of $I_{k_1,\dots,k_{n-1}}$, depending on the sign of τ. Similarly, the 'remaining gaps' in $T_{\alpha_{n-1}}^{(\tau_{n-1})}(I_{k_1,\dots,k_{n-1}})$ are the $|\tau|$ many intervals on the opposite end of $T_{\alpha_{n-1}}^{(\tau_{n-1})}(I_{k_1,\dots,k_{n-1}})$. Hence, we may apply assertion (iii) of Lemma 9.5 to conclude that

$$|\varphi^n(x) - \varphi^n(T_{\alpha_n}^{(\tau_n)}(x))| \leq c(M_1, \dots, M_{n-1}),$$

for those $x \in I_{k_1,\dots,k_{n-1}}$ where $\tau_n(x) \neq \tau_{n-1}(x)$. Summing over all 'good intervals' $I_{k_1,\dots,k_{n-1}}$, where $(k_1, \dots, k_{n-1}) \in J_{n-1}^g$, we conclude that the contribution to (9.55), with $n-1$ replaced by n, is controlled by the following factors: M_{n-1}, which is a bound for the number of elements in J_{n-1}^g, times M_{n-1}, which is a bound for $|\tau|$, times $\frac{1}{M_n}$, which is the length of the intervals $I_{k_1,\ldots,k_n}$, times the above-found constant $c(M_1,\ldots,M_{n-1})$. In total, this implies the estimate (9.55), with $n-1$ replaced by n.

We now turn to item (iv), i.e. to the *singular* indices: fix $k_1,\ldots,k_{n-1} \in J^s_{n-1}$ and let $\Delta\varphi$ denote the constant

$$\Delta\varphi := \varphi^{n-1}(T^{(\tau_{n-1})}_{\alpha_{n-1}}(x)) - \varphi^{n-1}(x), \quad x \in I_{k_1,\ldots,k_{n-1}},$$

and again τ the constant $\tau_{n-1}|_{I_{k_1,\ldots,k_{n-1}}}$, so that $0 \le \Delta\varphi \le |\tau| < M_{n-1}$.

Similarly, as for the case $n = 2$, define

$$J^{k_1,\ldots,k_{n-1},g,l} = \{k_n^l, k_n^l + 1 \ldots, \tfrac{m_n-1}{2}\}, \quad J^{k_1,\ldots,k_{n-1},g,r} = \{\tfrac{m_n+1}{2},\ldots, k_n^r\}.$$

Here, k_n^r is the largest number such that, for the orbit $(T^i_{\alpha_n}(x))^{\tau+\Delta\varphi M_{n-1}-1}_{i=\tau}$ and for $x \in I_{k_1,\ldots,k_{n-1},k_n^r}$, all its members lie in the right half of the respective intervals $I_{k'_1,\ldots,k'_{n-1}}$. In fact, we get as in the step $n = 2$ that $k_n^r = m_n - (\tau + \Delta\varphi M_{n-1})$.

Similarly, k_n^l is the smallest number such that, for the orbit $(T^i_{\alpha_n}(x))^{\tau-\Delta\varphi M_{n-1}+1}_{i=\tau}$ and for $x \in I_{k_1,\ldots,k_{n-1},k_n^l}$, all its members are in the left half of the respective intervals $I_{k'_1,\ldots,k'_{n-1}}$. We get $k_n^l = \tau - \Delta\varphi M_{n-1} + 1$.

Now we define τ_n as

$$\tau_n(x) = \tau + \Delta\varphi M_{n-1}, \quad \text{for } x \in I_{k_1,\ldots,k_{n-1},k_n}, k_n \in J^{k_1,\ldots,k_{n-1},g,r},$$

and

$$\tau_n(x) = \tau - \Delta\varphi M_{n-1}, \quad \text{for } x \in I_{k_1,\ldots,k_{n-1},k_n}, k_n \in J^{k_1,\ldots,k_{n-1},g,l}.$$

Similarly, as in (9.44) at step $n = 2$, we get for $k_n \in J^{k_1,\ldots,k_{n-1},g} := J^{k_1,\ldots,k_{n-1},g,l} \cup J^{k_1,\ldots,k_{n-1},g,r}$ and $x \in I_{k_1,\ldots,k_{n-1},k_n}$ that

$$\varphi^n(x) - \varphi^n(T^{(\tau_n)}_{\alpha_n}(x))$$
$$= [\varphi^n(x) - \varphi^n(T^{(\tau_{n-1})}_{\alpha_n}(x))] + [\varphi^n(T^{(\tau_{n-1})}_{\alpha_n}(x)) - \varphi^n(T^{(\tau_n)}_{\alpha_n}(x))]$$
$$= [\varphi^{n-1}(x) - \varphi^{n-1}(T^{(\tau_{n-1})}_{\alpha_{n-1}}(x))] + [\varphi^n(T^{(\tau_{n-1})}_{\alpha_n}(x)) - \varphi^n(T^{(\tau_n)}_{\alpha_n}(x))]$$
$$= -\Delta\varphi + \Delta\varphi = 0.$$

We still have to deal with the *singular* indices

$$J^{k_1,\ldots,k_{n-1},s} := \{1,\ldots,m_n\} \setminus J^{k_1,\ldots,k_{n-1},g} = \{1,\ldots, k_n^l - 1\} \cup \{k_n^r + 1,\ldots, m_n\},$$

which consists of $2\Delta\varphi M_{n-1}$ many indices. This number is bounded by $2M^2_{n-1}$ as $\Delta\varphi \le |\tau| < M_{n-1}$. These intervals have to be mapped onto the 'remaining gaps' in the interval $T^{(\tau_{n-1})}_{\alpha_{n-1}}(I_{k_1,\ldots,k_{n-1}})$. Make the crucial observation that, while the intervals $I_{k_1,\ldots,k_{n-1},k_n}$, for $k_n \in J^{k_1,\ldots,k_{n-1},s}$, are at the boundary of $I_{k_1,\ldots,k_{n-1}}$, the 'remaining gaps' are in the middle of the interval $T_{\alpha_{n-1}}^{(\tau_{n-1})}(I_{k_1,\ldots,k_{n-1}})$. This fact is analogous to the situation for $n = 1$ and $n = 2$.

Now define τ_n on the intervals $I_{k_1,\ldots,k_{n-1},k_n}$ for $k_n \in J^{k_1,\ldots,k_{n-1},s}$, in such a way that $T_{\alpha_n}^{(\tau_n)}$ maps these intervals onto the 'remaining gaps' in $T_{\alpha_{n-1}}^{(\tau_{n-1})}(I_{k_1,\ldots,k_{n-1}})$ and such that τ_n is constant on each of these intervals, takes values in $\{-M_n + 1, \ldots, M_n - 1\}$ and such that (9.52) (resp. (9.53)) is satisfied with $n - 1$ replaced by n. Applying Lemma 9.5, assertion (ii) as well as $2(M_{n-1} + 1)|\tau|$ many times assertion (i) we obtain, for $x \in I_{k_1,\ldots,k_{n-1},k_n}$ and $k_n \in J^{k_1,\ldots,k_{n-1},s}$,

$$\varphi^n(x) - \varphi^n(T_{\alpha_n}^{(\tau_n)}(x)) \leq -\frac{m_n}{2M_{n-1}} + c(M_1,\ldots,M_{n-1}).$$

Assuming that m_n is sufficiently large as compared with M_{n-1} we have that the right-hand side is negative.

Keeping in mind that there are $2\Delta\varphi M_{n-1}$ many indices in $J^{k_1,\ldots,k_{n-1},s}$, we may estimate the 'singular mass' on the interval $I_{k_1,\ldots,k_{n-1}}$ by

$$\sum_{k_n \in J^{k_1,\ldots,k_{n-1},s}} \int_{I_{k_1,\ldots,k_{n-1},k_n}} [\varphi^n(x) - \varphi^n(T_{\alpha_n}^{(\tau_n)}(x))]\,dx$$
$$\leq 2\Delta\varphi M_{n-1}[-\frac{m_n}{2M_{n-1}} + c(M_1,\ldots,M_{n-1})]\,\frac{1}{M_n} \qquad (9.58)$$
$$= -\frac{\Delta\varphi}{M_{n-1}}\,[1 - \frac{c(M_1,\ldots,M_{n-1})}{m_n}].$$

We have by the inductive hypothesis that

$$\sum_{k_1,\ldots,k_{n-1} \in J_{n-1}^s} \int_{I_{k_1,\ldots,k_{n-1}}} [\varphi^{n-1}(x) - \varphi^{n-1}(T_{\alpha_{n-1}}^{(\tau_{n-1})}(x))]\,dx$$

$$\leq -1 + \frac{3}{m_1} + \frac{c(M_1)}{m_2} + \cdots + \frac{c(M_1,\ldots,M_{n-2})}{m_{n-1}},$$

or, writing now $\Delta\varphi_{k_1,\ldots,k_{n-1}}$ for the above value of $\Delta\varphi$ on the interval $I_{k_1,\ldots,k_{n-1}}$,

$$\frac{1}{M_{n-1}} \sum_{k_1,\ldots,k_{n-1} \in J_{n-1}^s} \Delta\varphi_{k_1,\ldots,k_{n-1}} \leq -1 + \frac{3}{m_1} + \frac{c(M_1)}{m_2} + \cdots + \frac{c(M_1,\ldots,M_{n-2})}{m_{n-1}}.$$

Letting $J_n^s := \bigcup_{k_1,\ldots,k_{n-1} \in J_{n-1}^s} \{(k_1,\ldots,k_{n-1},k_n) : k_n \in J^{k_1,\ldots,k_{n-1},s}\}$ we obtain from (9.58)

$$\sum_{k_1,\ldots,k_n \in J_n^s} \int_{I_{k_1,\ldots,k_n}} [\varphi^n(x) - \varphi^n(T_{\alpha_n}^{(\tau_n)}(x))]\,dx$$
$$\leq (-1 + \frac{3}{m-1} + \cdots + \frac{c(M_1,\ldots,M_{n-2})}{m_{n-1}})(1 - \frac{c(M_1,\ldots,M_{n-1})}{m_n})$$
$$= -1 + \frac{3}{m_1} + \cdots + \frac{c(M_1,\ldots,M_{n-2})}{m_{n-1}} + \frac{c(M_1,\ldots,M_{n-1})}{m_n},$$

where we may have increased the constant $c(1, \ldots, M_{n-1})$ in the last line. This concludes the inductive step.

Construction of the Example. Let $\alpha = \lim_{n \to \infty} \alpha_n$ so that $T_\alpha = \lim_{n \to \infty} T_{\alpha_n}$ is the shift by the irrational number α.

The sequence $(\tau_n)_{n=1}^\infty$ of functions $\tau_n : [0, 1) \to \mathbb{Z}$ converges, by (9.54), almost surely to a $\mathbb{Z}$-valued function $\tau = \lim_{n \to \infty} \tau_n$. Hence the maps $(T_{\alpha_n}^{(\tau_n)})_{n=1}^\infty$ converge almost surely to a map

$$T_\alpha^{(\tau)} : \begin{cases} [0, 1) \to [0, 1) \\ x \mapsto T_\alpha^{(\tau)}(x) = T_\alpha^{\tau(x)}(x). \end{cases}$$

Using the fact that each $T_{\alpha_n}^{(\tau_n)}$ is a measure preserving almost sure bijection on $[0, 1)$, it is straightforward to check that $T_\alpha^{(\tau)}$ is so too.

Letting $\Gamma_\tau = \{(x, T_\alpha^{(\tau)}(x)), \ x \in [0, 1)\}$ in analogy to the notations $\Gamma_0 = \{(x, x), \ x \in [0, 1)\}$ and $\Gamma_1 = \{(x, T_\alpha(x)), \ x \in [0, 1)\}$, we define

$$c(x, y) = \begin{cases} h_+(x, y), & \text{if } (x, y) \in \Gamma_0 \cup \Gamma_1 \cup \Gamma_\tau, \\ \infty & \text{otherwise,} \end{cases}$$

where h is defined in (9.15). From this definition we deduce the almost sure identity, for $\tau(x) > 0$,

$$\begin{aligned} h(x, T_\alpha^{(\tau)}(x)) &= \#\{i \in \{0, \ldots, \tau(x) - 1\} : T_\alpha^i(x) \in [0, \tfrac{1}{2})\} \\ &\quad - \#\{i \in \{0, \ldots, \tau(x) - 1\} : T_\alpha^i(x) \in [\tfrac{1}{2}, 1)\} + 1 \qquad (9.59) \\ &= \lim_{n \to \infty} [\varphi^n(x) - \varphi^n(T_{\alpha_n}^{(\tau_n)}(x))] + 1, \end{aligned}$$

a similar formula holding true for $\tau(x) < 0$.

As regards the Borel functions $(\varphi_n, \psi_n)_{n=1}^\infty$ announced in (9.17), (9.18) and (9.19), we need to slightly modify the functions $(\varphi^n, \psi^n)_{n=1}^\infty$ constructed in the above induction to make sure that they satisfy the inequality

$$\varphi_n(x) + \psi_n(y) \leq c(x, y), \quad \text{for } x \in X, \ y \in Y. \qquad (9.60)$$

As $c = \infty$ outside of $\Gamma_0 \cup \Gamma_1 \cup \Gamma_\tau$ it is sufficient to make sure that the following inequalities hold true almost surely, for $x \in [0, 1)$:

(0) $\varphi_n(x) + \psi_n(x) \leq c(x, x) = 1$,

(1) $\varphi_n(x) + \psi_n(T_\alpha(x)) \leq c(x, T_\alpha(x)) = \begin{cases} 2, & \text{for } x \in [0, \tfrac{1}{2}), \\ 0, & \text{for } x \in [\tfrac{1}{2}, 1), \end{cases}$

(τ) $\varphi_n(x) + \psi_n(T_\alpha^{(\tau)}(x)) \leq c(x, T_\alpha^{(\tau)}(x))$.

The above-constructed $(\varphi^n, \psi^n)_{n=1}^{\infty}$ only satisfy condition (0). We still have to pass from φ^n to a smaller function φ_n – while leaving $\psi_n := \psi^n$ unchanged – to satisfy (1) and (τ) too. Let

$$
\begin{aligned}
\varphi_n(x) := \varphi^n(x) &- [\varphi^n(x) + \psi^n(T_\alpha(x)) - c(x, T_\alpha(x))]_+ \\
&- [\varphi^n(x) + \psi^n(T_\alpha^{(\tau)}(x)) - c(x, T_\alpha^{(\tau)}(x))]_+ .
\end{aligned}
\tag{9.61}
$$

Clearly $\varphi_n \leq \varphi^n$ and the functions (φ_n, ψ_n) satisfy the inequality (9.60).

We have to show that the functions φ_n defined in (9.61) satisfy that $\varphi^n - \varphi_n$ is small in the norm of $L^1(\mu)$, as $n \to \infty$, that is

$$
\lim_{n \to \infty} \int_{[0,1)} (\varphi^n(x) - \varphi_n(x)) \, dx = 0,
\tag{9.62}
$$

provided that $(m_n)_{n=1}^{\infty}$ increases sufficiently fast to infinity.

We may estimate the first correction term in (9.61) by

$$
\begin{aligned}
&[\varphi^n(x) + \psi^n(T_\alpha(x)) - c(x, T_\alpha(x))]_+ \\
&\leq [\psi^n(T_\alpha(x)) - \psi^n(T_{\alpha_n}(x))]_+ \\
&+ [\varphi^n(x) + \psi^n(T_{\alpha_n}(x)) - c(x, T_\alpha(x))]_+ .
\end{aligned}
$$

The second term above is dominated by $\mathbb{1}_{I^n_{\text{middle}}}$, which is harmless as $\|\mathbb{1}_{I^n_{\text{middle}}}\|_{L^1(\mu)} = \frac{1}{M_n}$. As regards the first term, note that $T_\alpha(x) \ominus T_{\alpha_n}(x) = \alpha - \alpha_n = \sum_{j=n+1}^{\infty} \frac{1}{M_j}$, which we may bound by $\frac{2}{M_{n+1}}$ by assuming that $(m_n)_{n=1}^{\infty}$ increases sufficiently fast to infinity. As ψ^n is constant on each of the M_n many intervals $I_{k_1,\dots,k_n}$ we get

$$
\mu\{x \in [0,1) : \psi^n(T_\alpha(x)) \neq \psi^n(T_{\alpha_n}(x)\} \leq M_n(\alpha - \alpha_n) < \frac{2}{m_{n+1}}.
$$

On this set we may estimate, using only the obvious bound $|\psi_n(x)| < M_n$, that

$$
|\psi^n(T_\alpha(x)) - \psi^n(T_{\alpha_n}(x))| \leq 2M_n, \quad x \in [0,1),
$$

to obtain

$$
\|\psi^n(T_\alpha(x)) - \psi^n(T_{\alpha_n}(x))\|_{L^1(\mu)} < \tfrac{4M_n}{m_{n+1}}.
$$

Hence for $(m_n)_{n=1}^{\infty}$ growing sufficiently fast to infinity, the first correction term in (9.61) is also small in L^1-norm.

To estimate the second correction term in (9.61) note that

$$
\varphi^n(x) + \psi^n(T_\alpha^{(\tau)}(x)) = \varphi^n(x) + \psi^n(T_{\alpha_n}^{(\tau_n)}(x)), \quad \text{for } x \in [0,1).
\tag{9.63}
$$

Indeed, $T_{\alpha_n}^{(\tau_n)}$ induces a permutation between the intervals $I_{k_1,\dots,k_n}$ and, by assertion (i) preceding the formula (9.52), we have that $T_{\alpha_{n+j}}^{(\tau_{n+j})}$ maps the intervals

$I_{k_1,\ldots,k_n}$ onto the intervals $T_{\alpha_n}^{(\tau_n)}(I_{k_1,\ldots,k_n})$, for each $j \geq 0$. Noting that ψ^n is constant on each of the intervals $I_{k_1,\ldots,k_n}$ we obtain (9.63), by letting j tend to infinity.

By (9.47), $\varphi^n(x) + \psi^n(T_{\alpha_n}^{(\tau_n)}(x))$ is the number of visits to L^n minus the number of visits to R^n plus one, of the orbit $(T_{\alpha_n}^j)_{j=0}^{\tau_n(x)-1}$. Similarly, by (9.15), $h(x, T_{\alpha}^{\tau(x)}(x))$ is the number of visits to L minus the number of visits to R plus one, of the orbit $(T_{\alpha}^j)_{j=0}^{\tau(x)-1}$. We have to show that the positive part of the difference

$$f_n(x) := [\varphi^n(x) + \psi^n(T_{\alpha_n}^{(\tau_n)}(x)) - h_+(x, T_{\alpha}^{(\tau)}(x))]_+, \quad x \in [0,1), \quad (9.64)$$

is small in L^1-norm, as $n \to \infty$. To do so, we argue separately on $I_{middle}^1 = [\frac{1}{2} - \frac{1}{2M_1}, \frac{1}{2} + \frac{1}{2M_1}]$, on the union of the *good* intervals at level n: $G_n = \bigcup_{(k_1,\ldots,k_n) \in J_n^g} I_{k_1,\ldots,k_n}$, and the union of the *singular* intervals at level n, $S_n = \bigcup_{(k_1,\ldots,k_n) \in J_n^s} I_{k_1,\ldots,k_n}$.

- For $x \in I_{middle}^1$, the correction term $f_n(x)$ in (9.64) simply equals zero as $\tau_n(x) = \tau(x) = 0$.
- For $x \in S_n$, we have by (9.56) that $\varphi^n(x) + \psi^n(T_{\alpha_n}^{\tau_n(x)}(x)) \leq 1$ so that $f_n(x) \leq 1$ too; hence $\lim_{n\to\infty} \|f_n \mathbb{1}_{S_n}\|_{L^1(\mu)} = 0$.
- For $x \in G_n$, we use

$$f_n(x) \leq [\varphi^n(x) + \psi^n(T_{\alpha_n}^{(\tau_n)}(x)) - h(x, T_{\alpha}^{\tau(x)}(x))]_+$$

$$\leq \sum_{k=n+1}^{\infty} [(\varphi^{k-1}(x) + \psi^{k-1}(T_{\alpha_{k-1}}^{(\tau_{k-1})}(x))) - (\varphi^k(x) + \psi^k(T_{\alpha_k}^{(\tau_k)}(x)))]_+$$

and (9.55) to conclude that

$$\lim_{n\to\infty} \|f_n \mathbb{1}_{G_n}\|_{L^1(\mu)} \leq \lim_{n\to\infty} \sum_{k=n+1}^{\infty} \frac{c(M_1,\ldots,M_{k-1})}{m_k} = 0.$$

This proves (9.62).

Hence (9.17), (9.18) and (9.19) are satisfied.

As regards assertion (9.16), let us verify that π_0 and π_1 are optimal transport plans. Indeed, it follows from (9.17) and (9.18) that the dual value of the present transport problem is greater than or equal to one, which implies that $\langle c, \pi_0 \rangle = \langle c, \pi_1 \rangle = 1$ is the optimal primal value.

The fact that $\langle c, \pi_\tau \rangle > 1$ should be rather obvious to a reader who has made it up to this point of the construction. It follows from rough estimates. The set $\{[0, \frac{1}{2}) \cap \{\tau = -1\}\} \cup \{[\frac{1}{2}, 1) \cap \{\tau = 1\}\}$ has measure bigger than $1 - \frac{3}{M_1} + \sum_{i=2}^{\infty} \frac{c(M_1,\ldots,M_{i-1})}{m_i}$, which is bigger than, say, $\frac{3}{4}$, for $(m_n)_{n=1}^{\infty}$ tending sufficiently quick to infinity. As $c(x, T_\alpha^{(\tau)}(x))$ equals 2 on this set we get

$$\langle c, \pi_\tau \rangle \geq \tfrac{3}{2} > 1.$$

A slightly more involved argument, whose verification is left to the energetic reader, shows that, for $\varepsilon > 0$, we may choose $(m_n)_{n=1}^\infty$ such that

$$\langle h, \pi_\tau \rangle \geq 2 - \varepsilon. \tag{9.65}$$

Finally, we show assertion (iv) at the beginning of this section (see (9.20)). Let $\widehat{h} \in L^1(\pi)^{**}$ be a dual optimizer in the sense of [BLS09, Theorem 4.2]. We know from this theorem that there is a sequence $(\varphi_n, \psi_n)_{n=1}^\infty$ of bounded Borel functions[5] such that

$(\alpha)\ \lim\limits_{n\to\infty} \| [\varphi_n \oplus \psi_n - c]_+ \|_{L^1(\pi)} = 0 \tag{9.66}$

$(\beta)\ \lim\limits_{n\to\infty} \left(\int_X \varphi_n(x)\,d\mu(x) + \int_Y \psi_n(y)\,d\nu(y) \right) = 1, \tag{9.67}$

$(\gamma)\ \lim\limits_{n\to\infty} \varphi_n \oplus \psi_n = \widehat{h}^r, \quad \pi\text{-a.s.}, \tag{9.68}$

$(\delta)\ \hat{h}$ is a $\sigma(L^1(\pi)^{**}, L^\infty(\pi))$ cluster point of $(\varphi_n \oplus \psi_n)_{n=1}^\infty$. $\tag{9.69}$

Here, $\widehat{h} = \widehat{h}^r + \widehat{h}^s$ is the decomposition of $\widehat{h} \in L^1(\pi)^{**}$ into its regular part $\widehat{h}^r \in L^1(\pi)$ and into its purely singular part $\widehat{h}^s \in L^1(\pi)^{**}$.

We shall show that $\widehat{h}^r$ equals h, π-almost surely. Indeed, by assertions (9.66) and (9.67) we have that, for $x \in [0, 1)$,

$$\lim_{n\to\infty} (\varphi_n(x) + \psi_n(x)) = c(x, x) = h(x, x) = 1,$$

and

$$\lim_{n\to\infty} (\varphi_n(x) + \psi_n(T_\alpha(x))) = c(x, T_\alpha(x)) = h(x, T_\alpha(x)) = \begin{cases} 2, & \text{for } x \in [0, \tfrac{1}{2}), \\ 0, & \text{for } x \in [\tfrac{1}{2}, 1), \end{cases}$$

the limit holding true in $L^1([0, 1], \mu)$ as well as for μ-a.e. $x \in [0, 1)$, possibly after passing to a subsequence. As in the discussion following [BLS09, Theorem 4.2], this implies that, for each fixed $i \in \mathbb{Z}$,

$$\lim_{n\to\infty} (\varphi_n(x) + \psi_n(T_\alpha^i(x))) = h(x, T_\alpha^i(x)), \quad i \in \mathbb{Z},$$

the limit again holding true in $L^1(\mu)$ and μ-a.s., after possibly passing to a diagonal subsequence. Whence, we obtain with (9.68) that

$$\lim_{n\to\infty} (\varphi_n(x) + \psi_n(T_\alpha^{(\tau)}(x))) = h(x, T_\alpha^{(\tau)}(x)) = \widehat{h}^r(x, T_\alpha^{(\tau)}(x)),$$

convergence now holding true for μ-a.e. $x \in [0, 1]$.

---

[5] The (φ_n, ψ_n) need not be the same as the special sequence constructed above; still, we find it convenient to use the same notation.

As $x \to T_\alpha^{(\tau)}(x)$ is a measure-preserving bijection we get

$$\int_{[0,1)} [\varphi_n(x) + \psi_n(T_\alpha^{(\tau)}(x))]\,dx = \int_{[0,1)} (\varphi_n(x) + \psi_n(x))\,dx = 1,$$

so that using (9.65) we get

$$\lim_{n\to\infty} \int_{[0,1)} [\varphi_n(x) + \psi_n(T_\alpha^{(\tau)}(x))]\mathbb{1}_{\{\varphi_n(x)+\psi_n(T_\alpha^{(\tau)}(x))<h(x,T_\alpha^{(\tau)}(x))\}}(x)\,dx$$

$$= 1 - \lim_{n\to\infty} \int_{[0,1)} [\varphi_n(x) + \psi_n(T_\alpha^{(\tau)}(x))]\mathbb{1}_{\{\varphi_n(x)+\psi_n(T_\alpha^{(\tau)}(x))\geq h(x,T_\alpha^{(\tau)}(x))\}}(x)\,dx$$

$$= 1 - \langle h, \pi_\tau \rangle$$

$$< 0.$$

From $\lim_{n\to\infty} \mu\{x : \varphi_n(x) + \psi_n(T_\alpha^{(\tau)}(x)) < h(x, T_\alpha^{(\tau)}(x))\} = 0$ we conclude that each σ^*-cluster point of $([\varphi_n(\cdot) + \psi_n(T_\alpha^{(\tau)}(\cdot))]_-)_{n=1}^\infty$ is a purely singular element of $L^1(\pi)^{**}$ of norm equal to $\langle h, \pi_\tau \rangle - 1$.

Finally, we still have to specify the prime numbers $(m_n)_{n=1}^\infty$ in the above induction. It is now clear what we need: apart from satisfying the conditions of Lemma 9.3 as well as the requirements whenever we wrote 'for m_n tending sufficiently fast to infinity', we choose the $(m_n)_{n=1}^\infty$ inductively such that in (9.54) we have $\frac{M_{n-2}}{m_{n-1}} < 2^{-n}$, that in (9.55) we have $\frac{c(M_1,\dots,M_{n-2})}{m_{n-1}} < 2^{-n}$ and in (9.57) we have $\frac{3}{m_1} < \frac{1}{4}$ as well as again $\frac{c(M_1,\dots,M_{n-2})}{m_{n-1}} < 2^{-n}$.

Hence, we have shown all the assertions (i)–(iv) of Example 1 and the construction of the example is complete. $\qquad\square$

9.4 A relaxation of the dual problem

As in [BLS09, Remark 3.4], for a given cost function $c : X \times Y \to [0, \infty]$, we consider the family of pairs of functions

$$\Psi^{\mathrm{rel}}(\mu, v) = \left\{ (\varphi, \psi) : \begin{array}{l} \varphi, \psi \text{ Borel, integrable and} \\ \varphi(x) + \psi(y) \leq c(x, y), \ \pi\text{-a.s,} \\ \text{for each finite transport plan } \pi \in \Pi(\mu, v, c) \end{array} \right\}$$

and define the relaxed value of the dual problem as

$$D^{\mathrm{rel}} = \sup\left\{ \int_X \varphi\,d\mu + \int_Y \psi\,dv : (\varphi, \psi) \in \Psi^{\mathrm{rel}}(\mu, v) \right\}. \tag{9.70}$$

Using the notation of [BLS09] it is obvious that $D \leq D^{\mathrm{rel}}$ and it is straightforward to verify that the trivial duality inequality $D^{\mathrm{rel}} \leq P$ still is satisfied. One might conjecture – and the present authors did so for some time – that $D^{\mathrm{rel}} = P$

holds true in full generality, i.e. for arbitrary Borel measurable cost functions $c : X \times Y \to [0, \infty]$, defined on the product of two Polish spaces X and Y. In this section we construct a counterexample showing that this is not the case, i.e. it may happen that we have a duality gap $P - D^{\text{rel}} > 0$. The example will be a variant of the example in the previous section, i.e. the $(n + 1)$th variation of [AP03, Example 3.2].

In Section 9.3 we constructed a measure-preserving bijection $T_\alpha^{(\tau)} : [0, 1) \to [0, 1)$ having certain properties; we now shall construct a sequence $(T_\alpha^{(\tau_n)})_{n=0}^\infty$ of such maps and consider as cost function the restriction of h_+, where h is defined in (9.15) to the graphs $(\Gamma_n)_{n=0}^\infty$ of the maps $(T_\alpha^{(\tau_n)})_{n=0}^\infty$. This sequence also "builds up a singular mass", which now is positive as opposed to the negative singular mass in the previous section, but it does so in a different way. We summarize the properties of these maps which we shall construct in the following proposition.

Proposition 9.7. *With the notation of Section 9.3 there is an irrational* $\alpha \in [0, 1)$ *and a sequence* $(\tau_n)_{n=0}^\infty$ *of maps* $\tau_n : [0, 1) \to \mathbb{Z}$, *with* $\tau_0 = 0$ *and* $\tau_1 = 1$, *such that the transformations* $T_\alpha^{(\tau_n)} : [0, 1) \to [0, 1)$, *defined by*

$$T_\alpha^{(\tau_n)}(x) = T_\alpha^{\tau_n(x)}(x), \qquad x \in [0, 1),$$

have the following properties.

(i) *Each* τ_n *is constant on a countable collection of disjoint, half-open intervals in* $[0, 1)$ *whose union has full measure. For* $n \geq 0$, *the map* $T_\alpha^{(\tau_n)}$ *defines a measure-preserving almost sure bijection of* $([0, 1), \mu)$ *onto itself, where* $\mu = \nu$ *denotes Lebesgue measure on* $[0, 1)$. *We have, for each* $n \geq 0$,

$$\int_{[0,1)} h(x, T_\alpha^{(\tau_n)}(x)) \, dx = 1. \tag{9.71}$$

(ii) *The function*

$$f_n(x) := h(x, T_\alpha^{(\tau_n)}(x)), \qquad x \in [0, 1),$$

where h *is defined in* (9.15), *satisfies*

$$\| f_n - g_n \|_{L^1(\mu)} < 2^{-n}, \tag{9.72}$$

where g_n *is a Borel function on* $[0, 1)$ *such that*

$$\mu\{g_n = 0\} = 1 - \eta_n, \qquad \mu\{g_n = \tfrac{1-\eta_n}{\eta_n}\} = \eta_n \tag{9.73}$$

for some sequence $(\eta_n)_{n=1}^\infty$ *tending to zero.*

(iii) *There is a sequence* $(\varphi_n, \psi_n)_{n=1}^{\infty}$ *of bounded Borel functions such that, for every fixed* $n \in \mathbb{N}$,

$$\lim_{m \to \infty} \|h(x, T_{\alpha}^{(\tau_n)}(x)) - [\varphi_m(x) + \psi_m(T_{\alpha}^{(\tau_n)}(x))]\|_{L^1(\mu)} = 0,$$

and

$$\lim_{n \to \infty} \left[\int_{[0,1)} \varphi_n(x)\, dx + \int_{[0,1)} \psi_n(y)\, dy \right] = 1.$$

(iv) *The sequence* $(T_{\alpha}^{(\tau_n)})_{n=1}^{\infty}$ *converges to the identity map in the following sense:*

$$\delta(x, T_{\alpha}^{(\tau_n)}(x)) < 2^{-n}, \qquad x \in [0, 1),\ n \geq 1, \qquad (9.74)$$

where $\delta(\cdot, \cdot)$ *denotes the Riemannian metric on* $\mathbb{T} = [0, 1)$.

We postpone the proof of the proposition and first draw some consequences. Suppose that α as well as $(T_{\alpha}^{(\tau_n)})_{n=0}^{\infty}$ have been defined and satisfy the assertions of Proposition 9.7.

Proposition 9.8. *Fix* $M \geq 2$ *and define the cost function* $c_M : [0, 1) \times [0, 1) \to [0, \infty]$ *by*

$$c_M(x, y) = \begin{cases} h_+(x, y), & \text{for } (x, y) \text{ in the graph of } T_{\alpha}^0, T_{\alpha}^1, T_{\alpha}^{(\tau_2)}, T_{\alpha}^{(\tau_3)}, \dots, T_{\alpha}^{(\tau_M)}, \\ \infty, & \text{otherwise.} \end{cases}$$

For this cost function c_M *we find that the primal value, denoted by* P^M, *as well as the dual value, denoted by* D^M, *of the Monge–Kantorovich problem both are equal to 1.*

In addition, there is $\beta = \beta(M) > 0$, *such that, for every partial transport*

$$\sigma \in \Pi^{\text{part}}(\mu, \nu) := \{\sigma : \mathcal{M}(X \times Y) : p_X(\pi) \leq \mu,\ p_Y(\pi) \leq \nu\}$$

with

$$\|\sigma\| \geq \tfrac{2}{3} \text{ and } \int_{X \times Y} c_M(x, y)\, d\sigma(x, y) \leq \tfrac{1}{2},$$

there is no partial transport $\varrho \in \Pi^{\text{part}}(\mu, \nu)$ *with*

$$\|\sigma + \varrho\| = 1 \text{ and } \sigma + \varrho \in \Pi(\mu, \nu)$$

with the property that ϱ *is supported by*

$$\Delta^{\beta} = \{(x, y) \in [0, 1)^2 : \delta(x, y) < \beta\}.$$

Proof. First note that there is an open and dense subset $G \subseteq [0, 1)$ of full measure $\mu(G) = 1$ such that c_M, restricted to $G \times G$, is lower semi-continuous.

This follows from assertion (i) of Proposition 9.7 by replacing the half-open intervals by their open interior. Noting that G is Polish we may apply the general duality theory [Kel84] to the cost function c_M restricted to $G \times G$ to conclude that there is no duality gap for the cost function $c_M|_{G \times G}$. It follows that there is also no duality gap for the original setting of c_M, defined on $[0, 1) \times [0, 1)$, either.

We claim that, for every $M \geq 0$, the value D^M of the dual problem equals 1. Indeed, let $(\varphi_n, \psi_n)_{n=1}^{\infty}$ be a sequence as in Proposition 9.7 (iii). Defining

$$\tilde{\varphi}_n := \varphi_n - \sum_{j=0}^{M} [\varphi_n(x) + \psi_n(T_\alpha^{(\tau_j)}(x)) - h(x, T_\alpha^{(\tau_j)}(x))]_+$$

and $\tilde{\psi}_n = \psi_n$, we have that

$$\tilde{\varphi}_n(x) + \tilde{\psi}_n(y) \leq h(x, y) \leq h_+(x, y),$$

for all (x, y) in the graph of $T_\alpha^0, T_\alpha^1, T_\alpha^{(\tau_2)}, \ldots, T_\alpha^{(\tau_M)}$, and

$$\lim_{n \to \infty} \left[\int_X \tilde{\varphi}_n(x)\, dx + \int_Y \tilde{\psi}_n(y)\, dy \right] = 1,$$

showing that $D^M \geq 1$. It follows that $D^M = P^M = 1$.

Now suppose that the final assertion of the proposition is wrong to find a sequence $(\sigma_n)_{n=1}^{\infty} \in \Pi^{\text{part}}(\mu, \nu)$ with $\|\sigma_n\| \geq \frac{2}{3}$ and $\int_{X \times Y} c_M(x, y)\, d\sigma_n(x, y) \leq \frac{1}{2}$, as well as a sequence $(\varrho_n)_{n=1}^{\infty} \in \Pi^{\text{part}}(\mu, \nu)$ with $\|\pi_n + \varrho_n\| = 1$ and $\pi_n + \varrho_n \in \Pi(\mu, \nu)$ such that ϱ_n is supported by

$$\Delta^{1/n} = \{(x, y) \in [0, 1)^2 : \delta(x, y) < \tfrac{1}{n}\}. \tag{9.75}$$

Considering $(\sigma_n)_{n=1}^{\infty}$ as measures on the product $G \times G$ of the Polish space G, we then can find by Prokhorov's theorem a subsequence $(\sigma_{n_k})_{k=1}^{\infty}$ converging weakly on $G \times G$ to some $\sigma \in \Pi^{\text{part}}(\mu, \nu)$, for which we find $\|\sigma\| \geq \frac{2}{3}$ and $\int_{X \times Y} c(x, y)\, d\sigma(x, y) \leq \frac{1}{2}$. By passing once more to a subsequence, we may also suppose that $(\varrho_{n_k})_{k=1}^{\infty}$ weakly converges (as measures on $G \times G$ or $[0, 1) \times [0, 1)$; here it does not matter) to some $\varrho \in \Pi^{\text{part}}(\mu, \nu)$ for which we get $\|\sigma + \varrho\| = 1$ and $\sigma + \varrho \in \Pi(\mu, \nu)$. By (9.75) we conclude that ϱ induces the identity transport from its marginal $p_X(\varrho)$ onto its marginal $p_Y(\varrho) = p_X(\varrho)$. As $c_M(x, x) = 1$, for $x \in [0, 1)$ we find that $\int_{X \times Y} c_M(x, y)\, d\varrho(x, y) = \|\varrho\| \leq \frac{1}{3}$, which implies that

$$\int c_M(x, y)\, d(\pi + \varrho)(x, y) \leq \tfrac{1}{2} + \tfrac{1}{3},$$

a contradiction to the fact that $P^M = 1$, which finishes the proof. $\qquad\square$

We now can proceed to the construction of the example.

Proposition 9.9. *Assume the setting of Proposition 9.7. For a subsequence $(i_j)_{j=2}^{\infty}$ of $\{2, 3, \ldots\}$ we define the cost function $c : [0, 1) \times [0, 1) \to [0, \infty]$ by*

$$c(x, y) = \begin{cases} h_+(x, y), & \text{for } (x, y) \text{ in the support of } T_\alpha^0, T_\alpha^1, T_\alpha^{(\tau_{i_2})}, T_\alpha^{(\tau_{i_3})}, \ldots, T_\alpha^{(\tau_{i_j})}, \ldots \\ \infty, & \text{otherwise.} \end{cases}$$

(9.76)

If $(i_j)_{j=2}^{\infty}$ tends sufficiently fast to infinity we have that, for this cost function c, the primal value P is strictly positive, while the relaxed primal value P^{rel} (see [BLS09, Example 4.3]) as well as the dual value D and the relaxed dual value D^{rel} (see (9.70)) all are equal to 0.

In particular, there is a duality gap $P - D^{\text{rel}} > 0$, disproving the conjecture mentioned at the beginning of this section.

Proof. We proceed inductively: let $j \geq 2$ and suppose that $i_0 = 0, i_1 = 1, i_2, \ldots, i_j$ have been defined. Apply Proposition 9.8 to

$$c_j(x, y) = \begin{cases} h_+(x, y), & \text{for } (x, y) \text{ in the support of } T_\alpha^0, T_\alpha^1, T_\alpha^{(\tau_{i_2})}, T_\alpha^{(\tau_{i_3})}, \ldots, T_\alpha^{(\tau_{i_j})}, \\ \infty, & \text{otherwise,} \end{cases}$$

to find $\beta_j > 0$ satisfying the conclusion of Proposition 9.8. We may and do assume that $\beta_j \leq \min(\beta_1, \ldots, \beta_{j-1})$. Now choose i_{j+1} such that

$$\delta(x, T_\alpha^{(\tau_{i_{j+1}})}(x)) < \beta_j, \quad x \in [0, 1).$$ (9.77)

This finishes the inductive step and well-defines the cost function $c(x, y)$ in (9.76).

By (9.71) each $T_\alpha^{(\tau_{i_j})}$ induces a Monge transport $\pi_{i_j} \in \Pi(\mu, \nu)$ which satisfies

$$\int_{X \times Y} h(x, y) \, d\pi_{i_j}(x, y) = \int_X h(x, T_\alpha^{(\tau_{i_j})}) \, dx = 1.$$

The fact that the relaxed primal value P^{rel} for the cost function c equals zero directly follows from the definition of P^{rel} [BLS09, Section 1.1], (9.72) and (9.73) by transporting the measure $\mu\mathbb{1}_{\{g_n = 0\}}$, which has mass $1 - \eta_n$, via the Monge transport map $T_\alpha^{(\tau_n)}$ where n is a large element of the sequence $(i_j)_{j=1}^{\infty}$. Hence we conclude from [BLS09, Theorem 1.2] that the dual value D of the Monge–Kantorovich problem for the cost function c defined in (9.76) also equals zero.

Finally, observe that we have $D = D^{\text{rel}}$ in the present example: indeed, the set $\{(x, y) \in [0, 1)^2 : c(x, y) < \infty\}$ is the countable union of the supports of the finite-cost Monge transport plans $T_\alpha^0, T_\alpha^1, T_\alpha^{(\tau_{i_1})}, T_\alpha^{(\tau_{i_2})}, \ldots, T_\alpha^{(\tau_{i_j})}, \ldots$, so that the requirements $\varphi(x) + \psi(y) \leq c(x, y)$, for all $(x, y) \in [0, 1)^2$, and $\varphi(x) + \psi(y) \leq c(x, y)$, π-a.s., for each finite transport plan $\pi \in \Pi(\mu, \nu)$, coincide (after possibly modifying $\varphi(x)$ on a μ-null set).

What remains to prove is that the primal value P satisfies $P > 0$. We shall show that, for every transport plan $\pi \in \Pi(\mu, \nu)$, we have $\int_{X \times Y} c(x, y) \, d\pi(x, y) \geq \frac{1}{2}$. Assume to the contrary that there is $\pi \in \Pi(\mu, \nu)$ such that

$$\int_{X \times Y} c(x, y) \, d\pi(x, y) < \frac{1}{2}.$$

Denoting by σ_j the restriction of π to the union of the graphs of the maps T_α^0, $T_\alpha^1, T_\alpha^{(\tau_{i_1})}, T_\alpha^{(\tau_{i_2})}, \ldots, T_\alpha^{(\tau_{i_j})}$, each σ_j is a partial transport in $\Pi^{\text{part}}(\mu, \nu)$ and the norms $(\|\sigma_j\|)_{j=1}^\infty$ increase to one. Choose j such that

$$\|\sigma_j\| > \tfrac{2}{3}.$$

We apply Proposition 9.8 to conclude that there is no partial transport plan ϱ_j such that $\pi_j + \varrho_j \in \Pi(\mu, \nu)$, and such that ϱ_j is supported by Δ^{β_j}. But this is a contradiction as $\varrho_j = \pi - \sigma_j$ has precisely these properties by (9.77). □

Proof of Proposition 9.7. The construction of the example described by Proposition 9.7 will be an extension of the construction in the previous section from which we freely use the notation.

We shall proceed by induction on $j \in \mathbb{N}$ and define a double-indexed family of maps $\tau_{n,j} : [0, 1) \to \mathbb{Z}$, where $1 \leq n \leq j$.

Step $j = 1$: Define

$$\tau_{1,1} : [0, 1) \to \mathbb{Z}$$

as

$$\tau_{1,1} = -\tau_1,$$

where we have $m_1 = M_1, \alpha_1 = \frac{1}{M_1}$ and τ_1 as in (9.28). At this stage the only difference to the previous section is that we change the sign of τ_1 as we now shall build up a 'positive singular mass', as opposed to the 'negative singular mass' which we constructed in the previous section. More precisely, defining

φ^1, ψ^1 as in (9.27), we obtain, similarly as in (9.29),

$$
\varphi^1(x) + \psi^1(T_{\alpha_1}^{(\tau_{1,1})}(x)) = \begin{cases} 0, & \text{for } x \in I_{k_1}, k_1 \in \{2, \ldots, (M_1 - 1)/2, \\ & \quad (M_1 + 3)/2, \ldots, M_1 - 1\}, \\ (M_1 - 1)/2, & \text{for } x \in I_{k_1}, k_1 = 1, M_1, \\ 1, & \text{for } x \in I_{(M_1+1)/2}. \end{cases}
$$

This finishes the inductive step for $j = 1$.

Step $j = 2$: Let m_2 and $M_2 = M_1 m_2$ be as in Section 9.3, where m_2 satisfies the requirements of Lemma 9.3, and still is free to be eventually specified. To define $\tau_{1,2} : [0, 1) \to \mathbb{Z}$ we want to make sure that the map $T_{\alpha_2}^{(\tau_{1,2})}$ maps the intervals I_{k_1} bijectively onto $T_{\alpha_1}^{(\tau_{1,1})}(I_{k_1})$. Using the notation of the previous section, we consider *all* the intervals I_{k_1} as *good* intervals so that we do not have to take extra care of some *singular* intervals.

More precisely, fix $1 \le k_1 \le M_1$, and write τ for $\tau_{1,1}|_{I_{k_1}}$. If $\tau > 0$, define $J^{k_1,c}$ as $\{m_2 - \tau + 1, \ldots, m_2\}$, i.e. the set of those indices k_2 such that the interval I_{k_1,k_2} is not mapped into $T_{\alpha_1}^{(\tau_{1,1})}(I_{k_1})$ under $T_{\alpha_2}^{(\tau_{1,1})}$. If $\tau < 0$, we define $J^{k_1,c}$ as $\{1, \ldots, |\tau|\}$, and if $\tau = 0$, we define $J^{k_1,c}$ as the empty set. The complement $\{1, \ldots, m_2\} \backslash J^{k_1,c}$ is denoted by $J^{k_1,u}$.

Define $\tau_{1,2} := \tau_{1,1} = \tau$ on the intervals I_{k_1,k_2}, for $k_2 \in J^{k_1,u}$. On the remaining intervals I_{k_1,k_2} with $k_2 \in J^{k_1,c}$ we define $\tau_{1,2}$ such that it takes constant values in $\{-M_2 + 1, \ldots, M_2 - 1\}$ on each of these intervals, such that (9.37) (resp. (9.38) is satisfied, and such that these intervals I_{k_1,k_2} are mapped onto the 'remaining gaps' in $T_{\alpha_1}^{(\tau_{1,1})}(I_{k_1})$.

Using again Lemma 9.3 we resume the properties of the thus constructed map $T_{\alpha_2}^{(\tau_{1,2})} : [0, 1) \to [0, 1)$.

(i) The measure-preserving bijection $T_{\alpha_2}^{(\tau_{1,2})}$ maps each interval I_{k_1} onto $T_{\alpha_1}^{(\tau_{1,1})}(I_{k_1})$. It induces a permutation of the intervals I_{k_1,k_2}, where $1 \le k_1 \le M_1, 1 \le k_2 \le m_2$.

(ii) Defining φ^2, ψ^2 as in (9.32) we get, for each $1 \le k_1 \le M_1$, similarly as in (9.39) and (9.40)

$$
\mu[I_{k_1} \cap \{\tau_{1,2} \ne \tau_{1,1}\}] \le \frac{M_1}{m_2} \mu[I_{k_1}],
$$

as well as

$$
\sum_{k_1=1}^{M_1} \int_{I_{k_1}} |(\varphi^1(x) - \varphi^1(T_{\alpha_1}^{(\tau_{1,1})}(x))) - (\varphi^2(x) - \varphi^2(T_{\alpha_2}^{(\tau_{1,2})}(x))| dx < \frac{4M_1^2}{m_2}.
$$

(iii) On the middle interval $I_{\text{middle}}^1 = I_{\frac{M_1+1}{2}}$ we have $\tau_{1,2} = \tau_{1,1} = 0$.

We now pass to the construction of the map $\tau_{2,2} : [0, 1) \to \mathbb{Z}$. We define, for each $1 \le k_1 \le M_1$, and $x \in I_{k_1,k_2}$,

$$\tau_{2,2}(x) = \begin{cases} a_2(k_2), & \text{for } k_2 \in \{1, \ldots, M_1\} \\ -M_1, & \text{for } k_2 \in \{M_1 + 1, \ldots, (m_2 - 1)/2\}, \\ 0, & \text{for } k_2 = (m_2 + 1)/2, \\ M_1, & \text{for } k_2 \in \{(m_2 + 3)/2, \ldots, m_2 - M_1\}, \\ a_2(k_2), & \text{for } k_2 \in \{m_2 - M_1 + 1, \ldots, m_2\}. \end{cases}$$

The definition of the function a_2 on the *singular* intervals I_{k_1,k_2}, where $k_2 \in \{1, \ldots, M_1\} \cup \{m_2 - M_1 + 1, \ldots, m_2\}$ is done such that $T_{\alpha_2}^{(\tau_{2,2})}$ maps these intervals onto 'remaining gaps' I_{k_1,l_2}, where l_2 runs through the set

$$\{(m_2 - 1)/2 - M_1 + 1, \ldots, (m_2 - 1)/2\}$$
$$\cup \{(m_2 + 3)/2, \ldots, (m_2 + 3)/2 + M_1 - 1\}$$

in the middle region of the interval I_{k_1}. As above, we require in addition that a_2 on each I_{k_1,k_2} takes constant values in $\{-M_2 + 1, \ldots, M_2 - 1\}$ and that (9.37) (resp. (9.38)) is satisfied.

The function $\tau_{2,2}$ mimics the construction of $\tau_{1,1}$ above, with the role of $[0, 1)$ replaced by each of the intervals I_{k_1}, for $1 \le k_1 \le M_1$. The idea is that, $T_{\alpha_1}^{M_1}$ being the identity map, we have that $T_{\alpha_2}^{M_1}$ satisfies $T_{\alpha_2}^{M_1}(x) = x \oplus \frac{M_1}{M_2}$ and $\frac{M_1}{M_2} = \frac{1}{m_2}$ is small. Hence, the role of T_{α_1} in the previous section now is taken by $T_{\alpha_2}^{M_1}$.

More precisely, we have, for each $k_1 = 1, \ldots, M_1$, and $x \in I_{k_1,k_2}$

$$\varphi^2(x) + \psi^2(T_{\alpha_2}^{(\tau_{2,2})}(x))$$

$$= \begin{cases} 0, & \text{for } k_2 \in \{M_1 + 1, \ldots, (m_2 - 1)/2, \\ & \quad (m_2 + 3)/2, \ldots, m_2 - M_1\}, \\ \frac{m_2}{2M_1} + \gamma(M_1), & \text{for } k_2 \in \{1, \ldots, M_1\} \cup \{m_2 - M_1 + 1, \ldots, m_2\}, \\ 1, & \text{for } k_2 = (m_2 + 1)/2. \end{cases}$$

$$(9.78)$$

The notation $\gamma(M_1)$ denotes a quantity verifying $|\gamma(M_1)| \le c(M_1)$ for some constant $c(M_1)$, depending only on M_1. The verification of (9.78) uses Lemma 9.3 and is analogous as in Section 9.3.

As $T_{\alpha_2}^{(\tau_{2,2})}$ defines a measure-preserving bijection on $[0, 1)$, we get

$$\int_0^1 (\varphi^2(x) + \psi^2(T_{\alpha_2}^{(\tau_{2,2})}(x))) \, dx = \int_0^1 (\varphi^2(x) + \psi^2(x)) \, dx = 1. \qquad (9.79)$$

This finishes the inductive step for $j = 2$.

General inductive step: For prime numbers $m_1, \ldots, m_{j-1}$ as in the previous section suppose that we have defined, for $1 \leq n \leq j-1$ maps $\tau_{n,j} : [0, 1) \to \mathbb{Z}$ such that the following inductive hypotheses are satisfied.

(i) For $1 \leq n \leq j-1$, the measure-preserving bijection $T_{\alpha_i}^{(\tau_{n,j-1})} : [0, 1) \to [0, 1)$ maps the intervals $I_{k_1,\ldots,k_{n-1}}$ onto themselves. It induces a permutation of the intervals $I_{k_1,\ldots,k_{j-1}}$, where $1 \leq k_1 \leq m_1, \ldots, 1 \leq k_{j-1} \leq m_{j-1}$.

(ii) For $1 \leq n < j-1$ we have, for $1 \leq k_1 \leq m_1, \ldots, 1 \leq k_{j-2} \leq m_{j-2}$,

$$\mu[I_{k_1,\ldots,k_{j-2}} \cap \{\tau_{n,j-2} \neq \tau_{n,j-1}\}] \leq \tfrac{M_{j-2}}{m_{j-1}} \mu[I_{k_1,\ldots,k_{j-2}}], \tag{9.80}$$

and

$$\sum_{1 \leq k_1 \leq m_1, \ldots, 1 \leq k_{j-2} \leq m_{j-2}} \int_{I_{k_1,\ldots,k_{j-2}}} \left| \left(\varphi^{j-2}(x) - \varphi^{j-2}(T_{\alpha_{j-2}}^{(\tau_{n,j-2})}(x)) \right) \right. \tag{9.81}$$
$$\left. - \left(\varphi^{j-1}(x) - \varphi^{j-1}(T_{\alpha_{j-1}}^{(\tau_{n,j-1})}(x)) \right) \right| dx < \tfrac{c(M_1,\ldots,M_{j-2})}{m_{j-1}}.$$

We now shall define $\tau_{n,j} : [0, 1) \to \mathbb{Z}$, for $1 \leq n \leq j$ and $\tau_{j,j} : [0, 1) \to \mathbb{Z}$.

Fix $1 \leq n \leq j-1$ as well as $1 \leq k_1 \leq m_1, \ldots, 1 \leq k_{j-1} \leq m_{j-1}$. Denote by τ the constant value $\tau_{n,j-1}|_{I_{k_1,\ldots,k_{j-1}}}$. If $\tau > 0$ define $J^{k_1,\ldots,k_{j-1},c}$ as $\{m_j - \tau + 1, \ldots, m_j\}$, similarly as for the case $j = 2$ above. If $\tau \leq 0$ define $J^{k_1,\ldots,k_{j-1},c}$ as $\{1, \ldots, |\tau|\}$ which, for $\tau = 0$, equals the empty set. On the intervals $I_{k_1,\ldots,k_{j-1},k_j}$, where k_j lies in the complement $J^{k_1,\ldots,k_{j-1},u} = \{1, \ldots, m_j\} \setminus J^{k_1,\ldots,k_{j-1},c}$, we define $\tau_{n,j} := \tau_{n,j-1}$. On the remaining intervals $I_{k_1,\ldots,k_{j-1},k_j}$, where $k_j \in J^{k_1,\ldots,k_{j-1},c}$, we define $\tau_{n,j}$ in such a way that it takes constant values in $\{-M_j + 1, \ldots, M_j - 1\}$ on each of these intervals, such that (9.37) (resp. (9.38)) is satisfied, and such that these intervals $I_{k_1,\ldots,k_{j-1},k_j}$ are mapped onto the 'remaining gaps' in $T_{\alpha_{j-1}}^{(\tau_{n,j-1})}(I_{k_1,\ldots,k_{j-1}})$.

Similarly as in the previous section we thus well-define the function $\tau_{n,j}$ which then verifies (9.80) and (9.81), with $j-1$ replaced by j.

We still have to define $\tau_{j,j} : [0, 1) \to \mathbb{Z}$. For $1 \leq k_1 \leq m_1, \ldots, 1 \leq k_{j-1} \leq m_{j-1}$, we define $\tau_{j,j}(x)$ on the intervals $I_{k_1,\ldots,k_{j-1},k_j}$ by

$$\tau_{j,j}(x) = \begin{cases} a_j(k_j), & \text{for } k_j \in \{1, \ldots, M_{j-1}\} \\ -M_j, & \text{for } k_j \in \{M_{j-1} + 1, \ldots, (m_j - 1)/2\}, \\ 0, & \text{for } k_j = (m_j + 1)/2, \\ M_j, & \text{for } k_j \in \{(m_j + 3)/2, \ldots, m_j - M_{j-1}\}, \\ a_j(k_j), & \text{for } k_j \in \{m_j - M_{j-1} + 1, \ldots, m_j\}. \end{cases}$$

Similarly, as in step $j = 2$ the $\{-M_j + 1, \ldots, M_j - 1\}$-valued function $a_j(k_j)$ is defined in such a way that $T_{\alpha_j}^{(\tau_j)}$ maps the intervals $I_{k_1,\ldots,k_{j-1},k_j}$ with

$k_j \in \{1, \ldots, M_{j-1}\} \cup \{m_j - M_{j-1} + 1, \ldots, m_j\}$ to the intervals $I_{k_1, \ldots, k_{j-1}, k_j}$, where k_j runs through the 'middle region'

$$\{(m_j - 1)/2 - M_{j-1} + 1, \ldots, (m_j - 1)/2\}$$
$$\cup \{(m_j + 3)/2, \ldots, (m_j + 3)/2 + M_{j-1} - 1\}.$$

We now deduce from Lemma 9.3 that, for $x \in I_{k_1, \ldots, k_{j-1}, k_j}$

$$\varphi^j(x) + \varphi^j(T_{\alpha_j}^{(\tau_j, j)}(x))$$

$$= \begin{cases} 0, & \text{for } k_j \in \{M_{j-1} + 1, \ldots, (m_j - 1)/2\} \\ & \qquad \cup \{(m_j + 3)/2, \ldots, m_j - M_{j-1}\}, \\ \frac{m_j}{2M_{j-1}} + \gamma(M_1, \ldots, M_{j-1}), & \text{for } k_j \in \{1, \ldots, M_{j-1}\} \\ & \qquad \cup \{m_j - M_{j-1} + 1, \ldots, m_j\}, \\ 1, & \text{for } k_j = (m_j + 1)/2, \end{cases}$$

where $\gamma(M_1, \ldots, M_{j-1})$ denotes a quantity which is bounded in absolute value by a constant $c(M_1, \ldots, M_{j-1})$ depending only on $M_1, \ldots, M_{j-1}$.

This completes the inductive step.

We now define $\tau_0 = 0$, $\tau_1 = 1$ and, for $n \geq 2$

$$\tau_n = \lim_{j \to \infty} \tau_{n-1, j}. \tag{9.82}$$

It follows from (9.80) that, for each $n \geq 2$, the limit (9.82) exists almost surely provided the sequence $(m_n)_{n=1}^{\infty}$ converges sufficiently fast to infinity, similarly as in Section 9.3. The $(\tau_n)_{n=0}^{\infty}$ and the above-constructed functions $(\varphi_n, \psi_n)_{n=1}^{\infty}$ satisfy the assertions of Proposition 9.7. The verification of items (i), (ii) and (iii) is analogous to the arguments of Section 9.3 and therefore skipped. As regards assertions (iv), note that, for $1 \leq n \leq j$, the function $T_{\alpha_j}^{(\tau_n, j)}$ maps the intervals $I_{k_1, \ldots, k_{n-1}}$ onto themselves. It follows that $T_{\alpha}^{(\tau_n)}$ does so too, whence

$$\delta(x, T_{\alpha}^{(\tau_n)}(x)) < M_n^{-1},$$

which readily shows (9.74). □

Acknowledgments

The first author acknowledges financial support from the Austrian Science Fund (FWF) under grant P21209. The third author acknowledges support from the Austrian Science Fund (FWF) under grant P19456, from the Vienna Science and Technology Fund (WWTF) under grant MA13 and by the Christian

Doppler Research Association (CDG). All authors thank A. Pratelli for helpful discussions on the topic of this paper. We also thank L. Summerer for his advice.

References

[AE06] J.-P. Aubin and I. Ekeland. *Applied Nonlinear Analysis*. Dover Publications Inc., Mineola, NY, 2006. Reprint of the 1984 original.

[AP03] L. Ambrosio and A. Pratelli. Existence and stability results in the L^1 theory of optimal transportation. In *Optimal Transportation and Applications (Martina Franca, 2001)*, volume 1813 of Lecture Notes in Mathematics, pages 123–160. Springer, Berlin, 2003.

[BLS09] M. Beiglböck, C. Léonard and W. Schachermayer. A general duality theorem for the Monge–Kantorovich transport problem. *Studia Math.* 209(2):151–167, 2012.

[BS09] M. Beiglböck and W. Schachermayer. Duality for Borel measurable cost functions. *Trans. Am. Math. Soc.*, 363(8):4203–4224, 2011.

[ET99] I. Ekeland and R. Témam. *Convex Analysis and Variational Problems*, volume 28 of Classics in Applied Mathematics. Society for Industrial and Applied Mathematics (SIAM), Philadelphia, PA, English edition, 1999. Translated from the French.

[GM96] W. Gangbo and R.J. McCann. The geometry of optimal transportation. *Acta Math.*, 177(2):113–161, 1996.

[Kel84] H.G. Kellerer. Duality theorems for marginal problems. *Z. Wahrsch. Verw. Gebiete*, 67(4):399–432, 1984.

[Pra08] A. Pratelli. On the sufficiency of c-cyclical monotonicity for optimality of transport plans. *Math. Z.*, 258:677–690, 2008.

[PSU88] A.L. Peressini, F.E. Sullivan, and J.J. Uhl, Jr. *The Mathematics of Nonlinear Programming*. Undergraduate Texts in Mathematics. Springer-Verlag, New York, 1988.

[Roc66] R. T. Rockafellar. Characterization of the subdifferentials of convex functions. *Pacific J. Math.*, 17:497–510, 1966.

[RR96] D. Ramachandran and L. Rüschendorf. Duality and perfect probability spaces. *Proc. Am. Math. Soc.*, 124(7):2223–2228, 1996.

[Rüs96] L. Rüschendorf. On c-optimal random variables. *Statist. Probab. Lett.*, 27(3):267–270, 1996.

[ST08] W. Schachermayer and J. Teichmann. Characterization of optimal transport plans for the Monge–Kantorovich problem. *Proc. Am. Math. Soc.*, 137:519–529, 2008.

[Vil03] C. Villani. *Topics in Optimal Transportation*, volume 58 of Graduate Studies in Mathematics. American Mathematical Society, Providence, RI, 2003.

[Vil09] C. Villani. *Optimal Transport. Old and New*, volume 338 of Grundlehren der mathematischen Wissenschaften. Springer, 2009.

10

Optimal coupling for mean field limits

FRANÇOIS BOLLEY

Abstract

We review recent quantitative results on the approximation of mean field
diffusion equations by large systems of interacting particles, obtained
by optimal coupling methods. These results concern a larger range of
models, more precise senses of convergence and links with the long time
behaviour of the systems to be considered.

Let us consider a Borel probability distribution $f_t = f_t(X)$ on $\mathbb{R}^d$ evolving
according to the McKean–Vlasov equation

$$\frac{\partial f_t}{\partial t} = \sum_{i,j=1}^{d} a_{ij} \frac{\partial^2 f_t}{\partial X_i \partial X_j} + \sum_{i=1}^{d} \frac{\partial}{\partial X_i} \big(b_i[X, f_t] f_t\big), \qquad t > 0, \; X \in \mathbb{R}^d.$$

(10.1)

Here, $a = (a_{ij})_{1 \le i,j \le d}$ is a nonnegative symmetric $d \times d$ matrix; moreover,
given X in $\mathbb{R}^d$ and a Borel probability measure p on $\mathbb{R}^d$,

$$b_i[X, p] = \int_{\mathbb{R}^d} b_i(X, Y) \, dp(Y), \qquad 1 \le i \le d,$$

where $b(X, Y) = (b_i(X, Y))_{1 \le i \le d}$ is a vector of $\mathbb{R}^d$. Equation (10.1) has the
following natural probabilistic interpretation: if f_0 is a distribution on $\mathbb{R}^d$, the
solution f_t of (10.1) is the law at time t of the $\mathbb{R}^d$-valued process $(X_t)_{t \ge 0}$
evolving according to the mean field stochastic differential equation

$$dX_t = \sigma \, dB_t - b[X_t, f_t] \, dt.$$

(10.2)

Ceremade, Umr Cnrs 7534, Université Paris-Dauphine, Place du Maréchal de Lattre de Tassigny,
F-75775 Paris cedex 16. E-mail address: bolley@ceremade.dauphine.fr.
Keywords: mean field limits, transportation inequalities, concentration inequalities, optimal
coupling.
MSC 2010: 82C22, 65C35, 35K55, 90C08.

Here, the $d \times d$ matrix σ satisfies $\sigma\sigma^* = 2a$, $(B_t)_{t\geq 0}$ is a Brownian motion in $\mathbb{R}^d$ and f_t is the law of X_t in $\mathbb{R}^d$. It is a mean field equation in the sense that the evolution of X_t is obtained by averaging the contributions $b(X_t, Y)$ over the system, according to the distribution $df_t(Y)$. Existence and uniqueness of solutions to (10.1) and (10.2) are proven in [11] for globally Lipschitz drifts b and initial data f_0 with finite second moment. Non-globally Lipschitz drifts are discussed in Section 10.1.

Two instances of such evolutions are particularly interesting. First of all, when $\mathbb{R}^d$ is the phase space of positions $x \in \mathbb{R}^{d'}$ and velocities $v \in \mathbb{R}^{d'}$ with $d = 2d'$, one is interested in the Vlasov–Fokker–Planck equation

$$\frac{\partial f_t}{\partial t} + v \cdot \nabla_x f_t - \left(\nabla_x U *_x \rho[f_t]\right) \cdot \nabla_v f_t = \Delta_v f_t + \nabla_v \cdot ((A(v) + B(x))f_t),$$

$$t > 0, \ x, v \in \mathbb{R}^{d'}. \tag{10.3}$$

Here, $a \cdot b$ denotes the scalar product of two vectors a and b in $\mathbb{R}^{d'}$, whereas ∇_v, $\nabla_v\cdot$ and Δ_v respectively stand for the gradient, divergence and Laplace operators with respect to the velocity variable $v \in \mathbb{R}^{d'}$. Moreover, $\rho[f_t](x) = \int_{\mathbb{R}^{d'}} f_t(x, v)\, dv$ is the macroscopic density in the space of positions $x \in \mathbb{R}^{d'}$, or the space marginal of f_t; $U = U(x)$ is an interaction potential in the position space and $*_x$ stands for the convolution with respect to the position variable $x \in \mathbb{R}^{d'}$; finally, $A(v)$ and $B(x)$ are respectively friction and position confinement terms. This equation is used in the modelling of diffusive stellar matter (see [8] for instance).

We are also concerned with the space homogeneous equation

$$\frac{\partial f_t}{\partial t} = \Delta_v f_t + \nabla_v \cdot ((\nabla_v V + \nabla_v W *_v f_t)f_t), \qquad t > 0, \ v \in \mathbb{R}^d, \tag{10.4}$$

with $d = d'$. Here, V and W are respectively exterior and interaction potentials in the velocity space, and this equation is used in the modelling of space homogeneous granular media (see [2]).

The *particle approximation* of (10.1) consists in introducing N processes $(X_t^{i,N})_{t\geq 0}$, with $1 \leq i \leq N$, which evolve no more according to the drift $b[X_t, f_t]$ generated by the distribution f_t as in (10.2), but according to its discrete counterpart, namely the empirical measure

$$\hat{\mu}_t^N = \frac{1}{N} \sum_{i=1}^{N} \delta_{X_t^{i,N}}$$

of the particle system $(X_t^{1,N}, \ldots, X_t^{N,N})$, In other words, we let the processes $(X_t^{i,N})_{t\geq 0}$ solve

$$dX_t^{i,N} = \sigma\, dB_t^i - \frac{1}{N}\sum_{i=1}^{N} b(X_t^{i,N}, X_t^{j,N})\, dt, \qquad 1 \leq i \leq N. \qquad (10.5)$$

Here, the $(B_t^i)_{t\geq 0}$ are N independent standard Brownian motions on $\mathbb{R}^d$ and we assume that the initial data $X_0^{i,N}$ for $1 \leq i \leq N$ are independent variables with given law f_0.

The mean field force $b[X_t, f_t]$ in (10.2) is replaced in (10.5) by the pairwise actions $\frac{1}{N}b(X_t^{i,N}, X_t^{j,N})$ of particle j on particle i. In particular, even in this case when the initial data $X_0^{i,N}$ are independent, the particles get correlated at all $t > 0$. But, since this interaction is of order $1/N$, it may be reasonable that two of these interacting particles (or a fixed number k of them) become less and less correlated as N gets large.

In order to state this *propagation of chaos* property we let, for each $i \geq 1$, $(\bar{X}_t^i)_{t\geq 0}$ be the solution of

$$\begin{cases} d\bar{X}_t^i = \sigma\, dB_t^i - b[\bar{X}_t^i, f_t]\, dt, \\ \bar{X}_0^i = X_0^{i,N}, \end{cases} \qquad (10.6)$$

where f_t is the distribution of $\bar{X}_t^i$. The processes $(\bar{X}_t^i)_{t\geq 0}$ with $i \geq 1$ are independent since the initial conditions and driving Brownian motions are independent. Moreover, they are identically distributed and their common law at time t evolves according to (10.1), so is the solution f_t of (10.1) with initial datum f_0. In this notation, and as N gets large, we expect the N processes $(X_t^{i,N})_{t\geq 0}$ to look more and more like the N independent processes $(\bar{X}_t^i)_{t\geq 0}$:

Theorem 10.1 ([11], [12]). *If b is a Lipschitz map on $\mathbb{R}^{2d}$ and f_0 a Borel distribution on $\mathbb{R}^d$ with finite second moment, then, in the above notation, for all $t \geq 0$ there exists a constant C such that*

$$\mathbb{E}|X_t^{i,N} - \bar{X}_t^i|^2 \leq \frac{C}{N}$$

for all N.

This results also holds in a more general setting when the diffusion matrix can depend on X and on the distribution in a Lipschitz way, and can be stated at the level of the paths of the processes on finite time intervals.

First of all, it ensures that the common law $f_t^{1,N}$ of any (by exchangeability) of the particles $X_t^{i,N}$ converges to f_t as N goes to infinity, according to

$$W_2^2(f_t^{1,N}, f_t) \leq \mathbb{E}|X_t^{i,N} - \bar{X}_t^i|^2 \leq \frac{C}{N}. \tag{10.7}$$

Here, the Wasserstein distance of order $p \geq 1$ between two Borel probability measures μ and v on $\mathbb{R}^q$ with finite moment of order $p \geq 1$ is defined by

$$W_p(\mu, v) = \inf \left(\mathbb{E}|X - Y|^p\right)^{1/p},$$

where the infimum runs over all couples of random variables (X, Y) with X having law μ and Y having law v (see [13] for instance).

Moreover, it proves a quantitative version of propagation of chaos: for all fixed k, the law $f_t^{k,N}$ of any (by exchangeability) k particles $X_t^{i,N}$ converges to the product tensor $(f_t)^{\otimes k}$ as N goes to infinity, according to

$$W_2^2(f_t^{k,N}, (f_t)^{\otimes k}) \leq \mathbb{E}|(X_t^{1,N}, \dots, X_t^{k,N}) - (\bar{X}_t^1, \dots, \bar{X}_t^k)|^2 \leq \frac{kC}{N}.$$

It finally gives the following first result on the convergence of the empirical measure $\hat{\mu}_t^N$ of the particle system to the distribution f_t: if φ is a Lipschitz map on $\mathbb{R}^d$, then

$$\mathbb{E}\left|\frac{1}{N}\sum_{i=1}^N \varphi(X_t^{i,N}) - \int_{\mathbb{R}^d} \varphi \, df_t\right|^2 \leq 2\mathbb{E}|\varphi(X_t^{i,N}) - \varphi(\bar{X}_t^i)|^2$$

$$+ 2\mathbb{E}\left|\frac{1}{N}\sum_{i=1}^N \varphi(\bar{X}_t^i) - \int_{\mathbb{R}^d} \varphi \, df_t\right|^2 \leq \frac{C}{N} \tag{10.8}$$

by Theorem 10.1 and a law of large numbers argument on the independent variables $\bar{X}_t^i$.

Recent attention has been brought to improve these classical results in three directions:

1. enlarging the setting to non-Lipschitz drift terms;
2. providing more precise estimates on the approximation of the solution to (10.1) by the empirical measure of the particle system;
3. providing time uniform estimates when possible, in connection with the long time behaviour of the solutions.

10.1 Non-Lipschitz drifts

The interest for such mean field limits with locally but non-globally Lipschitz drifts has been renewed by kinetic models, in which the interaction may become larger for larger relative velocities.

For instance, for space homogeneous models such as (10.4) with $W(z) = |z|^3$ and $d = 1$, the difficulty brought by the non-Lipschitz drift has been solved by convexity arguments, first in dimension one in [1], then more generally in any dimension in [7], [9].

Space inhomogeneous biological collective behaviour models for large groups of small animals have been considered in [3], for which convexity or compact support arguments do not apply and have been replaced by moment arguments.

10.2 Deviation bounds for the empirical measure

The averaged estimate (10.8) ensures that the particle system is an appropriate approximation to solutions to (10.1). However, when the particle system is used for numerical simulations, one may wish to establish estimates making sure that the numerical method has a very small probability to give wrong results.

This was achieved in [4], [7], [9], [10] by the use of (Talagrand) transportation inequalities. It is proved in these studies, under diverse hypotheses on the initial data and in diverse contexts, that for all t the law $f_t^{N,N}$ of the particle system at time t satisfies a transportation inequality

$$W_1(\nu, f_t^{N,N})^2 \le \frac{1}{c} H(\nu | f_t^{N,N})$$

for all measures ν on $\mathbb{R}^{dN}$. Here, c may depend on t (but neither on N nor on ν), and

$$H(\nu | f_t^{N,N}) = \int_{\mathbb{R}^{dN}} \frac{d\nu}{df_t^{N,N}} \ln \frac{d\nu}{df_t^{N,N}} \, df_t^{N,N}$$

is the relative entropy of ν with respect to $f_t^{N,N}$, to be interpreted as $+\infty$ if ν is not absolutely continuous with respect to $f_t^{N,N}$. Then an argument by S. Bobkov and F. Götze, based on the Kantorovich–Rubinstein dual formulation

$$W_1(\nu, f_t^{N,N}) = \sup \left\{ \iint_{\mathbb{R}^{dN}} \Phi \, d\nu - \int_{\mathbb{R}^{dN}} \Phi \, df_t^{N,N}, \ \Phi \ \text{1-Lipschitz on } \mathbb{R}^{dN} \right\}$$

$$(10.9)$$

and the dual formulation of the entropy, ensures that

$$\mathbb{P}\left[\frac{1}{N}\sum_{i=1}^{N}\varphi(X_t^{i,N}) - \int_{\mathbb{R}^d}\varphi\,df_t^{1,N} > r\right] \le e^{-cNr^2}$$

for all $N \ge 1, r > 0$ and all 1-Lipschitz maps φ on $\mathbb{R}^d$ (see [13, Chapter 22] for instance). Since, moreover,

$$\left|\frac{1}{N}\sum_{i=1}^{N}\varphi(X_t^{i,N}) - \int_{\mathbb{R}^d}\varphi\,df_t\right| \le \left|\frac{1}{N}\sum_{i=1}^{N}\varphi(X_t^{i,N}) - \int_{\mathbb{R}^d}\varphi\,df_t^{1,N}\right| + W_1(f_t^{1,N}, f_t)$$

$$\le \left|\frac{1}{N}\sum_{i=1}^{N}\varphi(X_t^{i,N}) - \int_{\mathbb{R}^d}\varphi\,df_t^{1,N}\right| + \sqrt{\frac{C}{N}}$$

by (10.9), (10.7) and the bound $W_1 \le W_2$, this ensures one-observable error bounds like

$$\mathbb{P}\left[\left|\frac{1}{N}\sum_{i=1}^{N}\varphi(X_t^{i,N}) - \int_{\mathbb{R}^d}\varphi\,df_t\right| > \sqrt{\frac{C}{N}} + r\right] \le 2\,e^{-cNr^2} \qquad (10.10)$$

for all N, r and all 1-Lipschitz maps φ on $\mathbb{R}^d$. In this argument we see how well adapted to this issue are the Wasserstein distances: from their definition they can easily be bounded from above by simple estimates on the processes, as in (10.7), and in turn they lead to straightforward estimates on Lipschitz observables, by (10.9).

Uniformly on Lipschitz observables, and for the Wasserstein distance W_1 which, up to moment conditions, metrizes the narrow topology on measures, estimates like

$$\mathbb{P}\left[W_1(\hat{\mu}_t^N, f_t) > r\right] \le e^{-\lambda Nr^2}, \qquad (10.11)$$

or even

$$\mathbb{P}\left[\sup_{0\le t\le T} W_1(\hat{\mu}_t^N, f_t) > r\right] \le e^{-cNr^2},$$

were reached in [5], provided N is larger than an explicit $N_0(r)$, hence ensuring that the probability of observing any significant deviation during a whole time period $[0, T]$ is small. Also, bounds were obtained on the *pointwise* deviation of a mollified empirical measure around the solution f_t.

10.3 Time uniform estimates and long time behaviour

Under convexity assumptions on the potentials V and W, the solution to the space homogeneous granular media equation (10.4) has been proven to converge algebraically or exponentially fast to a unique steady state (see [6] for an entropy dissipation proof based on interpreting (10.4) as a gradient flow in the Wasserstein space, and also [7], [9], [10]). In this setting one can hope for time uniform constants in estimates (10.8)–(10.10)–(10.11), which were obtained in [5], [7], [9], [10].

Also, convergence to equilibrium (through a contraction argument in W_2 distance) and time uniform deviation bounds were obtained in [4] for solutions to the (now space inhomogeneous) Vlasov–Fokker–Planck equation (10.3).

References

[1] S. Benachour, B. Roynette, D. Talay and P. Vallois. Nonlinear self-stabilizing processes. I. Existence, invariant probability, propagation of chaos. *Stoch. Proc. Appl. 75*, 2 (1998), 173–201.

[2] D. Benedetto, E. Caglioti, J.A. Carrillo and M. Pulvirenti. A non-Maxwellian steady distribution for one-dimensional granular media. *J. Statist. Phys. 91*, 5–6 (1998), 979–990.

[3] F. Bolley, J.A. Cañizo and J.A. Carrillo. Stochastic mean-field limit: non-Lipschitz forces and swarming. *Math. Mod. Meth. Appl. Sci. 21*, 11 (2011), 2179–2210.

[4] F. Bolley, A. Guillin and F. Malrieu. Trend to equilibrium and particle approximation for a weakly selfconsistent Vlasov–Fokker–Planck equation. *Math. Mod. Num. Anal. 44*, 5 (2010) 867–884.

[5] F. Bolley, A. Guillin and C. Villani. Quantitative concentration inequalities for empirical measures on non-compact spaces. *Prob. Theor. Rel. Fields 137*, 3–4 (2007), 541–593.

[6] J.A. Carrillo, R.J. McCann and C. Villani. Kinetic equilibration rates for granular media and related equations: entropy dissipation and mass transportation estimates. *Rev. Mat. Iberoamericana 19*, 3 (2003), 971–1018.

[7] P. Cattiaux, A. Guillin and F. Malrieu. Probabilistic approach for granular media equations in the non uniformly case. *Prob. Theor. Rel. Fields 140*, 1–2 (2008), 19–40.

[8] J. Dolbeault. Free energy and solutions of the Vlasov–Poisson–Fokker–Planck system: external potential and confinement (large time behavior and steady states). *J. Math. Pures Appl. 9*, 78, 2 (1999), 121–157.

[9] F. Malrieu. Logarithmic Sobolev inequalities for some nonlinear PDE's. *Stoch. Proc. Appl. 95*, 1 (2001), 109–132.

[10] F. Malrieu. Convergence to equilibrium for granular media equations and their Euler schemes. *Ann. Appl. Probab. 13*, 2 (2003), 540–560.

[11] S. Méléard. *Asymptotic Behaviour of Some Interacting Particle Systems; McKean–Vlasov and Boltzmann models*. Lecture Notes in Mathematics 1627, Springer, Berlin, 1996.

[12] A.-S. Sznitman. *Topics in Propagation of Chaos*. Lecture Notes in Mathematics 1464, Springer, Berlin, 1991.

[13] C. Villani. *Optimal Transport, Old and New*. Grundlehren der mathematischen Wissenschaften 338, Springer, Berlin, 2009.

11

Functional inequalities via Lyapunov conditions

PATRICK CATTIAUX[1] AND ARNAUD GUILLIN[2]

[1] *Université de Toulouse*
[2] *Université Blaise Pascal*

Abstract

We review here some recent results by the authors, and various coauthors, on (weak, super) Poincaré inequalities, transportation-information inequalities or logarithmic Sobolev inequality via a quite simple and efficient technique: Lyapunov conditions.

11.1 Introduction and main concepts

Lyapunov conditions appeared a long time ago. They were particularly well fitted to deal with the problem of convergence to equilibrium for Markov processes; see [23, 38–40] and references therein. They also appeared earlier in the study of large and moderate deviations for empirical functionals of Markov processes (for examples, see Donsker and Varadhan [21, 22], Kontoyaniis and Meyn [33, 34], Wu [47, 48], Guillin [28, 29]), for solving the Poisson equation [24].

Their use to obtain functional inequalities is however quite recent, even if one may afterwards find hint of such an approach in Deuschel and Stroock [19] or Kusuocka and Stroock [35]. The present authors and coauthors have developed a methodology that has been successful for various inequalities: Lyapunov–Poincaré inequalities [4], Poincaré inequalities [3], transportation inequalities for Kullback information [17] or Fisher information [32], super Poincaré inequalities [16], weighted and weak Poincaré inequalities [13], or [18] for super weighted Poincaré inequalities. We finally refer to the

Keywords: Lyapunov condition, Poincaré inequality, transportation information inequality, logarithmic Sobolev inequality.
MSC 2000: 26D10, 47D07, 60G10, 60J60.

forthcoming book [15] for a complete review. For more references on the various inequalities introduced here we refer to [1, 2, 36, 46]. The goal of this short review is to explain the methodology used in these papers and to present various general sets of conditions for this panel of functional inequalities. The proofs will of course be only schemed and we will refer to the original papers for complete statements.

Let us first describe the framework of our study. Let E be some Polish state space, μ a probability measure, and a μ-symmetric operator L. The main assumption on L is that there exists some algebra $\mathcal{A}$ of bounded functions, containing constant functions, which is everywhere dense (in the $\mathbb{L}_2(\mu)$ norm) in the domain of L. It enables us to define a "carré du champ" Γ, i.e. for $f, g \in \mathcal{A}$, $L(fg) = fLg + gLf + 2\Gamma(f, g)$. We will also suppose that Γ is a derivation (in each component), i.e. $\Gamma(fg, h) = f\Gamma(g, h) + g\Gamma(f, h)$; that is, we are in the standard "diffusion" case in [2] and we refer to the introduction of [12] for more details. For simplicity we set $\Gamma(f) = \Gamma(f, f)$. Also, since L is a diffusion, we have the following chain rule formula: $\Gamma(\Psi(f), \Phi(g)) = \Psi'(f)\Phi'(g)\Gamma(f, g)$.

In particular, if $E = \mathbb{R}^n$, $\mu(dx) = e^{-V(x)}dx$, where V is smooth and $L = \Delta - \nabla V.\nabla$, we may consider the C^∞ functions with compact support (plus the constant functions) as the interesting subalgebra $\mathcal{A}$, and then $\Gamma(f, g) = \nabla f \cdot \nabla g$. It will be our main object.

Now we define the notion of ϕ-Lyapunov function. Let $W \geq 1$ be a smooth enough function on E and ϕ be a $\mathcal{C}^1$ positive increasing function defined on $\mathbb{R}^+$. We say that W is a ϕ-Lyapunov function if there is an increasing family of exhausting sets $(A_r)_{r \geq 0} \subset E$ and some $b \geq 0$ such that for some $r_0 > 0$

$$LW \leq -\phi(W) + b\,\mathbb{I}_{A_{r_0}}. \tag{11.1}$$

One has very different behavior depending on ϕ: if ϕ is linear then results of [39, 40] assert that the associated semigroup converges to equilibrium with exponential speed (in total variation or with some weighted norm) so that it is legitimate to hope for a Poincaré inequality to be valid.

When ϕ is superlinear (or more generally in the form $\phi \times W$ where ϕ tends to infinity) we may hope for stronger inequalities (super Poincaré, ultracontractivity, ...).

Finally if ϕ is sublinear, as asserted in [23], only subexponential convergence to equilibrium is valid, so we should be in the regime of weak Poincaré inequalities. We will see a class of examples in the next section.

Note, however, that the results of [23, 39, 40] are valid even in a fully degenerate hypoelliptic setting (kinetic Fokker–Planck equation for example (see [4, 47])), whereas we cannot hope for a (weak, normal, or super) Poincaré inequality, due to the "degeneracy" of the Dirichlet form.

We thus have to impose another condition which will often be a "local inequality" such as a local Poincaré inequality (i.e., Poincaré inequality restricted to a ball, or a particular set) or local super Poincaré inequality, preventing the degeneracy case but quite easy to verify for general locally bounded measures.

We will present now the main lemma, which will show how the Lyapunov condition is used in our setting.

Lemma 11.1. *Let* $\psi : \mathbb{R}^+ \to \mathbb{R}^+$ *be a* $\mathcal{C}^1$ *increasing function. Then, for any* $f \in \mathcal{A}$ *and any positive* $h \in D(\mathcal{E})$,

$$\int \frac{-Lh}{\psi(h)} f^2 d\mu \le \int \frac{\Gamma(f)}{\psi'(h)} d\mu.$$

In particular,

$$\int \frac{-Lh}{h} f^2 d\mu \le \int \Gamma(f) d\mu.$$

Proof. Since L is μ-symmetric, using that Γ is a derivation and the chain rule formula, we have

$$\int \frac{-Lh}{\psi(h)} f^2 d\mu = \int \Gamma\left(h, \frac{f^2}{\psi(h)}\right) d\mu = \int \left(\frac{2f\Gamma(f,h)}{\psi(h)} - \frac{f^2\psi'(h)\Gamma(h)}{\psi^2(h)}\right) d\mu.$$

Since ψ is increasing and according to the Cauchy–Schwarz inequality we get

$$\frac{f\Gamma(f,h)}{\psi(h)} \le \frac{f\sqrt{\Gamma(f)\Gamma(h)}}{\psi(h)} = \frac{\sqrt{\Gamma(f)}}{\sqrt{\psi'(h)}} \cdot \frac{f\sqrt{\psi'(h)\Gamma(h)}}{\psi(h)}$$

$$\le \frac{1}{2}\frac{\Gamma(f)}{\psi'(h)} + \frac{1}{2}\frac{f^2\psi'(h)\,\Gamma(h)}{\psi^2(h)}.$$

The result follows.   □

Remark 1. In fact the conclusion of the preceding lemma, in the case where ψ is the identity, holds in a more general setting and requires only the reversibility assumption. It is thus valid for some Markov jump case ($M/M/\infty$, Levy process, ...); see [32] where the proof follows from a large deviations argument, or on more general Riemanian manifolds.

11.2 Examples of Lyapunov conditions

Before stating the results achievable by our method, let us present some examples of Lyapunov conditions.

We will restrict ourselves to the framework described before: $E = \mathbb{R}^n$, $\mu(dx) = Ze^{-V(x)}dx$, and $L = \Delta - \nabla V.\nabla$. The Lyapunov conditions may be quite different: first because of the very nature of V itself, and second because of the choice of the Lyapunov function W. Let us illustrate it in the Gaussian case $V(x) = |x|^2$:

(1) Choose first $W_1(x) = 1 + |x|^2$, so that

$$LW_1(x) = \Delta W_1(x) - 2x.\nabla W_1(x) = 2n - 2|x|^2$$
$$\leq -W_1(x) + (2n + 1)1_{|x|^2 \leq 2n+1}.$$

(2) Choose now $W_2(x) = e^{a|x|^2}$ for $0 < a < 1$,

$$LW_2(x) = (2n + 4a(a - 1)|x|^2)W_2(x) \leq -\lambda|x|^2W_2(x) + b1_{|x|^2 \leq R}$$

for some λ, b, R.

We may now consider to usual examples: let $U > c > 0$ be convex (convexity and positivity outside a large compact is sufficient if the measure is properly defined).

- Exponential-type measures: $V(x) = U^p$ for some positive p. Then, there exists $b, c, R > 0$ and $W \geq 1$ such that

$$LW \leq -\phi(W) + b1_{B(0,R)}$$

with $\phi(u) = u \log^{2(p-1)/p}(c + u)$ increasing. Furthermore, one can choose $W(x) = e^{\gamma|x|^p}$ for x large and γ small enough.
- Cauchy-type measures: $V(x) = (n + \beta)\log(U)$ for some positive β. Then, there exists $k > 2, b, R > 0$ and $W \geq 1$ such that

$$LW \leq -\phi(W) + b1_{B(0,R)}$$

with $\phi(u) = cu^{(k-2)/k}$ for some constant $c > 0$. Furthermore, one can choose $W(x) = |x|^k$ for x large. k has to be chosen so that there exists $\epsilon > 0$ such that $k + n\epsilon - 2 - \beta(1 - \epsilon) < 0$.

The details can be found in [13] for example.

11.3 Poincaré-like inequalities

The prototype of inequalities we will consider in this section is the following Poincaré inequality: for every nice function f there exists $C > 0$ such that

$$Var_\mu(f) := \int f^2 d\mu - \left(\int f d\mu \right)^2 \leq C \int |\nabla f|^2 d\mu.$$

Poincaré inequalities have attracted a lot of attention due to their beautiful properties: they are equivalent to the exponential L_2 decay of the associated semigroup, they give an exponential dimension-free concentration of measure,... We refer to [1, 36] for historical and mathematical references. Weak and super Poincaré inequalities will be variant (weaker or stronger) of this inequality. As we will see, this inequality may be proved very quickly using Lyapunov conditions and local inequalities.

11.3.1 Poincaré inequality

Let us begin with

Theorem 11.2. *Suppose that the following Lyapunov condition holds: there exists $W \geq 1$ in the domain of L, $\lambda > 0$, $b > 0$, $R > 0$ such that*

$$LW \leq -\lambda W + b 1_{\{|x| \leq R\}}. \tag{11.2}$$

Assume in addition that the following local Poincaré inequality holds: there exists κ_R such that for all nice functions f

$$\int_{|x| \leq R} f^2 d\mu \leq \kappa_R \int \Gamma(f) d\mu + \mu(\{|x| \leq R\})^{-1} \left(\int_{|x| \leq R} f d\mu \right)^2. \tag{11.3}$$

Then we have the following Poincaré inequality: for all nice f

$$Var_\mu(f) \leq \frac{b\kappa_R + 1}{\lambda} \int \Gamma(f) d\mu.$$

As the proof is very simple, it will be quite the only one we will write completely:

Proof. Denote $c_R = \int_{|x| \leq R} f d\mu$. Remark now that we may rewrite the Lyapunov condition as

$$1 \leq -\frac{LW}{\lambda W} + \frac{b}{\lambda} 1_{\{|x| \leq R\}}$$

so that by Lemma 11.1 and the local Poincaré inequality we have

$$
\begin{aligned}
Var_\mu(f) &\le \int (f - c_R)^2 d\mu \\
&\le \int (f - c_R)^2 \frac{-LW}{\lambda W} d\mu + \frac{b}{\lambda} \int_{|x| \le R} (f - c_R)^2 d\mu \\
&\le \frac{1}{\lambda} \int \Gamma(f) d\mu + \frac{b}{\lambda} \kappa_R \int |\nabla f|^2 d\mu,
\end{aligned}
$$

which is the desired result. $\qquad\qquad\square$

Remark 2. Using this theorem combined with the examples provided above, we may recover very simply the nice results of Bobkov [8] asserting that every log-concave measures (i.e. V convex) satisfies a Poincaré inequality ([3]).

One may also easily verifies that the following two sufficient conditions for the Poincaré inequality, are inherited from Lyapunov condition:

(1) There exist $0 < a < 1, c > 0$ and $R > 0$ such that for all $|x| \ge R$, we have $(1 - a)|\nabla V|^2 - \Delta V \ge c$.
(2) There exist $c > 0$ and $R > 0$ such that for all $|x| \ge R$, we have $x.\nabla V(x) \ge c|x|$.

Note that the first one was known with $a = 1/2$ for a long time but with quite a hard proof.

Remark 3. We will not develop it here, but in fact quite the same may be done for the Cheeger inequality (L_1-Poincaré); see [3].

11.3.2 Weighted and weak Poincaré inequality

We will now consider weaker inequalities: weighted Poincaré inequalities as introduced recently by Bobkov and Ledoux [11] or [13], i.e. with an additional weight in the Dirichlet form or in the variance, or weak Poincaré inequalities introduced by Röckner and Wang [42] (see also [6] or [13]), useful to establish sub-exponential concentration inequalities or algebraic rate of decay to equilibrium for the associated Markov process. We shall state here

Theorem 11.3. *Suppose that the following ϕ-Lyapunov condition holds: there exist some sublinear $\phi : [1, \infty[\to \mathbb{R}^+$ and $W \ge 1, b > 0, R > 0$ such that*

$$
LW \le -\phi(W) + b1_{\{|x| \le R\}}. \tag{11.4}
$$

Suppose also that μ satisfies a local Poincaré inequality (11.3) then:

(1) For all nice f, the following weighted Poincaré inequality holds:

$$Var_\mu(f) \le \max\left(\frac{b\kappa_R}{\phi(1)}, 1\right) \int \left(1 + \frac{1}{\phi'(W)}\right) \Gamma(f)d\mu.$$

(2) For all nice f, the following converse weighted Poincaré inequality holds:

$$\inf_c \int (f - c)^2 \frac{\phi(W)}{W} d\mu \le (1 + b\kappa_R) \int \Gamma(f)d\mu.$$

(3) Define $F(u) = \mu(\phi(W) < uW)$ and for $s < 1$, $F^{-1}(s) := \inf\{u; F(u) > s\}$ then the following weak Poincaré inequality holds:

$$Var_\mu(f) \le \frac{C}{F^{-1}(s)} \int \Gamma(f)d\mu + s\, Osc_\mu(f)^2.$$

Proof. The proof of the first two points may be easily derived using the proof for the usual Poincaré inequality. For the weak Poincaré inequality, start with the variance, divide the integral with respect to large or small values of $\phi(W)/W$ and use the converse Poincaré inequality established previously; see details in [13]. □

Remark 4. Using $V(x) = 1 + |x|^2$ in the examples of the previous section, one gets a weighted inequality with weight $1 + |x|^2$, and converse inequality with weight $(1 + |x|^2)^{-1}$ recovering results of Bobkov and Ledoux [11] (with worse constants, however). Note also that it enables us to get the correct order for the weak Poincaré inequality (as seen in dimension 1 in [6]). For this weak Poincaré inequality, one can find another approach in [4] based on weak Lyapunov–Poincaré inequality.

11.3.3 Super Poincaré inequality

Our next inequality was considered first by Wang [45] to study the essential spectrum of Markov operators. It is also useful for concentration of measures [45] or isoperimetric inequalities [7]. Wang also showed that they are, under Poincaré inequalities, equivalent to an F-Sobolev inequality (in particular one specific super Poincaré inequality is equivalent to the logarithmic Sobolev inequality), so that the results we will present now enable us to consider a very large class of inequalities stronger than the Poincaré inequality.

Theorem 11.4. *Suppose that there is an increasing family of exhausting sets $(A_r)_{r\ge0}$, an $r_0 > 0$ and a superlinear ϕ such that for some $b > 0$ the following Lyapunov condition holds:*

$$LW \le -\phi(W) + b1_{A_{r_0}}. \tag{11.5}$$

Assume in addition that a local super Poincaré inequality holds, i.e. there exists β_{loc} decreasing in s (for all r) such that for all s and nice f

$$\int_{A_r} f^2 d\mu \le s \int \Gamma(f) d\mu + \beta_{loc}(r, s) \left(\int_{A_r} |f| d\mu \right)^2. \tag{11.6}$$

Then, if $G(r) := (\inf_{A_r^c} \phi(W)/W)^{-1})$ tends to 0 as $r \to \infty$, μ satisfies for all positive s

$$\int f^2 d\mu \le 2s \int \Gamma(f) d\mu + \tilde{\beta}(s) \left(\int |f| d\mu \right)^2, \tag{11.7}$$

where

$$\tilde{\beta}(s) = c_{r_0} \beta_{loc}(G^{-1}(s), s/c_{r_0})$$

and $c_{r_0} = 1 + b \dfrac{\sup_{A_{r_0}} W}{\inf_{A_{r_0}^c} \phi(W)/W}$.

Proof. In fact, one just has to play with the extra strength provided by the Lyapunov condition (i.e., superlinear) and set A_r; that is,

$$\int f^2 d\mu \le \int_{A_r} f^2 d\mu + \int_{A_r^c} f^2 d\mu$$

$$\le \int_{A_r} f^2 d\mu + \frac{1}{\inf_{A_r^c} \phi(W)/W} \int f^2 \frac{\phi(W)}{W} d\mu.$$

The first term is treated by using the local inequality, and for the second one use the Lyapunov condition, the crucial Lemma 11.1, and once again the local super Poincaré inequality. Optimize in r to get the conclusion; see details in [16]. $\square$

Remark 5. Note that if the Boltzmann measure μ is locally bounded, using the Nash inequality for Lebesgue measures on balls, it is quite easy to find a local super Poincaré inequality; see discussion in [16].

Remark 6. Using this approach, one may recover famous criteria for the logarithmic Sobolev inequality: convexity Bakry–Emery criterion [5] (with worse constants), Kusuocka–Stroock conditions [35], or pointwise Wang's criterion [46].

11.4 Transportation inequalities

We will consider here another type of inequality linking Wasserstein distance to various information forms, namely Kullback information or Fisher information defined respectively by: if f is a density of probability with respect to μ

$$H(fd\mu, d\mu) := \text{Ent}_\mu(f) := \int f \log(f) d\mu$$

$$I(fd\mu, d\mu) := \int \frac{|\nabla f|^2}{f} d\mu.$$

The Wasserstein distance is defined by: for all measure ν and μ

$$W_p(\nu, \mu) := \inf \left\{ \mathbb{E}(d^p(X, Y))^{1/p}; \ X \sim \nu, Y \sim \mu \right\}.$$

11.4.1 Transportation and Kullback information

First, let us consider the usual transportation inequalities: for all probability density f w.r.t μ

$$W_p(\nu, \mu) \leq \sqrt{c\, H(fd\mu, d\mu)}.$$

These types of inequalities were introduced by Marton [37] as they imply straightforwardly concentration of measure, and deviation inequality by a beautiful characterization of Bobkov and Götze [10]. The case $p = 1$ was proved to be equivalent to Gaussian integrability [20].

The case $p = 2$ is much more difficult: Talagrand established the inequality for Gaussian measure [43], whereas Otto and Villani [41] and Bobkov *et al.* [9] proved that a logarithmic Sobolev inequaliy is a sufficient condition. More recently (see [14]), the authors proved that the logarithmic Sobolev inequality is strictly stronger, and provided such an example in dimension one. We will prove here that one may give a nice Lyapunov condition to verify this transportation inequality. Let us finish by the beautiful characterization obtained by Gozlan [25] proving that the case $p = 2$ is in fact equivalent to the Gaussian dimension-free concentration of measure; see [44] or [26] for more on the subject. We will prove here

Theorem 11.5. *Suppose that there exists $W \geq 1$, some point X_0 and constants b, c such that*

$$LW \leq (-cd^2(x, x_0) + b)W, \tag{11.8}$$

then there exists $C > 0$ such that for all density f w.r.t. μ

$$W_2(fd\mu, \mu) \leq \sqrt{K\, H(fd\mu, d\mu)}.$$

Proof. Refining arguments of Bobkov *et al.* [9], the authors proved that it is in fact sufficient to get a logarithmic Sobolev inequality for a restricted class of function, i.e. functions f such that

$$\log(f^2) \leq \log \left(f^2 d\mu\right) + 2\eta(d^2(x, x_0) + \int d^2(x, x_0)d\mu).$$

Using truncation arguments, and mainly this class of function's property, one sees how Lemma 11.1 comes into play. We refer to [17] for the tedious technical details. ☐

Remark 7. It is not difficult to remark that for $V(x) = x^3 + 3x^2 \sin(x) + x$ near infinity the Lyapunov condition is verified. However, the logarithmic Sobolev inequality does not hold in this case, as shown in [14].

11.4.2 Transportation and Fisher information

Transportation–information inequalities with Fisher inequalities were only very recently studied in [30–32], because of their equivalence with deviation inequality for Markov processes due to large deviations estimation. The two main interesting ones are for $p = 1$ and $p = 2$: for all probability density f w.r.t. μ

$$W_p(fd\mu, d\mu) \leq \sqrt{C\,I(fd\mu, d\mu)}.$$

In [32], various criteria were studied, such as the Lipschitz spectral gap. In particular, if $p = 1$ and the distance is the trivial one, this in equality is in fact equivalent to a Poincaré inequality. These authors also proved:

Theorem 11.6. *Suppose that a Poincaré inequality holds, and that the following Lyapunov condition holds: there exists $W \geq 1, x_0, c, b > 0$ such that*

$$LW \leq -cd^2(x, x_0)W + b. \tag{11.9}$$

Then we have for all probability density f w.r.t. μ

$$W_1(fd\mu, d\mu) \leq \sqrt{C\,I(fd\mu, d\mu)}.$$

Proof. Let us scheme the proof: by [44]

$$W_1(fd\mu, d\mu) \leq \int d(x, x_0)|f - 1|d\mu$$

$$\leq \sqrt{\int |f - 1|d\mu} \sqrt{\int d^2(x, x_0)|f - 1|d\mu}.$$

For the first term, we use the fact that the Poincaré inequality is equivalent to a control of the total variation by the square of the Fisher information, and for the second one the Lyapunov condition. One has of course to be careful as $|f - 1|$

is not in the domain of L, so that an approximation argument has to be done. We refer to [32] for details. $\square$

Remark 8. It is quite easy to remark that a logarithmic Sobolev inequality implies a transportation–information inequality with Fisher information in the case $p = 2$, but it is unknown if it is strictly weaker. *A fortiori*, no Lyapunov condition is known in the case $p = 2$.

11.5 Logarithmic Sobolev inequalities under curvature

Recall the classical logarithmic Sobolev inequality, i.e. for all nice f

$$Ent_\mu(f^2) \le c \int \Gamma(f)d\mu.$$

This inequality has a long history.

Initiated by Gross [27] to study hypercontractivity, it was largely studied by many authors due to its relationship with the study of decay to equilibrium, concentration of measure property, and efficacity in spin systems study; see [1, 2, 36, 44, 46] for further references. A breakthrough condition was the Bakry–Emery one: namely if $Hess(V) + Ric \ge \delta > 0$ then a logarithmic Sobolev holds. Kusuocka–Stroock gave a Lyapunov-type condition (recovered by the study given in the super Poincaré case), and using Harnack inequalities Wang proved that in the lower bounded curvature case, i.e.

$$Hess(V) + Ric \ge \delta \qquad (11.10)$$

with δ maybe negative, a sufficient Gaussian integrability, i.e. $\int e^{((-\delta)_+/2+\epsilon)|x|^2}d\mu < \infty$, is enough to prove a logarithmic Sobolev inequality. We will prove here

Theorem 11.7. *Suppose that (11.8) and (11.10) hold then μ satisfies a logarithmic Sobolev inequality.*

Proof. Remark first that by (11.8), a Poincaré inequality holds due to the effort of Section 11.3. Remark also that by Lyapunov conditions (and maximization argument)

$$W_2^2(fd\mu, d\mu) \le 2 \int d^2(x, x_0)|f - 1|d\mu \le I(fd\mu, d\mu) + C.$$

Use now an HWI inequality of Otto and Villani [41] (see also [9]):

$$H(fd\mu, d\mu) \le 2\sqrt{I(fd\mu, d\mu)}W_2(fd\mu, d\mu) - \frac{\delta}{2}W_2^2(fd\mu, d\mu).$$

so that a defective logarithmic Sobolev inequality holds, that may be tightened via Rothaus' lemma due to the Poincaré inequality. We refer to [17] for details. □

Remark 9. Let us give here an example not covered by Wang's condition. On $\mathbb{R}^2$, take $V(x, y) = r^2 g(\theta)$ in polar coordinates with $g(\theta) = 2 + \sin(k\theta)$. It is not hard to remark that $Hess(V)$ is bounded and that the Lyapunov condition (11.8) is verified. However, Wang's integrability condition is not verified, despite the fact that a logarithmic Sobolev inequality does hold by our theorem.

Remark 10. One may also give Lyapunov conditions when δ is replaced by some unbounded function of the distance.

Acknowledgments

A.G. thanks the organizers of this beautiful Grenoble Summer School and wonderful conference.

References

[1] C. Ané, S. Blachère, D. Chafaï, P. Fougères, I. Gentil, F. Malrieu, C. Roberto, and G. Scheffer. *Sur les inégalités de Sobolev logarithmiques*, volume 10 of *Panoramas et Synthèses*. Société Mathématique de France, Paris, 2000.

[2] D. Bakry. L'hypercontractivité et son utilisation en théorie des semigroupes. In *Lectures on Probability theory. École d'été de Probabilités de St-Flour 1992*, volume 1581 of *Lecture Notes in Mathematics*, pages 1–114. Springer, Berlin, 1994.

[3] D. Bakry, F. Barthe, P. Cattiaux, and A. Guillin. A simple proof of the Poincaré inequality for a large class of probability measures. *Electron. Commun. in Prob.*, 13:60–66, 2008.

[4] D. Bakry, P. Cattiaux, and A. Guillin. Rate of convergence for ergodic continuous Markov processes: Lyapunov versus Poincaré. *J. Func. Anal.*, 254:727–759, 2008.

[5] D. Bakry and M. Émery. Diffusions hypercontractives. In *Séminaire de probabilités, XIX, 1983/84*, volume 1123 of *Lecture Notes in Mathematics*, pages 177–206. Springer, Berlin, 1985.

[6] F. Barthe, P. Cattiaux, and C. Roberto. Concentration for independent random variables with heavy tails. *AMRX*, 2005(2):39–60, 2005.

[7] F. Barthe, P. Cattiaux, and C. Roberto. Isoperimetry between exponential and Gaussian. *Electron. J. Prob.*, 12:1212–1237, 2007.

[8] S.G. Bobkov. Isoperimetric and analytic inequalities for log-concave probability measures. *Ann. Prob.*, 27(4):1903–1921, 1999.

[9] S.G. Bobkov, I. Gentil, and M. Ledoux. Hypercontractivity of Hamilton–Jacobi equations. *J. Math. Pu. Appl.*, 80(7):669–696, 2001.

[10] S.G. Bobkov and F. Götze. Exponential integrability and transportation cost related to logarithmic Sobolev inequalities. *J. Funct. Anal.*, 163(1):1–28, 1999.

[11] S.G. Bobkov and M. Ledoux. Weighted Poincaré-type inequalities for Cauchy and other convex measures. *Ann. Probab.*, 37(2):403–427, 2009.

[12] P. Cattiaux. A pathwise approach of some classical inequalities. *Potential Anal.*, 20:361–394, 2004.

[13] P. Cattiaux, N. Gozlan, A. Guillin, and C. Roberto. Functional inequalities for heavy tailed distributions and applications to isoperimetry. Preprint, 2008.

[14] P. Cattiaux and A. Guillin. On quadratic transportation cost inequalities. *J. Math. Pures Appl.*, 88(4):341–361, 2006.

[15] P. Cattiaux and A. Guillin. Long time behavior of Markov processes and functional inequalities: Lyapunov conditions approach. Book in preparation, 2014.

[16] P. Cattiaux, A. Guillin, F.Y. Wang, and L. Wu. Lyapunov conditions for super Poincaré inequalities. *J. Funct. Anal.*, 256(6):1821–1841, 2009.

[17] P. Cattiaux, A. Guillin, and L. Wu. A note on Talagrand transportation inequality and logarithmic Sobolev inequality. *Probab. Theory Relat. Fields*, 148:285–304, 2010.

[18] P. Cattiaux, A. Guillin, and L. Wu. Some remarks on weighted logarithmic Sobolev inequality. Indiana Univ. Math. J., 60(6):1885–1904, 2011.

[19] J.D. Deuschel and D.W. Stroock. *Large Deviations*, volume 137 of *Pure and Applied Mathematics*. Academic Press Inc., Boston, MA, 1989.

[20] H. Djellout, A. Guillin, and L. Wu. Transportation cost–information inequalities and applications to random dynamical systems and diffusions. *Ann. Probab.*, 32(3B):2702–2732, 2004.

[21] M.D. Donsker and S.R.S. Varadhan. Asymptotic evaluation of certain Markov process expectations for large time. III. *Comm. Pure Appl. Math.*, 29(4):389–461, 1976.

[22] M.D. Donsker and S.R.S. Varadhan. Asymptotic evaluation of certain Markov process expectations for large time. IV. *Comm. Pure Appl. Math.*, 36(2):183–212, 1983.

[23] R. Douc, G. Fort, and A. Guillin. Subgeometric rates of convergence of f-ergodic strong Markov processes. *Stochastic Process. Appl.*, 119(3):897–923, 2009.

[24] P.W. Glynn and S.P. Meyn. A Liapounov bound for solutions of the Poisson equation. *Ann. Probab.*, 24(2):916–931, 1996.

[25] N. Gozlan. Poincaré inequalities and dimension free concentration of measure. Preprint, 2007.

[26] N. Gozlan and C. Leonard. Transportation–information inequalities. Preprint, 2009.

[27] L. Gross. Logarithmic Sobolev inequalities. *Am. J. Math.*, 97(4):1061–1083, 1975.

[28] A. Guillin. Moderate deviations of inhomogeneous functionals of Markov processes and application to averaging. *Stochastic Process. Appl.*, 92(2):287–313, 2001.

[29] A. Guillin. Averaging principle of SDE with small diffusion: moderate deviations. *Ann. Probab.*, 31(1):413–443, 2003.

[30] A. Guillin, A. Joulin, C. Léonard, and L. Wu. Transportation–information inequalities for Markov processes III. Preprint, 2010.

[31] A. Guillin, C. Léonard, L. Wu, and F.Y. Wang. Transportation–information inequalities for Markov processes II. Preprint, 2009.

[32] A. Guillin, C. Léonard, L. Wu, and N. Yao. Transportation–information inequalities for Markov processes. *Probab. Theory Relat. Fields*, 144(3–4):669–695, 2009.

[33] I. Kontoyiannis and S.P. Meyn. Spectral theory and limit theorems for geometrically ergodic Markov processes. *Ann. Appl. Probab.*, 13(1):304–362, 2003.

[34] I. Kontoyiannis and S. P. Meyn. Large deviations asymptotics and the spectral theory of multiplicatively regular Markov processes. *Electron. J. Probab.*, 10(3):61–123 (electronic), 2005.

[35] S. Kusuoka and D. Stroock. Some boundedness properties of certain stationary diffusion semigroups. *J. Func. Anal.*, 60:243–264, 1985.

[36] M. Ledoux. *The Concentration of Measure Phenomenon*, volume 89 of *Mathematical Surveys and Monographs*. American Mathematical Society, Providence, RI, 2001.

[37] K. Marton. Bounding $\bar{d}$-distance by informational divergence: a method to prove measure concentration. *Ann. Prob.*, 24:857–866, 1996.

[38] S.P. Meyn and R.L. Tweedie. *Markov Chains and Stochastic Stability*. Communications and Control Engineering Series. Springer-Verlag London Ltd. London, 1993.

[39] S.P. Meyn and R.L. Tweedie. Stability of Markovian processes II: continuous-time processes and sampled chains. *Adv. Appl. Probab.*, 25:487–517, 1993.

[40] S.P. Meyn and R.L. Tweedie. Stability of Markovian processes III: Foster–Lyapunov criteria for continuous-time processes. *Adv. Appl. Probab.*, 25:518–548, 1993.

[41] F. Otto and C. Villani. Generalization of an inequality by Talagrand and links with the logarithmic Sobolev inequality. *J. Funct. Anal.*, 173(2):361–400, 2000.

[42] M. Röckner and F.Y. Wang. Weak Poincaré inequalities and L^2-convergence rates of Markov semigroups. *J. Funct. Anal.*, 185(2):564–603, 2001.

[43] M. Talagrand. Transportation cost for Gaussian and other product measures. *Geom. Funct. Anal.*, 6:587–600, 1996.

[44] C. Villani. *Optimal Transport: Old and New*, volume 338 of *Grundlehren der Mathematischen Wissenschaften [Fundamental Principles of Mathematical Sciences]*. Springer-Verlag, Berlin, 2009.

[45] F.Y. Wang. Functional inequalities for empty essential spectrum. *J. Funct. Anal.*, 170(1):219–245, 2000.

[46] F.Y. Wang. *Functional Inequalities, Markov Processes and Spectral Theory*. Science Press, Beijing, 2005.

[47] L. Wu. Large and moderate deviations and exponential convergence for stochastic damping Hamiltonian systems. *Stochastic Process. Appl.*, 91(2):205–238, 2001.

[48] L. Wu. Essential spectral radius for Markov semigroups. I. Discrete time case. *Probab. Theory Relat. Fields*, 128(2):255–321, 2004.

12

Size of the medial axis and stability
of Federer's curvature measures

Abstract

In this chapter, we study the $(d-1)$-volume and the covering numbers of the medial axis of a compact subset of $\mathbb{R}^d$. In general, this volume is infinite; however, the $(d-1)$-volume and covering numbers of a filtered medial axis (the μ-medial axis) that is at distance greater than ε from the compact set can be explicitly bounded. The behavior of the bound we obtain with respect to μ, ε and the covering numbers of K is optimal.

From this result we deduce that the projection function on a compact subset K of $\mathbb{R}^d$ depends continuously on the compact set K, in the L^1 sense. This implies in particular that Federer's curvature measures of a compact subset of $\mathbb{R}^d$ with positive reach can be reliably estimated from a Hausdorff approximation of this subset, regardless of any regularity assumption on the approximating subset.

12.1 Introduction

We are interested in the following question: given a compact subset K of $\mathbb{R}^d$ with positive reach, and a Hausdorff approximation P of this set, is it possible to approximate Federer's curvature measures of K (see [9] or Section 12.2.2 for a definition) from P only? A positive answer to this question has been given in [8] using convex analysis. In this chapter, we show that such a result can also be deduced from a careful study of the "size" – that is, the covering numbers – of the medial axis.

2000 *Mathematics Subject Classification.* 28A78; 52A39; 35F21.
Key words: cut locus, medial axis, distance function, Federer curvature measure.

The notion of medial axis, also known as ambiguous locus in Riemannian geometry, has many applications in computer science. In image analysis and shape recognition, the skeleton of a shape is often used as an idealized version of the shape, which is known to have the same homotopy type as the original shape [14]. In the reconstruction of curves and surfaces from point cloud approximations, the distance to the medial axis provides an estimation of the size of the local features that can be used to give sampling conditions for provably correct reconstruction [1]. The flow associated with the distance function d_K to a compact set K that flows away from K toward local maxima of d_K (that lie in the medial axis of K) can be used for shape segmentation. The reader that is interested by the computation and stability of the medial axis with some of these applications in mind can refer to the survey [2].

The main technical ingredient needed for bounding the covering numbers of the subsets of the medial axis that we consider is a Lipschitz regularity result for the so-called normal distance to the medial axis $\tau_K : \mathbb{R}^d \setminus K \to \mathbb{R}$. It is defined as follows: if x belongs to the medial axis of K, then $\tau_K(x) = 0$; otherwise, $\tau_K(x)$ is the infimum time t such that $x + t \nabla_x d_K$ belongs to the medial axis of K. When K is a compact submanifold of class $C^{2,1}$, this function is globally Lipschitz on any r-level set of the distance function to K, when the radius r is small enough [5, 11, 13]. When K is the analytic boundary of a bounded domain Ω of $\mathbb{R}^2$, the normal distance to the medial axis of $\partial \Omega$ is 2/3-Hölder on Ω [3]. However, without a strong regularity assumption on the compact set K, it is hopeless to obtain a global Lipschitz regularity result for τ_K on a parallel set of K. Indeed, such a result would imply the finiteness of the $(d-1)$-Hausdorff measure of the medial axis, which is known to be false – for instance, the medial axis of a generic compact set is dense.

We show however, that the normal distance to the medial axis is Lipschitz on a suitable subset of a parallel set. This enables us to prove the following theorem on the Lebesgue covering numbers of the μ-medial axis. The μ-medial of a compact subset K of $\mathbb{R}^d$, denoted $\mathrm{Med}_\mu(K)$, is the set of points where the distance function to K admits a supergradient with length at most μ. We refer the reader to Section 12.3.1 for precise definitions.

Theorem 12.1. *For any compact set $K \subseteq \mathbb{R}^d$, a parameter ε smaller than the diameter of K, and η small enough,*

$$\mathcal{N}\left(\mathrm{Med}_\mu(K) \cap (\mathbb{R}^d \setminus K^\varepsilon), \eta\right) \leqslant \mathcal{N}(\partial K, \varepsilon/2) \, O\left(\left[\frac{\mathrm{diam}(K)}{\eta\sqrt{1-\mu}}\right]^{d-1}\right).$$

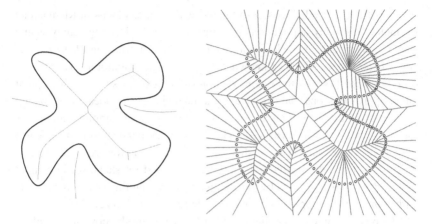

Figure 12.1 Medial axis of a curve C in the plane, and Voronoi diagram of a point cloud P sampled on the curve.

From this theorem, we deduce a quantitative Hausdorff-stability result for the projection function, which is the key to the stability of Federer's curvature measure (see (12.1)):

Theorem 12.2. *Let E be a bounded open set of $\mathbb{R}^d$. The application that maps a compact subset of $\mathbb{R}^d$ to the projection function $p_K \in L^1(E)$ is locally h-Hölder, for any exponent h smaller than $1/(4d - 2)$.*

Note that a similar result with optimal Hölder exponent $1/2$ has been obtained in [8], but we believe that the proofs in this chapter give a better geometric insight on the Hausdorff-stability of projection functions.

12.2 Boundary measures and medial axes

12.2.1 Distance, projection, boundary measures

Throughout this chapter, K will denote a compact set in the Euclidean d-space $\mathbb{R}^d$, with no additional regularity assumption unless specified otherwise. The *distance function* to K, denoted by $d_K : X \to \mathbb{R}^+$, is defined by $d_K(x) = \min_{p \in K} \| p - x \|$. A point p of K that realizes the minimum in the definition of $d_K(x)$ is called a *projection* of x on K. The set of such projections is denoted $\mathrm{proj}_K(x)$. The locus of the points with two or more distinct projections on K is called the *medial axis* of K, and denoted $\mathrm{Med}(K)$ (see Figure 12.1). The map $p_K : \mathbb{R}^d \setminus \mathrm{Med}(K) \to K$ such that $p_K(x) \in \mathrm{proj}_K(x)$ is called the *projection function* on K.

Let K be a compact subset and E be a Borel subset of $\mathbb{R}^d$. The *boundary measure* of K with respect to E is defined as the pushforward of the restriction of the Lebesgue measure to E by the projection function p_K, or more concisely $\mu_{K,E} = p_{K\#} \, \mathcal{H}^d \big|_E$. We will be especially interested in the case where E is of the form K^r, where K^r denotes the r-tubular neighborhood of K, i.e. $K^r = d_K^{-1}([0, r])$.

Example 1 (Steiner–Minkowski). If P is a convex solid polyhedron of $\mathbb{R}^3$, F its set of faces, E its set of edges, and V its set of vertices, then the following formula holds:

$$\mu_{P,P^r} = \mathcal{H}^3 \Big|_P + r \sum_{f \in F} \mathcal{H}^2 \Big|_f + \frac{1}{2}r^2 \sum_{e \in E} K(e)\mathcal{H}^1 \Big|_e + \frac{1}{3}r^3 \sum_{v \in V} K(v)\delta_v,$$

where $K(e)$ is the angle between the normals of the faces adjacent to the edge e, and $K(v)$ the solid angle formed by the normals of the faces adjacent to the vertex v. For a general convex polyhedron the measure μ_{K,K^r} can similarly be written as a sum of weighted Hausdorff measures supported on the i-skeleton of K, whose local density is the local external dihedral angle.

Example 2 (Weyl). Let M be a compact smooth hypersurface of $\mathbb{R}^d$, and denote by $\sigma_i(p)$ the ith elementary symmetric polynomial of the $(d-1)$ principal curvatures of M at a point p in M. Then, for any Borel subset B of $\mathbb{R}^d$, and r small enough, the μ_{K,K^r}-measure of B can be written as

$$\mu_{K,K^r}(B) = \sum_{i=0}^{d-1} \text{const}(i, d) \int_{B \cap M} \sigma_i(p)\mathrm{d}M(p).$$

This formula can be generalized to submanifolds of any codimension [18].

12.2.2 Federer curvature measures and reach

Following Federer [9], we call the *reach* of a compact subset K of $\mathbb{R}^d$ the minimum distance between K and its medial axis Med(K). Generalizing Steiner–Minkowski and Weyl tubes formulas, Federer proved that as long as r is smaller than the *reach* of K, the dependence in r of the boundary measure μ_{K,K^r} is polynomial, and the degree of the polynomial is bounded by the ambient dimension d. The coefficients of this polynomial are signed measures, and Federer calls them *curvature measures* of the compact set K.

Theorem 12.3 (Federer). *For any compact set $K \subseteq \mathbb{R}^d$ with reach greater than R, there exists $(d+1)$ uniquely defined (signed) measures $\Phi_K^0, \ldots, \Phi_K^d$*

supported on K such that for any $r \leqslant R$,

$$\mu_{K,K^r} = \sum_{i=0}^{d} \omega_{d-i}\, \Phi_K^i\, r^i,$$

where ω_k *is the volume of the k-dimensional unit sphere.*

12.2.3 Stability of boundary and curvature measures

The question of the stability of boundary measures is a particular case of the more general question of geometric inference [6,8]. Our goal here is to bound the Wasserstein distance between the boundary measures of two compact subsets as a function of their Hausdorff distance. Recall that the *Hausdorff distance* between two sets K and K' can be defined by $d_{\mathrm{H}}(K, K') := \|d_K - d_{K'}\|_{\infty}$. The *Wasserstein distance* with exponent one between two measures μ and ν with equal total mass is given by $\mathrm{W}(\mu, \nu) = \max_f |\int f\mathrm{d}\mu - \int f\mathrm{d}\nu|$, where the maximum is taken over any 1-Lipschitz function on $\mathbb{R}^d$. The following inequality holds for any bounded open set E and compact subsets K, K' of $\mathbb{R}^d$, and is proved in Proposition 3.1 in [8]:

$$\mathrm{W}_1\left(\frac{\mu_{K,E}}{\mathcal{H}^d(E)}, \frac{\mu_{K',E}}{\mathcal{H}^d(E)}\right) \leqslant \frac{1}{\mathcal{H}^d(E)} \|\mathrm{p}_K - \mathrm{p}_{K'}\|_{\mathrm{L}^1(E)}. \qquad (12.1)$$

It implies that, in order to obtain a Hausdorff stability result for boundary measures, one only needs to obtain a bound $\|\mathrm{p}_K - \mathrm{p}_{K'}\|_{\mathrm{L}^1(E)} = o(d_{\mathrm{H}}(K, K'))$. The possibility to estimate Federer's curvature measures follows from such an L^1 stability result for projection functions, as shown in [8, Section 4].

12.3 A first non-quantitative stability result

Intuitively, one expects that the two projections $p_K(x)$ and $p_{K'}(x)$ of a point x on two Hausdorff-close compact subsets can be far from each other only if x lies close to the medial axis of one of the compact sets. This makes it reasonable to expect an L^1 convergence property of the projections. Because the medial axis of a compact subset of $\mathbb{R}^d$ is generically dense (see [19] or Proposition I.2 in [15]), translating the above intuition into a proof is not completely straightforward. In this section, we prove the following non-quantitative L^1 convergence result for projections:

Proposition 12.4. *If* (K_n) *Hausdorff converges to a compact* $K \subseteq \mathbb{R}^d$, *then for any bounded open set* E, $\lim_{n \to +\infty} \|\mathrm{p}_{K_n} - \mathrm{p}_K\|_{\mathrm{L}^1(E)} = 0$.

12.3.1 Semi-concavity of d_K and μ-medial axis

The semi-concavity of the distance function to a compact set has been remarked and used in different contexts [4, 10, 14, 16]. We will use the fact that for any compact subset $K \subseteq \mathbb{R}^d$, the square of the distance function to K is 1-concave. This property is equivalent to the convexity of the function $\|.\|^2 - d_K^2$ is convex. This allows one to define a generalized gradient vector field for the distance function d_K everywhere on $\mathbb{R}^d$. A *supergradient* of the distance function to K at x is a vector v such that, for every h in $\mathbb{R}^d$,

$$d_K^2(x + h) \leqslant d_K^2(x) + 2d_K(x)\langle h|v \rangle + \|h\|^2 . \tag{12.2}$$

The *superdifferential* of d_K at a point x is denoted by $\partial_x^+ d_K$, and coincides with the convex hull of the set $\{(p - x)/\|p - x\| \; ; \; p \in \mathrm{proj}_K(x)\}$. The *generalized gradient* $\nabla_x d_K$ of the distance function d_K at a point x is the unique projection of the origin on the convex set $\partial_x^+ d_K$ [14, 16]. Denoting $\gamma_K(x)$ and $r_K(x)$ the center and the radius of the smallest ball enclosing the set of projections of x on K, the following formulas hold [14]:

$$\nabla_x d_K = \frac{x - \gamma_K(x)}{d_K(x)}$$

$$\|\nabla_x d_K\| = \left(1 - \frac{r_K^2(x)}{d_K^2(x)}\right)^{1/2} = \cos(\theta), \tag{12.3}$$

where θ is the (half) angle of the cone joining x to $B(\gamma_K(x), r_K(x))$.

12.3.2 μ-Medial axis of a compact set

The notion of μ-medial axes and μ-critical point of the distance function to a compact subset K of $\mathbb{R}^d$ was introduced for the purpose of geometric inference in [6]. A point x of $\mathbb{R}^d$ is called μ-*critical* for the distance function to K (with $\mu \geqslant 0$), or simply μ-critical for K, if for every vector h in $\mathbb{R}^d$,

$$d_K^2(x + h) \leqslant d_K^2(x) + \mu \|h\| d_K(x) + \|h\|^2 .$$

By definition of the superdifferential of d_K, a point x is μ-critical iff the norm of the generalized gradient $\|\nabla_x d_K\|$ is bounded by μ. The μ-*medial axis* $\mathrm{Med}_\mu(K)$ of a compact set $K \subseteq \mathbb{R}^d$ is the set of μ-critical points of the distance function. A point belongs to the medial axis of K iff $\|\nabla_x d_K\| < 1$, and therefore

$$\mathrm{Med}(K) = \bigcup_{0 \leqslant \mu < 1} \mathrm{Med}_\mu(K).$$

The lower semicontinuity of the map $x \mapsto \|\nabla_x d_K\|$ implies that the μ-medial axis is compact for μ less than one. We will use this in the following quantitative critical point stability theorem from [6]:

Theorem 12.5 (critical point stability theorem). *Let K, K' be two compact sets with $d_H(K, K') \leqslant \varepsilon$. For any point x in the μ-medial axis of K, there exists a point y in the μ'-medial axis of K' with $\mu' = \mu + 2\sqrt{\varepsilon/d_K(x)}$ and $\|x - y\| \leqslant 2\sqrt{\varepsilon d_K(x)}$.*

12.3.3 Proof of the non-quantitative stability result

For any pair of compact sets K and K' and $L > 0$, we let $\Delta_L(K, K')$ be the set of points of $\mathbb{R}^d \setminus (K \cup K')$ whose projections on K and K' are at distance at least L. More precisely,

$$\Delta_L(K, K') = \{x \in \mathbb{R}^d \setminus (\mathrm{Med}(K) \cup \mathrm{Med}(K')); \ \|p_K(x) - p_{K'}(x)\| \geqslant L\}.$$

Note that the points of the medial axes are removed for technical reasons, but this can be neglected since the Lebesgue measure of both medial axes is zero. The critical point stability theorem implies that $\Delta_L(K, K')$ lies close to the μ-medial axis of K for a certain value of μ. The following lemma is similar to [7, Theorem 3.1]:

Lemma 12.6. *Let $L > 0$ and K, K' be two compact sets and $\delta \leqslant L/2$ denote their Hausdorff distance. Then for any positive radius R, one has*

$$\Delta_L(K, K') \cap K^R \subseteq \mathrm{Med}_\mu(K)^{2\sqrt{R\delta}}$$

with

$$\mu = \left(1 + \left[\frac{L-\delta}{4R}\right]^2\right)^{-1/2} + 4\sqrt{\frac{\delta}{L}}.$$

Proof. Let x be a point in $\Delta_L(K, K')$ with $d_K(x) \leqslant R$, and denote by p and p' its projections on K and K' respectively. By assumption, $\|p - p'\|$ is at least L. We let q be the projection of p' on the sphere $\mathcal{S}(x, d_K(x))$, and let K_0 be the union of K and q. By hypothesis on the Hausdorff distance between K and K', there exists a point p'' in K such that $\|p'' - p'\| \leqslant \delta$. By definition of the distance to K, $\|x - p''\| \geqslant d_K(x)$: this means that $\|x - p'\| \geqslant d_K(x) - \delta$. Thus, because q is the projection of p' on the sphere $\mathcal{S}(x, d_K(x))$, the distance between p' and q is at most δ. Hence, $d_H(K, K_0)$ is at most 2δ.

By construction, the point x has two projections on K_0, and must belong to the μ_0-medial axis of K_0 for some value of μ. Letting m be the midpoint of the segment $[p, q]$, we are able to upper bound the value of μ_0:

$$\mu_0^2 := \|\nabla_x d_{K_0}\|^2 \leqslant \cos\left(\frac{1}{2}\angle(p - x, q - x)\right)^2 = \|x - m\|^2 / \|x - p\|^2 .$$

Since p, q belong to the sphere $B(x, d_K(x))$, one has $(p - q) \perp (m - x)$ and $\|x - p\|^2 = \|x - m\|^2 + \frac{1}{4}\|p - q\|^2$. This gives

$$\mu_0 \leqslant \left(1 + \frac{1}{4}\frac{\|p - q\|^2}{\|x - m\|^2}\right)^{-1/2} \leqslant \left[1 + \left(\frac{L - \delta}{2R}\right)^2\right]^{-1/2} .$$

To get the second inequality we used $\|x - m\| \leqslant R$ and $\|p - q\| \geqslant L - \delta$.

In order to conclude, one only needs to apply the critical point stability theorem (Theorem 12.5) to the compact sets K and K_0 with $d_H(K, K_0) \leqslant 2\delta$. Since x is in the μ_0-medial axis of K_0, there should exist a point y in $\mathrm{Med}_\mu(K)$ with $\|x - y\| \leqslant 2\sqrt{R\delta}$ and $\mu = \mu_0 + 4\sqrt{\delta/L}$.   □

Proof of Proposition 12.4. Fix $L > 0$, and suppose K and K' are given. One can decompose the set E between the set of points where the projections differ by at least L (i.e. $\Delta_L(K, K') \cap E$) and the remaining points. This gives the bound

$$\|p_{K'} - p_K\|_{L^1(E)} \leqslant L\mathcal{H}^d(E) + \mathcal{H}^d(\Delta_L(K, K') \cap E)\,\mathrm{diam}(K \cup K').$$

Now, take $R = \sup_E \|d_K\|$, so that E is contained in the tubular neighborhood K^R, and fix $L = \varepsilon/\mathcal{H}^d(E)$. Then, for $\delta = d_H(K, K')$ small enough (e.g., less than some δ_0), the value of μ given in Lemma 12.6 is smaller than one. Denote by μ_0 the value given by the lemma for δ_0. Then

$$\|p_{K'} - p_K\|_{L^1(E)} \leqslant \varepsilon + \mathcal{H}^d(\mathrm{Med}_{\mu_0}(K)^{2\sqrt{R\delta}})\,\mathrm{diam}(K \cup K'). \tag{12.4}$$

Being compact, $\mathrm{Med}_{\mu_0}(K)$ is the intersection of its tubular neighborhoods. Combining this with the outer-regularity of the Lebesgue measure gives

$$\lim_{\delta \to 0} \mathcal{H}^d(\mathrm{Med}_{\mu_0}(K)^{2\sqrt{R\delta}}) = \mathcal{H}^d(\mathrm{Med}_{\mu_0}(K)) = 0.$$

Putting this limit in Equation (12.4) concludes the proof.   □

12.4 Size and volume of the μ-medial axis

From the proof of Proposition 12.4, one can see that a way to get a quantitative stability of the projection functions is to control the volume of tubular neighborhoods of some part of the μ-medial axis. Recall that the *ε-covering*

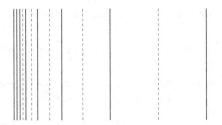

Figure 12.2 The "comb" and a part of its medial axis (dotted).

number of a subset $X \subseteq \mathbb{R}^d$ is the minimum number N of points $x_1, \ldots, x_N$ such that X is contained in the union of balls $\cup_{i=1}^{N} \overline{\mathrm{B}}(x_i, \varepsilon)$. This number is denoted $\mathcal{N}(X, \varepsilon)$. The following inequality is straightforward:

$$\mathcal{H}^d(X^\varepsilon) \leqslant \mathcal{H}(\mathrm{B}(0, \varepsilon)) \mathcal{N}(X, \varepsilon). \tag{12.5}$$

Our goal in this section is to show the following bound on the covering numbers of a certain part of the μ-medial axis, which will allow us to control the growth of the volume of its tubular neighborhoods.

Theorem 12.7. *For any compact set* $K \subseteq \mathbb{R}^d$, $\varepsilon \leqslant \mathrm{diam}(K)$, *and* η *small enough,*

$$\mathcal{N}\left(\mathrm{Med}_\mu(K) \cap (\mathbb{R}^d \setminus K^\varepsilon), \eta\right) \leqslant \mathcal{N}(\partial K, \varepsilon/2) \, \mathrm{O}\left(\left[\frac{\mathrm{diam}(K)}{\eta \sqrt{1-\mu}}\right]^{d-1}\right).$$

In particular, one can bound the $(d-1)$-*volume of the* μ-*medial axis*

$$\mathcal{H}^{d-1}\left(\mathrm{Med}_\mu(K) \cap (\mathbb{R}^d \setminus K^\varepsilon)\right) \leqslant \mathcal{N}(\partial K, \varepsilon/2) \, \mathrm{O}\left(\left[\frac{\mathrm{diam}(K)}{\sqrt{1-\mu}}\right]^{d-1}\right).$$

The bound on the covering numbers of $\mathrm{Med}_\mu(K) \cap (\mathbb{R}^d \setminus K^{2\varepsilon})$ given in this theorem is obtained by parameterizing this set by a suitable subset of the level set ∂K^ε using the *normal projection on the medial axis* $\ell : \mathbb{R}^d \setminus K \to \overline{\mathrm{Med}}(K)$ defined below. One cannot expect this map ℓ to be Lipschitz on the whole surface ∂K^ε. Our goal is therefore to construct a subset S_μ^ε that is both (i) large enough so that its image under ℓ covers the ε-away μ-medial axis, and (ii) small enough so that the restriction of ℓ to S_μ^ε is Lipschitz. Before turning to the proof of the theorem, we make two remarks.

Remark 1. Because of its compactness, one could expect that the μ-medial axis of a well-behaved compact set will have finite $\mathcal{H}^{d-1}$-measure for $\mu < 1$, but this not the case in general. Consider a "comb," as displayed in Figure 12.2, that is an infinite union of parallel segments of fixed length in $\mathbb{R}^2$, such as

$C = \cup_{i \in \mathbb{N}^*}[0, 1] \times \{2^{-i}\} \subseteq \mathbb{R}^2$. Then, the set of critical points of the distance sanction to C contains an imbricate comb, and the measure $\mathcal{H}^1(\mathrm{Med}_\mu(C))$ is therefore infinite for any $\mu \geqslant 0$. On the other hand, note that, for any positive ε, the part of the μ-medial axis that is ε-away from C, in other words the set $\mathrm{Med}_\mu(C) \cap (\mathbb{R}^d \setminus C^\varepsilon)$, consists of a finite union of segments, and has finite $\mathcal{H}^1$-measure. Theorem 12.7 shows quantitatively that this remains true for any compact set.

Remark 2. Let x, y be two points of $\mathbb{R}^d$ at distance D and $K = \{x, y\}$. The medial axis of K is the medial hyperplane H between x and y. A point m in this hyperplane belongs to $\mathrm{Med}_\mu(K)$ iff the cosine of the angle $\theta = \frac{1}{2}\angle(x - m, y - m)$ is at most μ, and $\mathrm{Med}_\mu(K)$ coincides with the intersection between H and a certain ball $\mathrm{B}(z, r)$, where z is the midpoint between x and y and the radius r is to be determined.

$$\cos^2(\theta) = 1 - \frac{\|x - y\|^2}{4\mathrm{d}_K^2(m)} = 1 - \frac{\mathrm{diam}(K)^2}{4\mathrm{d}_K^2(m)} \leqslant \mu^2$$
$$\Longleftrightarrow \|z - m\|^2 \leqslant \frac{1}{4}\frac{\mu^2}{1 - \mu^2}\,\mathrm{diam}(K)^2.$$

This implies $\mathcal{H}^{d-1}(\mathrm{Med}_\mu(K)) = \mathrm{const}(d)(\mathrm{diam}(K)\mu/(1 - \mu^2)^{1/2})^{d-1}$. Hence, the behavior of the bound in the theorem in $\mathrm{diam}(K)$ and μ is sharp as μ converges to one.

12.4.1 Covering numbers of the μ-medial axis

We now proceed to the proof of Theorem 12.7. For any point x in $\mathbb{R}^d$, we define the *normal distance of x to the medial axis* as $\tau_K(x) := \inf\{t \geqslant 0; x + t\nabla_x \mathrm{d}_K \in \mathrm{Med}(K)\}$, and we set $\tau_K(x)$ to zero at any point x that lies in K or in its medial axis. For any $t < \tau(x)$, we let $\Psi_K^t(x) := x + t\nabla_x \mathrm{d}_K$. For any point x in the complementary of K, *the normal projection of x on the medial axis of K* is the point $\ell_K(x) := x + \tau_K(x)\nabla_x \mathrm{d}_K$. Note that the point $\ell_K(x)$ always belongs to the closure of the medial axis of K, but not necessarily to the medial axis itself.

Lemma 12.8. *Let m be a point of the medial axis $\mathrm{Med}(K)$ with $\mathrm{d}(x, K) > \varepsilon$, and x be a projection of m on ∂K^ε. Then $\ell(x) = m$.*

Proof. By definition of K^ε, $d(m, K) = d(m, K^\varepsilon) + \varepsilon$, so that the projection p of x on K must also be a projection of m on K. Hence, m, x and p must be aligned. Since the open ball $B(m, d(m, p))$ does not intersect K, for any point $y \in]p, m[$ the ball $B(y, d(y, p))$ intersects K only at p. In particular, by definition of the gradient, $\nabla_x d_K$ must be the unit vector directing $]p, m[$, i.e. $\nabla_x d_K = (m - x)/d(m, x)$. Moreover, since $[x, p[$ is contained in the complement of the medial axis, $\tau(x)$ must be equal to $d(x, m)$. Finally, one gets $\Psi^{\tau(x)}(x) = x + d(x, m)\nabla_x d_K = m$. $\qquad\square$

This statement means in particular that ε-away medial axis – that is, the intersection $\mathrm{Med}(K) \cap (\mathbb{R}^d \setminus K^\varepsilon)$ – is contained in the image of the piece of hypersurface $\{x \in \partial K^\varepsilon \,;\, \tau_K(x) \geqslant \varepsilon\}$ by the map ℓ.

Recall that the radius of a set $K \subseteq \mathbb{R}^d$ is the radius of the smallest ball enclosing K, while the diameter of K is the maximum distance between two points in K. The following inequality between the radius and the diameter is known as Jung's theorem [12]: $\mathrm{radius}(K)\sqrt{2(1 + 1/d)} \leqslant \mathrm{diam}(K)$.

Lemma 12.9. *For any point m in the μ-medial axis $\mathrm{Med}_\mu(K)$, there exist two projections $x, y \in \mathrm{proj}_K(m)$ of m on K such that the cosine of the angle $\frac{1}{2}\angle(x - m, y - m)$ is smaller than $\left(\frac{1+\mu^2}{2}\right)^{1/2}$.*

Proof. We use the characterization of the gradient of the distance function given in Equation (12.3). If $B(\gamma_K(m), r_K(m))$ denotes the smallest ball enclosing $\mathrm{proj}_K(m)$, then $\mu^2 \geqslant 1 - r_K^2(m)/d_K^2(m)$. Using Jung's theorem and the definition of the diameter, there must exists two points x, y in $\mathrm{proj}_K(m)$ whose distance r' is larger than $\sqrt{2}r_K(m)$. The following bound on the cosine of the angle $\theta = \frac{1}{2}\angle(x - m, y - m)$ concludes the proof:

$$\cos^2(\theta) = 1 - \frac{(r'/2)^2}{d_K^2(m)} \leqslant 1 - \frac{1}{2}\frac{r_K^2(m)}{d_K^2(m)} \leqslant (1 + \mu^2)/2 \qquad (12.6)$$

$\qquad\square$

Lemma 12.10. *The maximum distance from a point in $\mathrm{Med}_\mu(K)$ to K is bounded by $\frac{1}{\sqrt{2}}\mathrm{diam}(K)/\left(1 - \mu^2\right)^{1/2}$.*

Proof. Let x, y be two projections of $m \in \mathrm{Med}_\mu(K)$ on K as given by the previous lemma. Then, using Equation (12.6), one obtains

$$1 - \frac{\|x - y\|^2/4}{d_K^2(m)} \leqslant (1 + \mu^2)/2.$$

Hence, $d_K^2(m) \leqslant \frac{1}{2}(1 - \mu^2)^{-1}\|x - y\|^2$, which proves the result. $\qquad\square$

Let us denote by S_μ^ε the set of points x of the hypersurface ∂K^ε that satisfies the three conditions below:

(i) the normal distance to the medial axis is bounded below: $\tau(x) \geqslant \varepsilon$;
(ii) the image of x by ℓ is in the μ-medial axis of K: $\ell(x) \in \mathrm{Med}_\mu(K)$;

(iii) there exists another projection y of $m = \ell(x)$ on ∂K^{ε} with

$$\cos\left(\frac{1}{2}\angle(p-m, q-m)\right) \leqslant \sqrt{\frac{1+\mu^2}{2}}.$$

A reformulation of Lemmas 12.9 and 12.8 is the following corollary:

Corollary 12.11. *The image of S_{μ}^{ε} by the map ℓ covers the whole 2ε-away μ-medial axis: $\ell(S_{\mu}^{\varepsilon}) = \mathrm{Med}_{\mu}(K) \cap (\mathbb{R}^d \setminus K^{2\varepsilon})$.*

12.4.2 Lipschitz estimations for the map ℓ

In this section, we bound the Lipschitz constants of the restriction of the maps ∇d_K, τ, and ℓ to the subset $S_{\mu}^{\varepsilon} \subseteq \partial K^{\varepsilon}$. The first step is to prove that the maps Ψ^t and $\nabla_x d_K$ are Lipschitz on the set $\partial K^{\varepsilon,t} := \{x \in \partial K^{\varepsilon}; \ \tau_K(x) \geqslant t\}$. The main ingredient of the proof is a particular case of the Monge–Mather shortening principle in the quadratic case [17, Eq. (8.2)].

Lemma 12.12.

 (i) *The restriction of Ψ^t to $\partial K^{\varepsilon,t}$ is $(1+t/\varepsilon)$-Lipschitz.*
 (ii) *The gradient of the distance function is $3/\varepsilon$-Lipschitz on $\partial K^{\varepsilon,\varepsilon}$.*

Proof. (i) Let x and x' be two points of ∂K^{ε} with $\tau(x), \tau(x') > t$, p and p' their projections on K and y and y' their image by Ψ^t. We let $u = 1 + t/\varepsilon$ be the scale factor between $x - p$ and $y - p$, i.e.:

$$y' - y = u(x' - x) + (1-u)(p' - p) \tag{12.7}$$

Using the fact that y projects to p, and the definition of u, we have

$$\|y - p\|^2 \leqslant \|y - p'\|^2 = \|y - p\|^2 + \|p - p'\|^2 + 2\langle y - p|p - p'\rangle$$

$$\text{i.e. } 0 \leqslant \|p - p'\|^2 + 2u\langle x - p|p - p'\rangle$$

$$\text{i.e. } \langle p - x|p - p'\rangle \leqslant \frac{1}{2}u^{-1}\|p - p'\|^2.$$

Summing this last inequality, the same inequality with primes and the equality $\langle p' - p|p - p'\rangle = -\|p' - p\|^2$ gives

$$\langle x' - x|p' - p\rangle \geqslant \left(1 - u^{-1}\right)\|p' - p\|^2. \tag{12.8}$$

Using (12.7) and (12.8) we get the desired Lipschitz inequality

$$\|y - y'\|^2 = u^2\|x - x'\|^2 + (1-u)^2\|p' - p\|^2 + 2u(1-u)\langle x' - x|p' - p\rangle$$

$$\leqslant u^2\|x - x'\|^2 - (1-u)^2\|p' - p\|^2 \leqslant (1 + t/\varepsilon)^2\|x - x'\|^2.$$

(ii) If x belongs to $\partial K^{\varepsilon,\varepsilon}$, then $\nabla_x d_K = \frac{1}{\varepsilon}(\Psi^\varepsilon(x) - x)$. The result follows from the Lipschitz estimation of (i). □

Since S^ε_μ is contained in $\partial K^{\varepsilon,\varepsilon}$, Lemma 12.12 shows that the restriction of ∇d_K to S^ε_μ is Lipschitz. The second step is to prove that the restriction of τ to the set S^ε_μ is also Lipschitz. The technical core of the proof is contained in the following easy geometric lemma:

Lemma 12.13. *Let $t(x, v)$ denote the intersection time of the ray $x + tv$ with the medial hyperplane $H_{x,y}$ between x and another point y, and $t(x', v')$ the intersection time between the ray $x' + tv'$ and $H_{x'y}$. Then, assuming*

$$\alpha \|x - y\| \leqslant \langle v | x - y \rangle, \tag{12.9}$$
$$\|x' - y\| \leqslant D, \tag{12.10}$$
$$\|v' - v\| \leqslant \lambda \|x' - x\|, \tag{12.11}$$
$$\varepsilon \leqslant t(x, v), \tag{12.12}$$

one obtains the bound

$$t(x', v') \leqslant t(x, v) + \frac{6}{\alpha^2}(1 + \lambda D) \|x' - x\|$$

as soon as $\|x' - x\|$ is small enough (namely, smaller than $\varepsilon\alpha^2(1 + 3\lambda D)^{-1}$).

Proof. We search the time t such that $\|x' + tv' - x'\|^2 = \|x' + tv' - y\|^2$, i.e.

$$t^2 \|v'\|^2 = \|x' - y\|^2 + 2t\langle x' - y | v' \rangle + t^2 \|v'\|^2 .$$

Hence, the intersection time is $t(x', v') = \|x' - y\|^2 / 2\langle y - x' | v' \rangle$. The lower bound on $t(x, y)$ translates as

$$\varepsilon \leqslant \frac{1}{2} \frac{\|x - y\|^2}{\langle x - y | v \rangle} \leqslant \frac{1}{2\alpha} \|x - y\|.$$

If $\nabla_{x'} t$ and $\nabla_{v'} t$ denote the gradients of this function in the direction of v' and x', one has

$$\nabla_{v'} t(x', v') = \frac{1}{2} \frac{\|x' - y\|^2 (x' - y)}{\langle y - x' | v' \rangle^2}$$

$$\nabla_{x'} t(x', v') = \frac{1}{2} \frac{\|x' - y\|^2 v' + 2\langle y - x' | v' \rangle (x' - y)}{\langle y - x' | v' \rangle^2}.$$

Now, we bound the denominator of this expression:

$$\langle x' - y | v' \rangle = \langle x' - y | v' - v \rangle + \langle x' - x | v \rangle + \langle x - y | v \rangle$$
$$\geqslant \alpha \, \|x - y\| - (1 + \lambda \, \|x' - y\|) \, \|x' - x\|$$
$$\geqslant \alpha \, \|x' - y\| - (2 + \lambda D) \, \|x' - x\| \,.$$

The scalar product $\langle x' - y | v' \rangle$ will be larger than (say) $\frac{\alpha}{2} \|x' - y\|$ provided that

$$(2 + \lambda D) \, \|x' - x\| \leqslant \frac{\alpha}{2} \, \|x' - y\|$$

or, bounding from below $\|x' - y\|$ by $\|x - y\| - \|x - x'\| \geqslant 2\alpha\varepsilon - \|x - x'\|$, provided that

$$(3 + \lambda D) \, \|x' - x\| \leqslant \alpha^2 \varepsilon.$$

This is the case in particular if $\|x' - x\| \leqslant \alpha^2 \varepsilon (3 + \lambda D)^{-1}$. Under that assumption, we have the following bound on the norm of the gradient, from which the Lipschitz inequality follows:

$$\|\nabla_{x'} t(x', v')\| \leqslant 6/\alpha^2 \text{ and } \|\nabla_{v'} t(x', v')\| \leqslant 4D/\alpha^2.$$

$\square$

We are now ready to show that the function ℓ is locally Lipschitz S_μ^ε:

Proposition 12.14. *The restriction of τ to S_μ^ε is locally L-Lipschitz, in the sense that if $(x, y) \in S_\mu^\varepsilon$ are such that $\|x - y\| \leqslant \delta_0$, then $\|\ell(x) - \ell(y)\| \leqslant L \, \|x - y\|$ with*

$$L = O\left(\frac{1 + \mathrm{diam}(K)/\varepsilon}{(1 - \mu)^{1/2}} \right) \text{ and } \delta_0 = O(\varepsilon/L).$$

In order to simplify the proof of this proposition, we will make use of the following notation, where f is any function from $X \subseteq \mathbb{R}^d$ to $\mathbb{R}$ or $\mathbb{R}^d$:

$$\mathrm{Lip}_\delta \, f|_X := \sup\{\|f(x) - f(y)\| / \|x - y\| \, ; (x, y) \in X^2 \text{ and } \|x - y\| \leqslant \delta\}.$$

Proof. We start the proof by evaluating the Lipschitz constant of the restriction of τ to S_μ^ε, using Lemma 12.13 (Step 1), and then deduce the Lipschitz estimate for the function ℓ (Step 2).

Step 1. Thanks to Lemma 12.9, for any x in S_μ^ε, there exists another projection y of $m = \ell(x)$ on ∂K^ε such that the cosine of the angle $\theta = \frac{1}{2} \angle(x - m, y - m)$ is at most $\sqrt{(1 + \mu^2)/2}$. Let us denote by $v = \nabla_x d_K$ the unit vector from x to

m. The angle between $\overrightarrow{xy}$ and v is $\pi/2 - \theta$. Then,

$$\cos(\pi/2 - \theta) = \sin(\theta) = \sqrt{1 - \cos^2(\theta)} \geqslant \alpha := \left(\frac{1 - \mu^2}{2}\right)^{1/2}.$$

As a consequence, with the α introduced above, one has $\alpha \|x - y\| \leqslant |\langle v|x - y\rangle|$. Moreover, $\|x - y\|$ is smaller than $D = \mathrm{diam}(K^\varepsilon) \leqslant \mathrm{diam}(K) + \varepsilon$. For any other point x' in S_μ^ε, and $v' = \nabla_{x'} d_K$, one has $\|v - v'\| \leqslant \lambda \|x - x'\|$ with $\lambda = 3/\varepsilon$ (thanks to Lemma 12.12).

These remarks allow us to apply Lemma 12.13. Using the notations of this lemma, one sees that $t(x, v)$ is simply $\tau(x)$ while $t(x', v')$ is an upper bound for $\tau(x')$. This gives us

$$\tau(x') \leqslant \tau(x) + \frac{6}{\alpha^2}(1 + \lambda D) \|x - x'\|$$
$$\leqslant \tau(x) + M \|x - x'\|,$$
$$\text{where } M = O\left(\frac{1 + \mathrm{diam}(K)/\varepsilon}{\sqrt{1 - \mu^2}}\right),$$

as soon as x' is close enough to x. From the statement of Lemma 12.13, one sees that $\|x - x'\| \leqslant \delta_0$ with $\delta_0 = O(\varepsilon/M)$ is enough. Exchanging the role of x and x', one proves that $|\tau(x) - \tau(x')| \leqslant M \|x - x'\|$, provided that $\|x - x'\| \leqslant \delta_0$. As a conclusion,

$$\mathrm{Lip}_{\delta_0}\left[\tau|_{S_\mu^\varepsilon}\right] = O\left(\frac{1 + \mathrm{diam}(K)/\varepsilon}{\sqrt{1 - \mu^2}}\right). \tag{12.13}$$

Step 2. We can use the following decomposition of the difference $\ell(x) - \ell(x')$:

$$\ell(x) - \ell(x') = (x - x') + (\tau(x) - \tau(x'))\nabla_x d_K + \tau(x')(\nabla_x d_K - \nabla_{x'} d_K) \tag{12.14}$$

in order to bound the (local) Lipschitz constant of the restriction of ℓ to S_μ^ε from those computed earlier. One deduces from this equation that

$$\mathrm{Lip}_{\delta_0}\left[\ell|_{S_\mu^\varepsilon}\right] \leqslant 1 + \mathrm{Lip}_{\delta_0}\left[\tau|_{S_\mu^\varepsilon}\right] + \|\tau\|_\infty \mathrm{Lip}_{\delta_0}\left[\nabla d_K|_{S_\mu^\varepsilon}\right]. \tag{12.15}$$

Thanks to Lemma 12.10, one has $|\tau(x)| = O(\mathrm{diam}(K)/(1 - \mu)^{1/2})$; combining this with the estimate from Lemma 12.12 that $\mathrm{Lip}\, \nabla d_K|_{S_\mu^\varepsilon} \leqslant 3/\varepsilon$, this gives

$$\|\tau\|_\infty \mathrm{Lip}_{\delta_0}\left[\nabla d_K|_{S_\mu^\varepsilon}\right] = O(\mathrm{diam}(K)/[\varepsilon(1 - \mu)^{1/2}]). \tag{12.16}$$

Putting the estimates (12.13) and (12.16) into (12.15) concludes the proof. $\square$

12.4.3 Proof of Theorem 12.7

In order to deduce the theorem from Proposition 12.14 we need the following bound on the covering numbers of the level set ∂K^r:

$$\mathcal{N}(\partial K^r, \varepsilon) \leqslant \mathcal{N}(\partial K, r)\mathcal{N}(\mathcal{S}^{d-1}, \varepsilon/2r). \tag{12.17}$$

This bound holds for any compact set in $\mathbb{R}^d$ and is proved in Proposition 4.2 of [8]. Applying Proposition 12.14, we get the existence of

$$L = \mathrm{Lip}_{\delta_0}\left[\ell|_{S_\mu^{\varepsilon/2}}\right] = O(\mathrm{diam}(K)/(\varepsilon\sqrt{1-\mu})) \text{ and } \delta_0 = O(\varepsilon/L)$$

such that ℓ is locally L-Lipschitz. In particular, for any η smaller than δ_0,

$$\mathcal{N}\left(\mathrm{Med}_\mu(K) \cap (\mathbb{R}^d \setminus K^\varepsilon), \eta\right) = \mathcal{N}\left(\ell(S_\mu^{\varepsilon/2}), \eta\right) \leqslant \mathcal{N}\left(S_\mu^{\varepsilon/2}, \eta/L\right)$$
$$\leqslant \mathcal{N}(\partial K^{\varepsilon/2}, \eta/L). \tag{12.18}$$

The bound on the covering number of the boundary of tubular neighborhoods given in Equation (12.17) gives

$$\mathcal{N}(\partial K^{\varepsilon/2}, \eta/L) \leqslant \mathcal{N}(\partial K, \varepsilon/2)\mathcal{N}\left(\mathcal{S}^{d-1}, \frac{\eta}{L\varepsilon}\right). \tag{12.19}$$

Putting Equations (12.18) and (12.19), and the estimation $\mathcal{N}(\mathcal{S}^{d-1}, \rho) \sim \omega_{d-1}\rho^{d-1}$ together, we get

$$\mathcal{N}\left(\mathrm{Med}_\mu(K) \cap (\mathbb{R}^d \setminus K^\varepsilon), \eta\right) = \mathcal{N}(\partial K, \varepsilon/2)\,O\left(\left[\frac{\eta}{L\varepsilon}\right]^{d-1}\right).$$

It suffices to replace L by its value from Proposition 12.14 to finish the proof.

12.5 A quantitative stability result for boundary measures

In this section we show how to use the bound on the covering numbers of the ε-away μ-medial axis given in Theorem 12.7 in order to get a quantitative version of the L^1 convergence results for projections. Notice that the meaning of *locally* in the next statement could also be made quantitative using the same proof.

Theorem 12.15. *The map* $K \mapsto p_K \in L^1(E)$ *is locally* h-*Hölder for any exponent* $h < \frac{1}{2(2d-1)}$.

Proof. As in the previous proof, we will let $R = \|d_K\|_{E,\infty}$, so that E is contained in the tubular neighborhood K^R. Remark first that if a point x is such that $d_K(x) \leqslant \frac{1}{2}L - d_H(K, K')$, then by definition of the Hausdorff distance, $d_{K'}(x) \leqslant \frac{1}{2}L$. In particular, the projections of x on K and K' are at distance at most L. Said otherwise, the set $\Delta_L(K, K')$ is contained in the complementary of the $\frac{L}{2} - \delta$ tubular neighborhood of K, with $\delta := d_H(K, K')$. Using this fact and the result of Lemma 12.6, we have

$$\Delta_L(K, K') \cap K^R \subseteq \mathrm{Med}_\mu(K)^{2\sqrt{R\delta}} \cap (\mathbb{R}^d \setminus K^{\frac{L}{2}-\delta}) \tag{12.20}$$

$$\subseteq \left(\mathrm{Med}_\mu(K) \cap \left(\mathbb{R}^d \setminus K^{\frac{L}{2}-\delta-2\sqrt{R\delta}} \right) \right)^{2\sqrt{R\delta}}, \tag{12.21}$$

$$\text{where } \mu = \left(1 + \left[\frac{L-\delta}{4R} \right]^2 \right)^{-1/2} + 4\sqrt{\frac{\delta}{L}}. \tag{12.22}$$

We now choose L to be δ^h, where $h > 0$, and see for which values of h we are able to get a converging bound. For $h < 1/2$, the radius $\frac{1}{2}(L - \delta) - 2\sqrt{R\delta}$ will be greater than $L/3$ as soon as δ is small enough. For these values,

$$\Delta_L(K, K') \cap K^R \subseteq \left(\mathrm{Med}_\mu(K) \cap (\mathbb{R}^d \setminus K^{L/3}) \right)^{2\sqrt{R\delta}}. \tag{12.23}$$

The μ above, given by Lemma 12.6, can then be bounded as follows. Note that the constants in the "big O" will always be positive in the remainder of the proof. From Equation (12.22), one deduces

$$\mu = 1 + O(-\delta^{2h} + \delta^{1/2-h/2}).$$

This term will be asymptotically smaller than 1 provided that $2h < 1/2 - h/2$ i.e. $h < 1/5$, in which case $\mu = 1 - O(\delta^{2h})$. By definition of the covering number, one has

$$\mathcal{H}^d(\Delta_L(K, K') \cap K^R) \leqslant \mathcal{H}^d \left[\left(\mathrm{Med}_\mu(K) \cap \left(\mathbb{R}^d \setminus K^{L/3} \right) \right)^{2\sqrt{R\delta}} \right]$$

$$\leqslant \mathcal{N} \left(\mathrm{Med}_\mu(K) \cap \left(\mathbb{R}^d \setminus K^{L/3} \right), 2\sqrt{R\delta} \right) \times O(\delta^{d/2}). \tag{12.24}$$

The covering numbers of the intersection $\mathrm{Med}_\mu(K) \cap (\mathbb{R}^d \setminus K^{L/3})$ can be bounded using Theorem 12.7:

$$\mathcal{N} \left(\mathrm{Med}_\mu(K) \cap \left(\mathbb{R}^d \setminus K^{L/2} \right), 2\sqrt{R\delta} \right)$$

$$= \mathcal{N}(\partial K, L/4) O \left(\left[\frac{\mathrm{diam}(K)/\sqrt{R\delta}}{\sqrt{1-\mu^2}} \right]^{d-1} \right) \tag{12.25}$$

$$= \mathcal{N}(\partial K, L/4) O \left(\delta^{-(h+\frac{1}{2})(d-1)} \right).$$

Combining Equations (12.24) and (12.25), and using the (crude) estimation $\mathcal{N}(\partial K, L/4) = O(1/L^d) = O(\delta^{-hd})$,

$$\mathcal{H}^d(\Delta_L(K, K') \cap K^R) \leqslant \mathcal{N}(\partial K, L/4)\, O(\delta^{-h(d-1)-\frac{1}{2}(d-1)+\frac{1}{2}d})$$
$$\leqslant O\left(\delta^{\frac{1}{2}-h(2d-1)}\right).$$

Hence, following the proof of Proposition 12.4,

$$\|\mathrm{p}_{K'} - \mathrm{p}_K\|_{\mathrm{L}^1(E)} \leqslant L\mathcal{H}^d(E) + \mathcal{H}^d(\Delta_L(K, K') \cap E)\,\mathrm{diam}(K \cup K')$$
$$= O(\delta^h + \delta^{1/2-h(2d-1)}).$$

The second term converges to zero as $\delta = \mathrm{d}_H(K, K')$ does if $h < \frac{1}{2(2d-1)}$. This concludes the proof. $\qquad\square$

References

[1] N. Amenta and M. Bern, Surface reconstruction by Voronoi filtering, *Discrete and Computational Geometry* **22** (1999), no. 4, 481–504.

[2] D. Attali, J.-D. Boissonnat, and H. Edelsbrunner, Stability and computation of medial axes: a state of the art report, *Mathematical Foundations of Scientific Visualization, Computer Graphics, and Massive Data Exploration*, Springer, 2007.

[3] P. Cannarsa, P. Cardaliaguet, and E. Giorgieri, Hölder regularity of the normal distance with an application to a PDE model for growing sandpiles, *Transactions of the American Mathematical Society* **359** (2007), no. 6, 2741.

[4] P. Cannarsa and C. Sinestrari, *Semiconcave Functions, Hamilton–Jacobi Equations, and Optimal Control*, Birkhäuser, 2004.

[5] M. Castelpietra and L. Rifford, Regularity properties of the distance functions to conjugate and cut loci for viscosity solutions of Hamilton–Jacobi equations and applications in Riemannian geometry, *ESAIM: Control, Optimisation and Calculus of Variations* **16** (2010), no. 3, 695–718.

[6] F. Chazal, D. Cohen-Steiner, and A. Lieutier, A sampling theory for compact sets in Euclidean space, *Discrete and Computational Geometry* **41** (2009), no. 3, 461–479.

[7] F. Chazal, D. Cohen-Steiner, and A. Lieutier, Normal cone approximation and offset shape isotopy, *Computational Geometry: Theory and Applications* **42** (2009), no. 6–7, 566–581.

[8] F. Chazal, D. Cohen-Steiner, and Q. Mérigot, Boundary measures for geometric inference, *Journal of Foundations of Computational Mathematics* **10** (2010), 221–240.

[9] H. Federer, Curvature measures, *Transactions of the American Mathematical Society* **93** (1959), no. 3, 418–491.

[10] J.H.G. Fu, Tubular neighborhoods in Euclidean spaces, *Duke Mathematics Journal* **52** (1985), no. 4, 1025–1046.

[11] J. Itoh and M. Tanaka, The Lipschitz continuity of the distance function to the cut locus, *Transactions of the American Mathematical Society* **353** (2001), no. 1, 21–40.

[12] H.W.E. Jung, Über den kleinsten Kreis, der eine ebene Figur einschließt., *Journal für die Reine und Angewandte Mathematik* **1910** (1910), no. 137, 310–313.

[13] Y. Li and L. Nirenberg, The distance function to the boundary, Finsler geometry, and the singular set of viscosity solutions of some Hamilton–Jacobi equations, *Communications on Pure and Applied Mathematics* **58** (2005), 0085–0146.

[14] A. Lieutier, Any open bounded subset of $\mathbb{R}^n$ has the same homotopy type as its medial axis, *Computer Aided Geometric Design* **36** (2004), no. 11, 1029–1046.

[15] Q. Mérigot, Détection de structure géométrique dans les nuages de points, Thèse de doctorat, Université de Nice Sophia-Antipolis, 2009.

[16] A. Petrunin, Semiconcave functions in Alexandrov's geometry, *Surveys in Differential Geometry*. Vol. XI, International Press, Somerville, MA, 2007, pp. 137–201.

[17] C. Villani, *Optimal Transport: Old and New*, Springer, 2009.

[18] H. Weyl, On the volume of tubes, *American Journal of Mathematics* **61** (1939), no. 2, 461–472.

[19] T. Zamfirescu, On the cut locus in Alexandrov spaces and applications to convex surfaces, *Pacific Journal of Mathematics* **217** (2004), 375–386.

Printed in the United States
By Bookmasters